WILLIAM F. MAAG LIBRARY
YOUNGSTOWN STATE UNIVERSITY

# Electrochemistry
## Volume 7

A Specialist Periodical Report

# Electrochemistry
Volume 7

A Review of Recent Literature

Senior Reporter
**H. R. Thirsk** Department of Physical Chemistry
University of Newcastle upon Tyne

Reporters
**W. I. Archer** University of Newcastle upon Tyne
**R. D. Armstrong** University of Newcastle upon Tyne
**J. Grimshaw** The Queen's University of Belfast
**N. Lakshminarayanaiah** Thomas Jefferson University, Philadelphia, U.S.A.
**S. K. Rangarajan** Indian Institute of Science, Bangalore, India

The Chemical Society
Burlington House, London W1V 0BN

**British Library Cataloguing in Publication Data**
Electrochemistry.–
  (Chemical Society. Specialist periodical reports).
  Vol. 7
  1. Electrochemistry
  I. Thirsk, Harold Reginald   II. Series
  541′.37     QD552     72-23822

ISBN 0-85186-870-3
ISSN 0305-9979

Copyright © 1980
The Chemical Society

*All Rights Reserved*
*No part of this book may be reproduced or transmitted*
*in any form or by any means – graphic, electronic,*
*including photocopying, recording, taping or*
*information storage and retrieval systems – without*
*written permission from The Chemical Society*

Printed in Great Britain
at the Alden Press, Oxford

# Preface

To get the priorities right, the Senior Reporter again must express his thanks to the contributors who have found the time, in a busy professional life, to produce the chapters for Volume 7.

It is often commented in reviews of this series that sometimes the contents may not be up-to-date or that the coverage goes back over many years. There is without doubt a basis for this criticism but it would also seem to be inevitable. In recent years interest in electrochemistry has both widened and deepened; new areas require reporting, older fields continue to provide new material of value, but large monographs have become a formidable expense, and, if it is desirable that this and similar volumes should be a possible personal rather than a purely library acquisition, cost is of importance.

It would therefore seem inescapable, within a comparatively small book, to avoid the faults, but one would hope that this will continue to be recompensed by articles of depth on a variety of interests, even if only a few are covered each year but covered in the informed manner which hopefully is the mark of this series.

H R Thirsk

# Contents

**Chapter 1  Organic Electrochemistry – Synthetic Aspects    1**
*By J. Grimshaw*

1 General   1

2 Reduction   4
  General   4
  Hydrocarbons   5
  Activated Olefins   6
  Carbonyl Compounds   7
  Nitro- and Nitroso-compounds   9
  Other Nitrogen-containing Compounds   10
  Oxygen (except Carbonyl), Sulphur, and Phosphorus
    Compounds   13
  Halides   15
  Simple Inorganic Materials   19

3 Oxidation   20
  General   20
  Aliphatic Hydrocarbons and Ethers   20
  Aromatic Compounds   24
  Carboxylic Acids   29
  Boron Compounds   30
  Nitrogen, Phosphorus, and Arsenic Compounds   31
  Oxygen and Sulphur Compounds   34

4 Electrochemical Halogenation   36
  Fluorination   36
  Chlorination and Bromination   36

5 Organometallic Compounds   36
  General   36
  Synthesis from Electrode Material   37
  Induced Ligand Exchange   37
  Induced Ligand Reactivity   38

Chapter 2  **Membrane Phenomena**  40
By N. Lakshminarayanaiah

1 **Introduction**  40

2 **Theoretical Considerations**  43

3 **Equilibrium and other Phenomena due to Chemical and Electrochemical Gradients**  50
   Self Diffusion  52
   Diffusion of Electrolyte  53
      Solute and Solvent Fluxes  53
      Membrane Potential  58
      Membrane Electrodes  62
         General considerations  63
         Air-gap electrodes  72
         Coated wire electrodes  74
         Glass membrane electrodes  77
         CATION-SELECTIVE MEMBRANE ELECTRODES  82
            Ammonium- and ammonia-selective electrodes  82
            Barium-selective electrode  84
            Cadmium- and caesium-selective electrodes  84
            Calcium-selective electrode  84
            Copper-selective electrode  87
            Lead-selective electrode  89
            Lithium-selective electrode  89
            Mercury-selective electrode  90
            Molybdenum-selective electrode  90
            Potassium-selective electrode  91
            Selenium-selective electrode  93
            Silver-selective electrode  93
            Sodium-selective electrode  94
            Thallium- and antimony-selective electrodes  94
            Uranyl-ion-selective electrode  95
            Zinc-selective electrode  95
         ANION-SELECTIVE MEMBRANE ELECTRODES  95
            Carbonate- and carbon-dioxide-selective electrodes  95
            Chlorate-selective electrode  96
            Chromate-selective electrode  96
            Cyanide-selective electrode  96
            Fluoroborate-selective electrodes  97
            Halide-selective electrodes  97
            Nitrate-selective electrode  102
            Perchlorate-selective electrode  103
            Perrhenate- and tetrachloroaurate-selective electrodes  104
            Phosphate-selective electrode  104
            Sulphate-selective electrode  105

Sulphide-selective electrode 105
Sulphonate-selective electrode 106
MISCELLANEOUS MEMBRANE ELECTRODES 106
ENZYME MEMBRANE ELECTRODES 108
Diffusion of Non-electrolyte 114
Diffusion of Electrolyte and Non-electrolyte 118

4 **Phenomena due to an Applied Electric Field** 119
Electrical Conductance 119
Electro-osmosis 126
Electrodialysis 132
Oscillatory Phenomena 134

5 **Phenomena due to an Applied Pressure Gradient** 134
Streaming Potential 136
Hyperfiltration and Reverse Osmosis 137
Hydrodynamic Flow 144
Nature of Flow through Membranes 148

6 **Phenomena due to a Gradient of Temperature** 153

Chapter 3 The Application of A.C. Impedance Methods
to Solid Electrolytes 157
*By W. I. Archer and R. D. Armstrong*

1 **Introduction** 157

2 **Definitions and Derivation of Simple A.C. Impedance Theory** 157

3 **Theoretical Models for the Metal/Superionic Conductor Interphase** 161
The Point-charge Model 161
Completely Blocking Electrodes (0,0): No Specific
Adsorption 162
One Mobile Species: Blocked but Specifically Adsorbed (0) 164
Non-blocking Electrodes with One Mobile, Charged
Species ($\infty$) 164
One Mobile, Charged Species: Partially Blocked ($r_n$) 166
Two Mobile, Charged Species: One Non-blocked (0,$\infty$) 166
The Finite-ion-size Model 166
Blocking Electrodes with No Specific Adsorption (0,0) 168
Blocking Electrodes with Specific Adsorption (0,0) 169
Non-blocking Electrodes ($\infty$) 170
Situations where a Normal Warburg Impedance Can Arise 173
Complex Reaction Mechanisms 173
A Comparison of the Two Theoretical Models 174

4   The Effect of Sample State upon the Observed Impedance   174
    Single Crystal   174
        Perfectly Flat Electrodes   174
        Rough Electrodes   174
    Compressed Powder or Sinter   175
        Voidage   175
        Particle Orientation   175
        Inter-particle Impedance   177
        Grain-boundary Conduction   177

5   Instrumentation   178

6   Some Experimental Factors which Affect the Analysis of Impedance Measurements   182

7   Review of Some Experimental Results   184
    Stabilized Zirconia   184
    The $\beta$-Aluminas   188
    Silver Rubidium Iodide and Related Compounds   191
    Silver Halides   201

## Chapter 4   The Electrical Double Layer   203
### By S. K. Rangarajan

*Part I: Models for Solvent Structure at Interfaces*

1   What is the Information Sought?   203

2   The First Stages of Double-layer Modelling   205

3   Primary Model Reduction   207

4   Primary Data Reduction   208

5   Gouy Theory – Present Status   208

6   Gouy Theory – The Dielectric Constant   210

7   Gouy Theory – Further Phenomenological Modelling   211

8   Inactive Ions – Are there any?   212

9   Criticisms of 'the Model and Data Reduction'   213

10  Experimental Observations for Verifying Solvent Models   213

## Contents

11 A Critique of Two-state Models 216

12 Three- and Four-state Models 218

13 A Critique of Multi-state Models 219

14 Phenomenological Models and Correlations 220

15 Are Two-state Models Inadequate? 223

16 Perspectives 224

*Part II: Theories of Adsorption Isotherms*

17 Interactions 226

18 Electrode–Adsorbate Interactions 231

19 Adsorbate–Adsorbate Interactions – Size Effects 231

20 Particle–Particle Interactions – Coulombic 233

21 Phase Transitions 236
Orientational Transitions 237
Surface Crystallization 239

22 Medium Effects 240
Phenomena 240
Analysis of Mixtures 241

23 Multi-component Systems and the pseudo-Frumkin Isotherms 246

24 The Frumkin Isotherm, Congruence, and Electrosorption Valency 249
The Frumkin Isotherm 249
Congruence 252
Electrosorption Valency 254

25 Perspectives 255

Author Index 257

# Errata

In volume 6 of this series the equation numbered (129) on page 161 should have read:

$$\mathbf{H}_s = (2\pi/C)[P_{ir}^2 + \omega_s^{-2}\, \dot{P}_{ir}^2] \qquad (129)$$

and equation (196) on page 180 should have been:

$$\mathbf{H}_{s(\gamma)} = \mathbf{H}_s(\{\mathbf{R}_\gamma\}) \cong \mathbf{H}_s(\{\mathbf{R}_\gamma^0\}) + \sum_{n \in N, \xi} \omega_\xi^2 (\nabla_{\mathbf{R}_n} \bar{\Pi}_\xi) \cdot (\mathbf{R}_n - \mathbf{R}_n^0)(\Pi_\xi - \bar{\Pi}_\xi) \qquad (196)$$

# 1
# Organic Electrochemistry – Synthetic Aspects

BY J. GRIMSHAW

This Report covers material published during 1975. Papers dealing with physical organic chemistry, such as reaction mechanisms, which have a bearing on electrochemical synthesis are included. Studies of radical-ions by e.p.r. have been excluded, as have papers on electrochemically initiated polymerization, electrocoating, and related technical fields.

Abbreviations used throughout this chapter are as follows: AN, acetonitrile; DME, 1,2-dimethoxyethane; DMF, dimethylformamide; DMSO, dimethyl sulphoxide; HMPT, hexamethylphosphoric triamide; THF, tetrahydrofuran.

## 1 General

The coverage of electro-organic synthesis in the Techniques of Chemistry series was completed in 1975.[1] Anodic oxidation was surveyed in another book[2] and chapters on electrochemistry have appeared in textbooks on the chemistry of quinones[3] and of hydrazo-, azo-, and azoxy-groups.[4] Other books have covered electrode kinetics,[5] experimental electrochemistry,[6] and electrochemical data for organic, organometallic, and biochemical substances.[7] Reviews have appeared on general synthetic reactions,[8] the synthesis of cyclic compounds,[9] electroreduction,[10] oxidation,[11] the synthesis and reactions of organometallic compounds,[12] and industrial electrosynthesis,[13] including indirect electrochemical processes[14] and reactor design.[15] The use

---

[1] 'Techniques of Electro-organic Synthesis', ed. N. L. Weinberg, Wiley, New York, 1974.
[2] S. D. Ross, M. Finkelstein, and E. Rudd, 'Anodic Oxidation', Academic Press, New York, 1975.
[3] J. Q. Chambers in 'Chemistry of Quinonoid Compounds', ed. S. Patai, Wiley, Chichester, 1974, Part 2, p. 737.
[4] F. G. Thomas and K. G. Botoin, 'Chemistry of Hydrazo, Azo and Azoxy Groups', ed. S. Patai, Wiley, Chichester, 1975, Part I, p. 443.
[5] J. Albery, 'Electrode Kinetics', Oxford University Press, Oxford, 1975.
[6] D. J. Sawyer and J. L. Roberts, 'Experimental Electrochemistry for Chemists', Wiley, New York, 1974; E. Gileadi, E. Kirowa-Eisner, and J. Penciner, 'Interfacial Electrochemistry – An Experimental Approach', Addison–Wesley, Reading, Mass., 1975.
[7] L. Meites, P. Zuman, W. J. Scott, B. H. Campbell, and A. M. Kardos, 'Electrochemical Data. Vol. 1. Organic, Organometallic and Biochemical Substances. Pt A', Wiley, New York, 1974.
[8] T. Shono, *Denki Kagaku Oyobi Kogyo Butsuri Kagaku*, 1974, **42**, 542.
[9] M. Lacan and I. Tabakovic, *Kem. Ind.*, 1975, **24**, 227.
[10] M. Tarle, *Kem. Ind.*, 1974, **23**, 647; M. R. Rifi, *Tech. Chem. (N.Y.)*, 1975, **5**, 83.
[11] O. A. Petrii, *Usp. Khim.*, 1975, **44**, 2067; L. A. Mirkind, *ibid.*, p. 2088.
[12] G. A. Tedoradze, *J. Organomet. Chem.*, 1975, **88**, 1; W. J. Settineri and L. D. McKeever, *Tech. Chem. (N.Y.)*, 1975, **5**, 397.
[13] M. Fleischmann and D. Pletcher, *Chem. Br.*, 1975, **11**, 50; K. C. Narasimham and M. S. V. Pathy, *Chem. Eng. World*, 1975, **10**, 87.
[14] R. Clarke, A. Kuhn, and E. Okoh, *Chem. Br.*, 1975, **11**, 59.
[15] M. Fleischmann and R. E. W. Jansson, *Chem. Eng. (London)*, 1975, **302**, 603.

of ion-exchange membranes in electrochemical cells has been reviewed.[16] Electrochemistry in thin layers of solution is discussed in a critical review[17] and the application of electrochemistry to physical organic problems is discussed.[18] IUPAC have published recommendations for sign conventions and the plotting of electrochemical data.[19]

A process for the purification of HMPT by fractional freezing, vacuum distillation, and drying over calcium oxide has been described,[20] and the electrochemical properties of several other solvents have been evaluated. Oxydipropionitrile shows a large potential range for reductions at a mercury cathode but no reactions were studied in this solvent,[21] and a possible limitation is that electrogenerated bases will cause elimination to give acrylonitrile. Ethylene carbonate is liquid at 40°C and shows a good range for oxidation and reduction:[22] nitromethane[23] and 1,2-dichloroethane[24] are both satisfactory solvents for oxidation processes. Triethyl-n-hexylammonium triethyl-n-hexylboride is a new ambient-temperature molten-salt solvent with a useful working range for reduction, but the solvent readily undergoes oxidation.[25] A mixture of aluminium chloride (2 moles) and ethylpyridinium bromide (1 mole) is molten at ambient temperatures and forms a strong Lewis acid solvent that is useful for oxidation processes.[26] Quinones solubilized in micelles formed in aqueous sodium dodecylsulphate show well-defined diffusion waves on polarography.[27]

Advances in electronic apparatus for electrochemistry have been reviewed[28] and new designs proposed for function generators[29] and integrators.[30] New designs[31,32] for laboratory electrolysis cells are available. One of these is formed in a rolled sandwich construction of the two electrodes and a separator cloth.[32] The reaction solution is pumped through a tube packed with the sandwich, so that the substrate is in contact with both the cathode and anode. If the latter situation can be tolerated, then this cell design gives high flow rates and current densities. Porous Teflon has been proposed as a diaphragm material.[33]

Tungsten bronzes have been studied as electrode materials for use in both aqueous and aprotic solvents.[34] They have a large reduction and oxidation range.

[16] G. Richter, *Chem.-Ing.-Tech.*, 1975, **47**, 909.
[17] A. T. Hubbard, *Crit. Rev. Anal. Chem.*, 1973, **3**, 201.
[18] R. Breslow, *Pure Appl. Chem.*, 1974, **40**, 493.
[19] *Information Bull. IUPAC*, Appendix: *Provisional Nomenclature: Symbols, Units, Standardisation*, 1975, p. 42.
[20] M. G. Formicheva, Yu. M. Kessler, E. E. Zabusova, and N. M. Alpatova, *Elektrokhimiya*, 1975, **11**, 163.
[21] J. Y. Gal and M. Persin, *C.R. Hebd. Seances Acad. Sci., Ser. C*, 1975, **280**, 1305.
[22] J. Y. Cabon, M. L'Her, and M. LeDemezet, *Bull. Soc. Chim. Fr.*, 1975, 1020.
[23] M. Breant and G. Demange-Guerin, *Bull. Soc. Chim. Fr.*, 1975, 163.
[24] V. G. Mairanovskii, N. T. Ioffe, and A. A. Engovatov, *Elektrokhimiya*, 1975, **11**, 1303.
[25] W. T. Ford, *Anal. Chem.*, 1975, **47**, 1125.
[26] H. L. Chum, V. R. Koch, L. L. Miller, and R. A. Osteryoung, *J. Am. Chem. Soc.*, 1975, **97**, 3264.
[27] T. Erabi, H. Hiura, and M. Tanaka, *Bull. Chem. Soc. Jpn*, 1975, **48**, 1354.
[28] P. Gilgen, K. Kaempf, and R. Rach, *Chimia*, 1975, **29**, 232.
[29] F. Magno, G. Bontempelli, G. A. Mazzocchin, and I. Patane, *Chem. Instrum.*, 1975, **6**, 239; D. F. Unterker, W. G. Sherwood, G. A. Martinchek, T. M. Reidhammer, and S. Bruckenstein, *ibid.*, p. 259.
[30] M. Lindstrom and G. Sundholm, *Finn. Chem. Lett.*, 1975, 27.
[31] P. Jeroschewski, Ger. (East) P. 111 158 (*Chem. Abs.*, 1975, **83**, 185 519).
[32] P. M. Robertson, F. Schwager, and N. Ibl, *J. Electroanal. Chem. Interfacial Electrochem.*, 1975, **65**, 883.
[33] D. S. Riley and C. R. Vallance, Ger. P. 2 433 941 (*Chem. Abs.*, 1975, **83**, 68 083).
[34] M. Amjad and D. Pletcher, *J. Electroanal. Chem. Interfacial Electrochem.*, 1975, **59**, 61.

Attempts have been made to improve the qualities of graphite as cathode material by coating it with mercury[35] and by attaching, with chemical bonds, a surface layer of (S)-(−)-phenylalanine methyl ester, bonded through the amine nitrogen.[36] The latter forms a chiral electrode surface which promotes the reduction of ketones to carbinols with partial asymmetric induction. Acetophenone afforded 1-phenylethanol for which $[\alpha]_D$ ($c=3$, $CHCl_3$) was $-7\cdot2°$. However, other workers[37] were unable to repeat this claim of asymmetric reduction. The properties of platinized silica particles as a fluidized-bed electrode for the Kolbe reaction have been examined.[38]

Experimental and theoretical studies have been made on the effect of adsorption of neutral molecules on electrochemical reactions.[39] Cryptate complexes of alkali-metal ions are reduced at very negative potentials, but the potassium ion complex of kryptofix-[2,2,2] is strongly adsorbed at a mercury cathode from dilute solutions, which limits the use of this ion in conducting salts.[40]

General studies on the properties of redox reactions have included a study of the reversible oxidation and reduction of four polynuclear hydrocarbons at pressures up to 2000 atm.[41] The change in partial molar volume which accompanies the redox reaction can then be determined, and this gives information on the solvation changes which accompany electron transfer. When redox potential is determined by cyclic voltammetry, it is usually assumed that the ratio of diffusion coefficients for the redox species is sufficiently close to unity that its logarithm can be taken as zero. In a critical study of some aromatic radical-cations, the diffusion coefficient for the parent molecule was always found to be greater than for the cation, but the ratio could be taken as unity with sufficient accuracy.[42] A linear relationship has been shown between values of the electron affinity and the polarographic half-wave potentials for some cyclic anhydrides.[43] Polarography has been used to detect short-lived radicals and radical-cations that are generated by pulse radiolysis from anthracene, naphthalene, benzene, and acetone.[44]

Further papers have appeared on the use of convolution potential sweep voltammetry in the determination of electrochemical reaction mechanisms, including the acetophenone pinacolization, the intramolecular cyclization of 1,3-dibenzoylpropane, and the coupling of 4-methylbenzylidenemalononitrile.[45] The technique can also be used to determine standard electrode potentials where one part of the couple is unstable.[46]

35 H. G. Tennant, U.S. P. 3 914 500 (*Chem. Abs.*, 1976, **84**, 23 763).
36 B. F. Watkins, J. R. Belling, E. Kariv, and L. L. Miller, *J. Am. Chem. Soc.*, 1975, **97**, 3549.
37 L. Horner and W. Brich, *Justus Liebigs Ann. Chem.*, 1977, 1354.
38 S. Yoshizawa, Z. Takehara. Z. Ogumi, and M. Matsubara, *Denki Kagaku Oyobi Kogyo Butsuri Kagaku*, 1975, **43**, 526 (*Chem. Abs.*, 1976, **84**, 81 627).
39 J. Lipkowski and Z. Galus, *J. Electroanal. Chem. Interfacial Electrochem.*, 1975, **61**, 11; H. Fischer, *ibid.*, 1975, **62**, 163.
40 D. Britz and D. Knittel, *Electrochim. Acta*, 1975, **20**, 891; F. Peter and M. Gross, *J. Electroanal. Chem. Interfacial Electrochem.*, 1975, **61**, 245.
41 M. Fleischmann, W. B. Gara, and G. J. Hills, *J. Electroanal. Chem. Interfacial Electrochem.*, 1975, **60**, 313.
42 U. Svanholm and V. D. Parker, *J. Chem. Soc., Perkin Trans. 2*, 1975, 755.
43 O. B. Nagy, H. Lion, and J. B. Nagy, *Bull. Soc. Chim. Belg.*, 1975, **84**, 1053.
44 K. D. Asmus and M. Graetzel, *Nucl. Sci. Abs.*, 1975, **31**, 110.
45 J. M. Savéant and D. Tessier, *J. Electroanal. Chem. Interfacial Electrochem.*, 1975, **61**, 251; C. P. Andrieux, J. M. Savéant, and D. Tessier, *ibid.*, 1975, **63**, 429; L. Nadjo, J. M. Savéant, and D. Tessier, *ibid.*, 1975, **64**, 143.
46 J. M. Savéant and D. Tessier, *J. Electroanal. Chem. Interfacial Electrochem.*, 1975, **65**, 57.

Ultraviolet spectroscopy[47] has been used to study the intermediates in electrochemical reactions, and there is a developing interest in the application of resonance Raman spectroscopy to the detection of intermediates.[48,49] Thin carbon films deposited on germanium prisms form optically transparent electrodes suitable for i.r. spectroelectrochemistry.[50]

The contrasting colours of radical-ions and their neutral substrates have been made the basis of electrochromic display systems.[51] Electrochemiluminescence continues to be examined.[52]

## 2 Reduction

**General.**—Reduction of acetophenone in a chiral solvent, $(S)$-$(+)$-Me$_2$NCH$_2$-CH(OMe)CH(OMe)CH$_2$NMe, gives the same ratio of *meso* to ($\pm$)-pinacols and the same degree of asymmetric induction in the ($\pm$)-pinacol as is obtained by photo-reduction of acetophenone in the same solvent. This strongly suggests that dimerization occurs by the same step in the two reactions; *i.e.*, by combination of two radicals PhĊH(OH)CH$_3$.[53] Cobalt(III) trisacetonylacetonate is destroyed on cathodic reduction, and the reaction in the presence of trimethyl-($-$)-menthylammonium perchlorate as supporting electrolyte was found to exhibit enantioselectivity.[54] The magnitude of this enantioselectivity varies systematically with potential and with electrolyte concentration.

A series of papers on mechanistic electrochemistry in liquid ammonia has appeared.[55—57] Liquid ammonia has a low dielectric constant, very low acidity, and is a suitable medium for reduction. In the absence of added protonating agents, nitrobenzene and nitrosobenzene are reduced by two reversible one-electron steps to the radical-anion and the dianion. In the presence of isopropyl alcohol as a weak acid, the dianion of nitrobenzene adds one proton and rapidly decomposes to nitrosobenzene. The dianion of nitrosobenzene adds one proton to give an anionic species which can be reversibly oxidized to the parent nitrosobenzene. In the presence of strong acids such as ammonium ions, both compounds are reduced to phenylhydroxylamine.[55] Quinoline is reduced in two one-electron steps, and the radical-anion dimerizes to a dianion, which can be re-oxidized to the parent quinoline.[56] Diethyl fumarate, cinnamonitrile, and acrylonitrile show similar electrochemical behaviour in liquid ammonia to that in aprotic solvents. Dimerization

---

[47] W. Paatsch, *Metalloberflaeche – Angew. Elektrochem.*, 1974, **28**, 485; A. Bewick, G. J. Edwards, and J. M. Mellor, *Tetrahedron Lett.*, 1975, 4685; W. J. Plieth and K. Naegele, *Electrochim. Acta*, 1975, **20**, 421.
[48] M. R. Suchanski and R. P. Van Duyne, *J. Am. Chem. Soc.*, 1975, **97**, 1699; D. L. Jeanmaire and R. P. Van Duyne, *J. Electroanal. Chem. Interfacial Electrochem.*, 1975, **66**, 235.
[49] A. J. McQuillan, P. J. Hendra, and M. Fleischmann, *J. Electroanal. Chem. Interfacial Electrochem.*, 1975, **65**, 933.
[50] J. S. Mattson and C. A. Smith, *Anal. Chem.*, 1975, **47**, 1122.
[51] T. Kawata, M. Yamamoto, M. Yamana, M. Tajima, and T. Nakano, *Jpn. J. Appl. Phys.*, 1975, **14**, 725; I. F. Chang, B. L. Gilbert, and T. I. Sun, *J. Electrochem. Soc.*, 1975, **122**, 955; N. V. Philips, Fr. P. 2 228 262 (*Chem. Abs.*, 1975, **83**, 106 250).
[52] K. G. Boto and A. J. Bard, *J. Electroanal. Chem. Interfacial Electrochem.*, 1975, **65**, 945; N. Periasamy and K. S. V. Santhanam, *Proc. Indian Acad. Sci., Sect. A*, 1974, **80**, 194.
[53] D. Seabach and H. A. Oei, *Angew. Chem.*, 1975, **87**, 629. (*Angew. Chem. Int. Ed. Engl.*, 1975, **14**, 634).
[54] S. Mazur and K. Ohkubo, *J. Am. Chem. Soc.*, 1975, **97**, 2911.
[55] W. H. Smith and A. J. Bard, *J. Am. Chem. Soc.*, 1975, **97**, 5203.
[56] W. H. Smith and A. J. Bard, *J. Am. Chem. Soc.*, 1975, **97**, 6491.

# Organic Electrochemistry – Synthetic Aspects

occurs by combination of radical-anions, and the rate is increased by the presence of potassium ions.[57]

**Hydrocarbons.**—A patent has been issued for the reduction of aromatic steroids in a mixture of liquid ammonia and THF at a steel cathode (Scheme 1).[58] Trioxan has

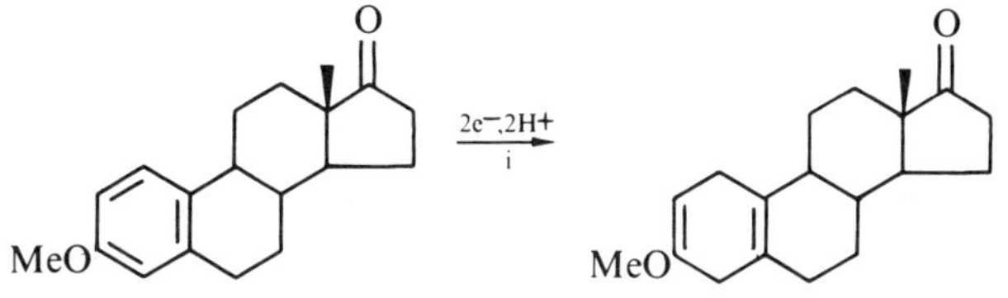

Reagents: i, liquid NH$_3$, THF

**Scheme 1**

been suggested as a very useful solvent for the related reduction of benzene to cyclohexadiene at a mercury cathode.[59] The reduction of naphthalene in AN to 1,4-dihydronaphthalene has been patented.[60] 3-Hydroxyphenalenone (1) behaves in a manner like that of naphthalene on reduction in an aqueous buffer at a mercury cathode (Scheme 2), to give a dihydro-derivative.[61] Related to the reduction of

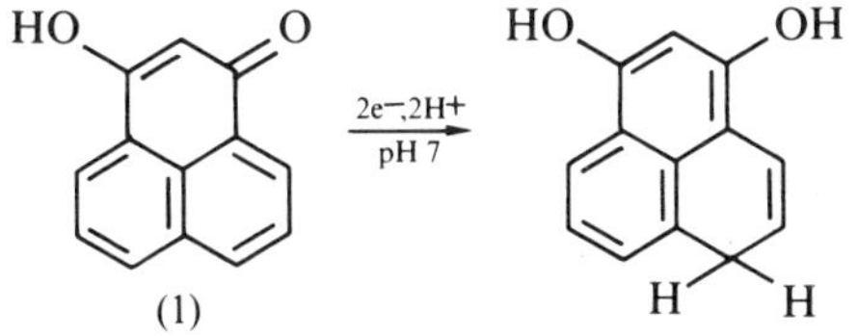

**Scheme 2**

benzenoid compounds is the electrosynthesis of 2,5-dihydrothiophen-2-carboxylic acid by the reduction, over a mercury cathode, of the lithium salt of thiophen-2-carboxylic acid.[62]

A full paper [63] has appeared describing the advantages of drying solvents over alumina actually in the electrolysis vessel, so as to stabilize the dianions from aromatic hydrocarbons. Under these conditions anthracene, benzanthracene, chrysene, coronene, and perylene show reversible behaviour on cyclic voltammetry due to the formation of radical-anions and dianions. Cyclo-octatetraene shows two reversible one-electron reduction steps under these conditions, and the rate of charge transfer for addition of the first electron depends on the supporting tetra-alkylammonium cation, being $10^3$ times faster for Me$_4$N$^+$ than for Bu$_4$N$^+$.[64] The

---

[57] I. Vartires, W. H. Smith, and A. J. Bard, *J. Electrochem. Soc.*, 1975, **122**, 894.
[58] K. Junghans, Ger. P. 2 337 155 (*Chem. Abs.*, 1975, **83**, 10 598).
[59] T. Hatayama, Y. Hamano, and K. Udo, Jpn. P. 74 41 192 (*Chem. Abs.*, 1975, **82**, 147 142).
[60] A. Misono, T. Cho, and H. Yamagishi, Jpn. P. 75 02 508 (*Chem. Abs.*, 1975, **82**, 161 873).
[61] J. C. Dufresne, *C.R. Hebd. Seances Acad. Sci., Ser. C.*, 1975, **280**, 1493.
[62] V. S. Mikhailov, V. P. Gul'tyai, S. G. Mairanovskii, S. Z. Taits, I. V. Proskurovskaya, and Yu. G. Dubovik, *Izv. Akad. Nauk SSSR, Ser. Khim.*, 1975, 888 (*Chem. Abs.*, 1975, **83**, 87 240).
[63] B. S. Jensen and V. D. Parker, *J. Am. Chem. Soc.*, 1975, **97**, 5211.
[64] S. Jensen, A. Ronlan, and V. D. Parker, *Acta Chem. Scand., Ser. B*, 1975, **29**, 394; A. J. Fry, C. S. Hutchins, and L. L. Chung, *J. Am. Chem. Soc.*, 1975, **97**, 591.

activation barrier for addition of the first electron, due to a conformational effect, is not as important as previously considered; electrolyte double-layer effects are greater than any conformational effects.

Polarography of 1-phenylhex-1-yne in DMF shows a single four-electron wave during which hexylbenzene is formed. However, under the conditions of preparative electrolysis, the isomerization of acetylene to allene becomes important, and the dominant process is reduction of the allene (Scheme 3).[65]

$$PhC{\equiv}C-CH_2C_4H_9 \longrightarrow PhCH{=}C{=}CHC_4H_9 \xrightarrow[2H^+]{2e^-} PhCH{=}CHCH_2C_4H_9$$
$$\downarrow \begin{array}{c} 2e^- \\ 2H^+ \end{array}$$
$$PhCH_2CH_2CH_2C_4H_9$$

**Scheme 3**

**Activated Olefins.**—A number of patents [66–68] have appeared on the conversion of acrylonitrile into adiponitrile. A mechanistic study of the hydrodimerization of ethyl cinnamate and diethyl fumarate in DMF at room temperature and lower shows that the reaction proceeds in both cases *via* a radical-anion dimerization step.[69] Alkali-metal ions ($Li^+$, $Na^+$, $K^+$) greatly increase the rate of dimerization of dialkyl fumarates, ethyl cinnamate, and cinnamonitrile in DMF due to ion pairing with the radical-anions and then rapid dimerization of the ion pairs.[70] The unsaturated nitriles (2; $R = H$) and (2; $R = Me$) undergo irreversible one-electron reduction, with dimerization and then cyclization, in DMF, with or without an added proton source (Scheme 4); reaction between two radical-anions is proposed as the dimerization step.[71, 72] The related nitriles (2; $R = Bu^t$) and (2; $R = Ph$) show reversible radical-anion formation in DMF and further reduction to the dianion (Scheme 5). On addition of a proton donor, two-electron reduction to the dihydro-compound occurs at the potential of the first wave, and an *ECE* mechanism has been pro-

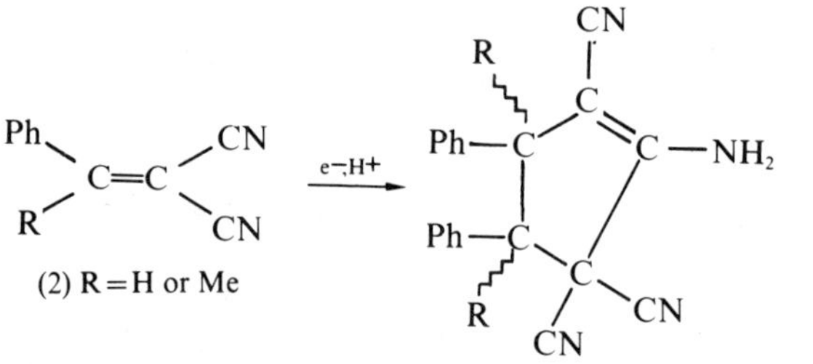

(2) $R = H$ or Me

cis- and trans-

**Scheme 4**

[65] M. W. Moore and D. G. Peters, *J. Am. Chem. Soc.*, 1975, **97**, 139.
[66] M. Seko, A. Tokuyama, T. Isotani, K. Inada, and N. Matsumoto, Jpn. P. 75 02 492 (*Chem. Abs.*, 1975, **82**, 161 872).
[67] J. F. Connolly, U. S. P. 3 871 976 (*Chem. Abs.*, 1975, **83**, 68 077).
[68] M. Seko, Y. Takahashi, S. Ogawa, H. Iwashita, A. Yamaguchi, and H. Ide, Jpn. P. 75 05 171 (*Chem. Abs.*, 1975, **83**, 50 050); M. Seko, S. Ogawa, M. Yoshida, N. Oishi, S. Hazama, and K. Okubo, Jpn. P. 75 05 172 (*Chem. Abs.*, 1975, **83**, 50 051).
[69] R. D. Grypa and J. T. Maloy, *J. Electrochem. Soc.*, 1975, **122**, 377, 509.
[70] M. J. Hazelrigg and A. J. Bard, *J. Electrochem. Soc.*, 1975, **122**, 211.
[71] L. A. Avaca and J. H. P. Utley, *J. Chem. Soc., Perkin Trans. 2*, 1975, 161.
[72] L. A. Avaca and J. H. P. Utley, *J. Chem. Soc., Perkin Trans. 1*, 1975, 971.

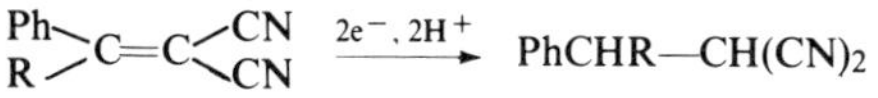

posed.[71] Electroreduction of $\alpha\beta$-unsaturated nitriles in acidic aqueous solution leads to the production of amines.[73] Co-electrodimerization of carbonyl compounds with acrylonitrile in aqueous buffers leads to $\gamma$-hydroxy-nitriles,[74, 75] while a similar reaction with acrylic acid leads to $\gamma$-lactones (Scheme 6).[76]

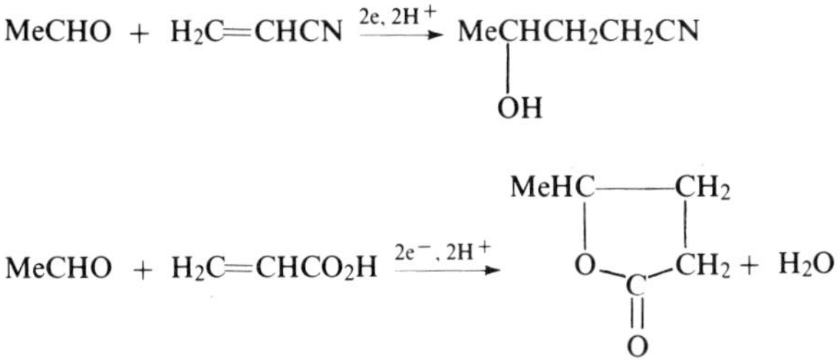

**Scheme 5**

**Scheme 6**

**Carbonyl Compounds.**—The mixed electrolytic reduction of 1,4-dimethylpyridinium methylsulphate and acetone leads to mixed coupling products (Scheme 7)

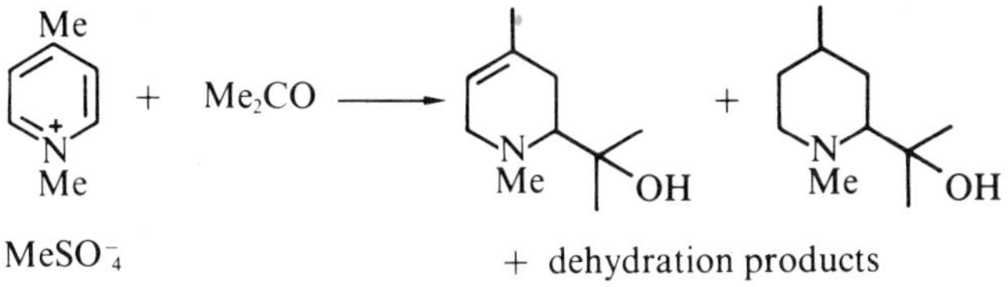

**Scheme 7**

along with products from reduction of the pyridine compound.[77] Patents[78, 79] have been issued for the electrolytic preparation of pinacols from simple aliphatic ketones, and in one process the corresponding secondary alcohol is used as the solvent.[79]

Two strikingly similar stereo- and enantio-selective hydrodimerization reactions have been described. Reduction of benzoin gives the racemic pinacol formed by *threo*-coupling between two molecules of the same enantiomeric configuration (Scheme 8).[80] Hydrodimerization of the racemic tricyclic enone (3) also gives the

---

[73] Yu. D. Smirnov, A. P. Tomilov, and S. K. Smirnov, *Zh. Org. Khim.*, 1975, **11**, 522.
[74] A. P. Tomilov, B. L. Klyuer, and V. D. Nechepurnoi, *Zh. Org. Khim.*, 1975, **11**, 1344.
[75] S. M. Makarochkina and A. P. Tomilov, *Zh. Obshch. Khim.*, 1974, **44**, 2566.
[76] A. P. Tomilov, B. L. Klyuer, and V. D. Nechepurnoi, *Zh. Org. Khim.*, 1975, **11**, 1984.
[77] M. Ferles, M. Lebl, P. Štern, and P. Trška, *Coll. Czech. Chem. Commun.*, 1975, **40**, 2183.
[78] H. Nohe, F. Beck, W. Dietmar, and E. J. Schier, Ger. P. 2 345 461 (*Chem. Abs.*, 1975, **83**, 9181).
[79] H. Hobe and F. Beck, Ger. P. 2 343 054 (*Chem. Abs.*, 1976, **84**, 10 498).
[80] R. E. Juday and W. L. Waters, *Tetrahedron Lett.*, 1975, 4321.

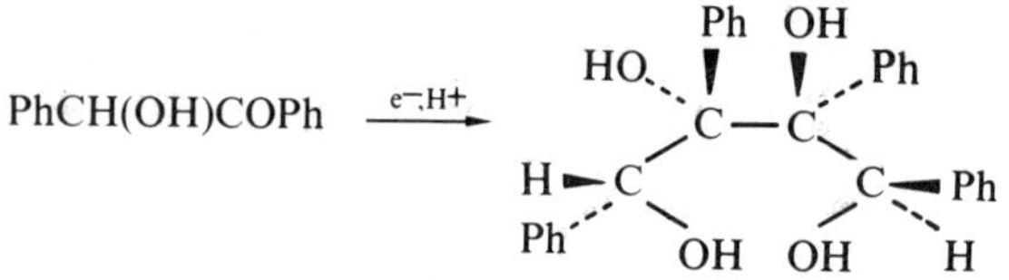

PhCH(OH)COPh  $\xrightarrow{e^-,H^+}$

**Scheme 8**

pinacol by *threo*-coupling, between two molecules of identical enantiomeric configuration (Scheme 9).[81] No other stereoisomers of the pinacol are formed in each

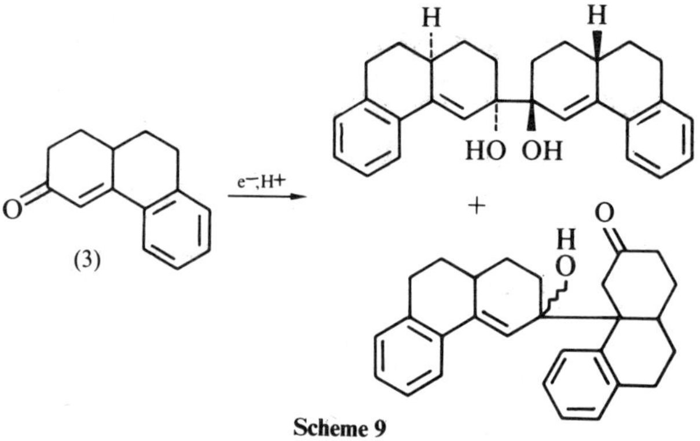

**Scheme 9**

case, although the enone also gives a mixture of ketols. Pinacol formation has been recorded during the reduction of thiophen-2,5-dicarboxaldehyde,[82] 2-benzoylthiophen,[83] 2-formylselenophen,[84] 2-acetylselenophen,[85] and acetylferrocene.[86]

Studies on the rate of the hydrodimerization of benzaldehyde in sulpholan, using a rotating ring-disc electrode, have been interpreted as showing that there is dimerization of the radical-anion.[87] The radical-anion of 4-nitrobenzaldehyde reacts too slowly for a rate constant to be determined using this technique. 4-Cyanobenzaldehyde undergoes dimerization by the same mechanism at high current densities but by an *ECE* mechanism at low current densities, where the chemical step is reaction between the radical-anion and a neutral molecule.

Reduction of amino-desoxybenzoins is dependent on the pH of the solution.[88] If the pH is sufficiently acid that the amino-function is protonated, then cleavage of the carbon–nitrogen bond occurs, as shown in Scheme 10. In more alkaline solutions this reaction is suppressed, and reduction of the carbonyl group to secondary alcohol occurs, giving a mixture of stereoisomers. Griseofulvin is reduced to dihydrogriseofulvin in aqueous buffer solutions.[89]

Examples have been given of the reduction of carboxylic acid to primary alcohol

---

[81] E. Touboul and G. Dana, *Tetrahedron*, 1975, **31**, 1925.
[82] J. P. Salaun, M. Salaun-Bouix, and C. Caullet, *C.R. Hebd. Seances Acad. Sci., Ser. C.*, 1975, **280**, 165.
[83] P. Foulatier, J. P. Salaun, and C. Caullet, *C.R. Hebd. Seances Acad. Sci., Ser. C*, 1974, **279**, 779.
[84] D. Guerout and C. Caullet, *C.R. Hebd. Seances Acad., Sci., Ser. C*, 1975, **281**, 643.
[85] D. Guerout and C. Caullet, *C.R. Hebd. Seances Acad. Sci., Ser. C*, 1975, **281**, 667.
[86] M. Lacan and Z. Ibrisagic, *Croat, Chem. Acta*, 1974, **46**, 107.
[87] N. R. Armstrong, N. E. Vanderborgh, and R. K. Quinn, *J. Electrochem. Soc.*, 1975, **122**, 615.
[88] J. Armand and L. Boulares, *Bull. Soc. Chim. Fr.*, 1975, 711.
[89] H. J. Baer and E. Beer, *Wiss. Z. Tech. Univ. Dresden*, 1975, **24**, 5 (*Chem. Abs.*, 1976, **84**, 10 440).

# Organic Electrochemistry – Synthetic Aspects

$$PhCOCHPh \atop | \atop X \quad \xrightarrow[pH < pK_a]{2e^-, 2H^+} \quad PhCOCH_2Ph + XH$$

X = NH$_2$, NHC$_6$H$_{11}$, NHPh, or NHNHCONH$_2$

**Scheme 10**

in acidic aqueous buffers,[90] reduction of pyridine-2-carboxylic acid and -2,6-dicarboxylic acid, and the reduction of an amide function (see Scheme 11).[91] The reduction of oxalic acid in aqueous solution to glyoxalic acid is the subject of a patent.[92]

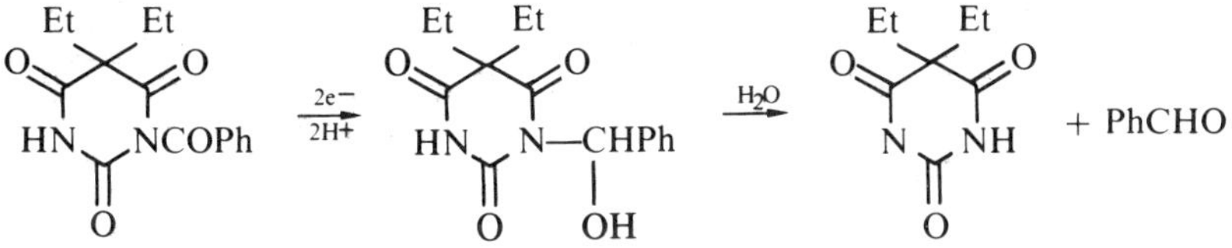

**Scheme 11**

**Nitro- and Nitroso-compounds.**—Reduction of the two nitro-groups in 2,4-dinitrophenol[93] and 2,4-dinitrotoluene[94] to amine has been examined. 3-Nitro-4-hydroxycoumarin is also smoothly reduced to the corresponding amino-compound.[95] Reduction of α-nitrocinnamic acid methyl ester in acid solution gives (±)-phenylalanine.[96] Reduction of 2-nitro-2'-isothiocyanatobiphenyl causes electrochemically initiated intramolecular cyclization (Scheme 12), and the product depends on the

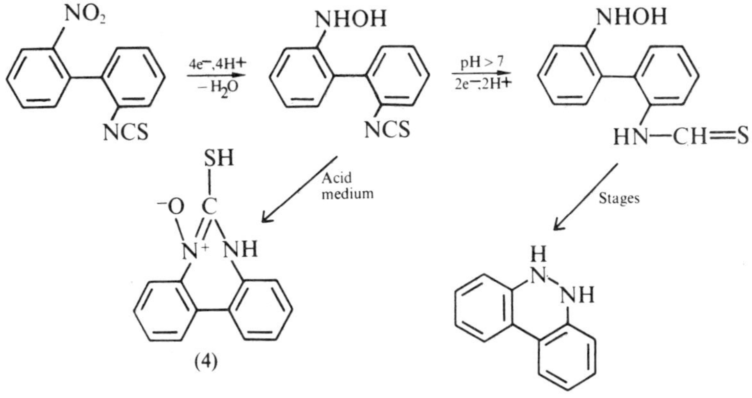

**Scheme 12**

[90] O. R. Brown, J. A. Harrison, and K. S. Sastry, *J. Electroanal. Chem. Interfacial Electrochem.*, 1975, **58**, 387.
[91] W. Pasek, J. Volke, and O. Manousek, *Coll. Czech. Chem. Commun.*, 1975, **40**, 819.
[92] F. Goodridge and K. Lister, Br. P. 1 411 371 (*Chem. Abs.*, 1976, **84**, 66 955).
[93] V. D. Bezuglyi, L. A. Kotok, E. K. Ostis, V. A. Ekel, and R. F. Ramakaeva, *Khim. Prom-st.* (*Moscow*), 1975, 17 (*Chem. Abs.*, 1975, **82**, 161 742).
[94] F. Goodridge and K. C. Nath, *Electrochim. Acta*, 1975, **20**, 685.
[95] M. Trkovnik, M. Lacan, Z. Stunic, and D. Nahmijuz, *Org. Prep. Proced. Int.*, 1975, **7**, 47.
[96] I. A. Avrutskaya, K. K. Babievskii, V. M. Belikov, E. V. Zaporozheto, V. T. Novikov, and M. Ya. Fioshin, *Elektrokhimiya*, 1975, **11**, 661.

pH of the solution.[97] In acidic solution, condensation between the generated hydroxylamino-function and the isothiocyanato-group to give (4) is rapid, but in neutral or alkaline solution this condensation is suppressed, and the isothiocyanato-group undergoes reduction. Further reactions then lead to the dihydrobenzo-[c]cinnoline (5).

A detailed mechanistic study of the reduction of nitrosobenzene in DMF is available.[98] The anion from 1,1-dinitroethane in aqueous alkaline solution undergoes reversible one-electron reduction. At more negative potentials an irreversible reduction process occurs.[99]

**Other Nitrogen-containing Compounds.**—X-Ray crystallography has been used to define the structure of the dihydroquinaldine dimer (6) obtained from cathodic reduction of quinaldine (Scheme 13).[100] The old process for reduction of indoles to their dihydro-derivatives at a lead cathode in 20% sulphuric acid has been revived in a recent patent.[101]

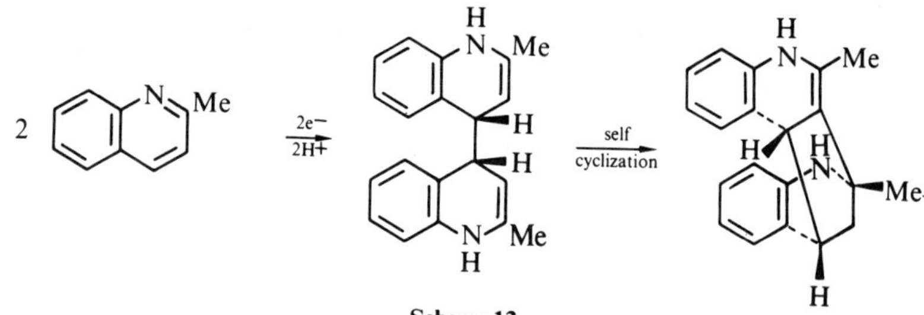

Scheme 13

(6)

Reduction of the C=N function in aqueous medium to its dihydro-derivative has been observed for a number of heterocyclic systems. Thus reduction of 7-methylguanosine (7) in acid medium leads to reduction of the imidazole ring (Scheme 14).[102] Reduction of 1,4-benzodiazepines leads to their dihydro-derivatives.[103–105]

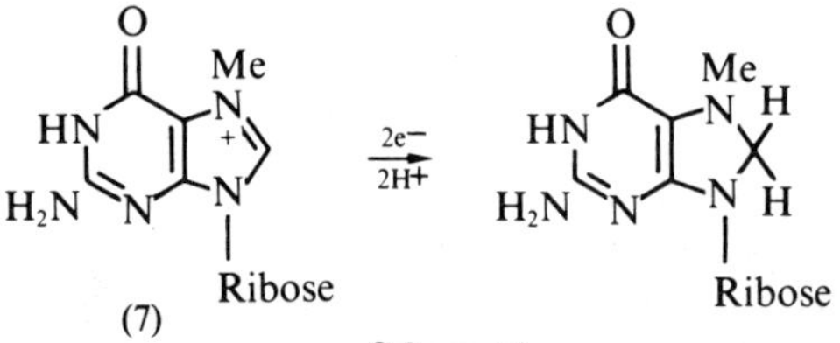

(7)

Scheme 14

[97] J. Hlewaty, J. Volke, and O. Manousek, *J. Electroanal. Chem. Interfacial Electrochem.*, 1975, **61**, 219.
[98] M. Lipsztajn, T. Krygowski, E. Laren, and Z. Galus, *J. Electroanal. Chem. Interfacial Electrochem.*, 1974, **57**, 339.
[99] I. P. Ryvkina, L. N. Nekrasov, V. A. Petrosyan, and V. I. Slovetakii, *Dokl. Akad. Nauk SSSR*, 1975, **220**, 1339.
[100] J. Bordner and I. W. Elliot, *Cryst. Struct. Commun.*, 1974, **3**, 689.
[101] H. Nohe and H. R. Müller, Ger. P. 2 403 446 (*Chem. Abs.*, 1975, **83**, 210 783).
[102] J. M. Sequaris and J. A. Reynaud, *J. Electroanal. Chem. Interfacial Electrochem.*, 1975, **63**, 207.
[103] S. A. Andronati, A. V. Bogatskii, V. P. Gul'tyai, T. A. Klyagul, S. P. Smul'skii, and Yu. I. Vikhlyaev, *Fiziol. Akt. Veshchestva*, 1975, **7**, 75 (*Chem. Abs.*, 1975, **83**, 123 150).
[104] H. Oelschläger and F. I. Senguen, *Arch. Pharm. (Weinheim)*, 1974, **307**, 909.
[105] H. Oelschläger and F. I. Senguen, *Chem. Ber.*, 1975, **108**, 3303.

Organic Electrochemistry – Synthetic Aspects

The cyclopropyl ring in prazepam (8) remains intact during this process, as shown in Scheme 15.[104] Reduction of lorazepam (9) in a buffer of pH 10.4 can be

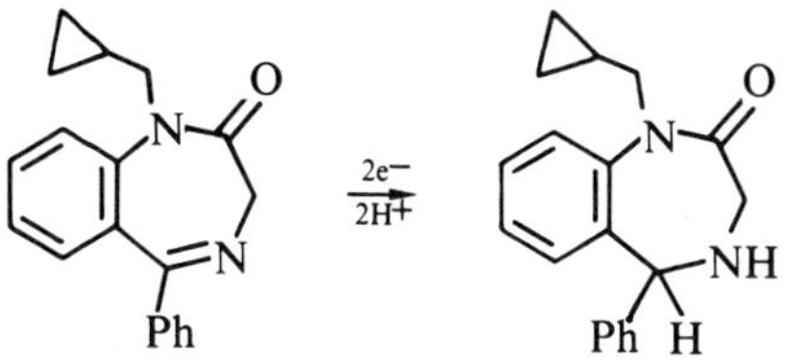

**Scheme 15**

terminated at the dihydro-stage, but the initial product loses water in a slow step (see Scheme 16), so that the product that is isolated corresponds to apparently simple replacement of a hydroxy-group by hydrogen.[105] This product will undergo further reduction of the C=N function.

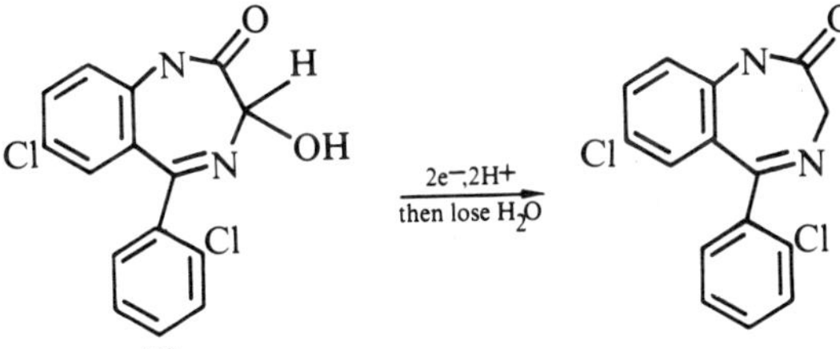

**Scheme 16**

Reduction of the C=N function is the electrochemical step in a potentially useful preparation of amines from amides.[106] The amide is first converted into its O-methyl ether (10), which is reduced in AN at a mercury cathode, an amine being the final product (Scheme 17).

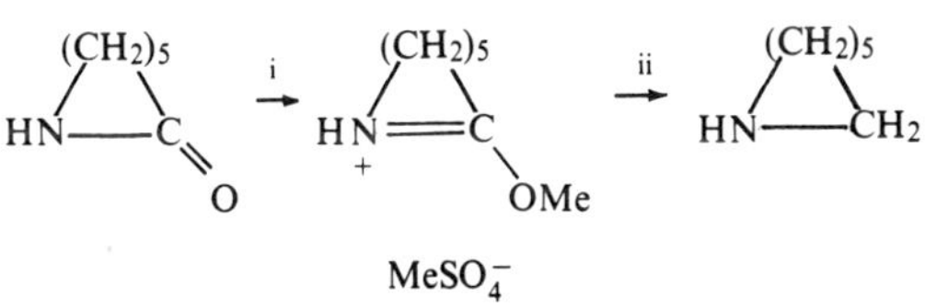

Reagents: i, $Me_2SO_4$; ii, AN solution, reduction at mercury cathode

**Scheme 17**

In contrast with the work just described, an extensive study of the reduction of N-benzylidene-4-toluidine in a solvent mixture of MeOH–MeOAc–$H_2O$ indicates that the products are the stereoisomeric hydro-dimers as well as the dihydro-compound.[107] Those benzodiazepines which give only the dihydro-compound can be

---

[106] H. Herbig, W. W. Wiersdorff, and D. Dagner, Ger. P. 2 408 532 (*Chem. Abs.*, 1976, **84**, 23 765).
[107] L. Horner and D. H. Skaletz, *Justus Liebigs Ann. Chem.* 1975, 1210.

regarded as Schiff's bases from benzophenone and an alkylamine; it would be useful to extend the study of Schiff's bases to detect any systematic variation of hydro-dimer, dihydro-product yields. The yield of hydro-dimer from $N$-benzylidene-4-toluidine depends upon the availability of protons in the double layer and is increased by using a hydrophobic supporting electrolyte. Typical results are 44% hydro-dimer using KOAc and 58% hydro-dimer using $Bu_4PBr$ as electrolyte and a mercury cathode, the yield decreasing at copper, lead, or glassy carbon. The ratio of $(\pm)$-:$meso$-hydro-dimer is in the range 0.9—1.1:1.

Benzyltrimethylammonium salts give bibenzyl as the principle product from reduction in HMPT at an aluminium cathode. The yield of bibenzyl falls when the water content of the solvent rises above 0.25 mol l$^{-1}$, and toluene becomes the principal product.[108] Use of a platinum cathode in HMPT leads to mixtures of bibenzyl and toluene; toluene is the only product from a platinum cathode in DMSO, DMF, or DME. Toluene is formed by reduction of the intermediate benzyl radical to the carbanion and then proton abstraction. Under these conditions the solution becomes sufficiently basic to promote the Sommelet–Hauser rearrangement of the substrate to form $NN'$-dimethyl-$o$-toluidine.

Reduction of azobenzene in HMPT at a mercury cathode with lithium chloride as electrolyte yields the dianion, which functions as a base and a nucleophile. In the presence of iodomethane, $NN'$-dimethylhydrazobenzene is formed.[109] The solvent must be rigorously dried, and this is best achieved by electrolysis using a platinum cathode until the blue colour due to solvated electrons is stable. The preparation of 4-methoxyphenylhydrazine by reduction of the benzenediazonium salt at a lead, tin, or amalgamated copper cathode in aqueous solution has been patented.[110] Benzonitrile is reduced to benzylamine at a platinum-coated graphite electrode, and this technique is proposed as being of general application for the synthesis of amines from nitriles.[111]

The redox interconversion of a series of $\alpha,\omega$-diaza-polyenes in AN (Scheme 18)

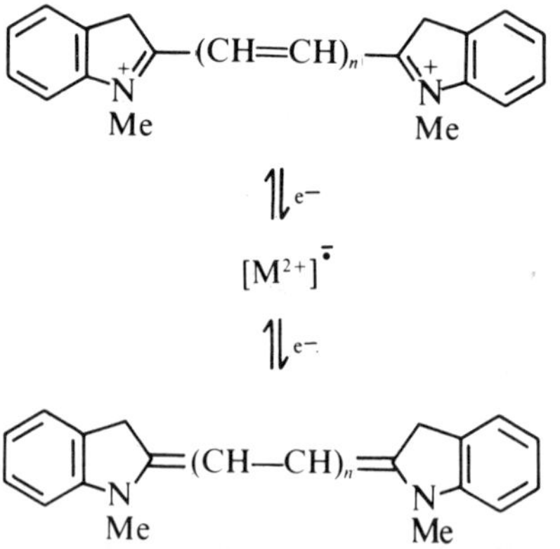

Scheme 18

[108] A. Desbene-Monvernay, Y. Robillard, P. C. Lacaze, and J. E. Dubois, *Bull. Soc. Chim. Fr.*, 1975, 548.
[109] T. Troll and M. M. Baizer, *Electrochim. Acta*, 1975, **20**, 33.
[110] T. A. Volodina, I. A. Avrutskaya, O. G. Ginsberg, E. F. Bystrikova, N. Ya. Fioshin, and N. N. Suvorov, U.S.S.R. P. 473 710 (*Chem. Abs.*, 1975, **83**, 78 852).
[111] V. Krishnan, K. Ragupathy, and H. V. K. Udupa, *J. Appl. Electrochem.*, 1975, **5**, 125.

shows two reversible one-electron steps when $n=1$. The formation constant ($K$) for the intermediate radical-ion, defined (at 25 °C) by $E_2-E_1=0.059 \log K$, falls from $7 \times 10^5$ when $n=1$ to $ca.$ 3 when $n=3$, showing a very marked dependence on the length of the polyene chain.[112] Benzazolylpyridinium salts also form highly stable delocalized radicals on reduction.[113] 2,5-Diaryl-1,3,4-oxadiazoles are reduced in DMF to radical-anions that have been characterized by e.s.r. spectroscopy. At more negative potentials a further one-electron irreversible reduction occurs.[114] Pyrazine mono-oxide reversibly forms the radical-anion in DMF, and the irreversible addition of a second electron is followed by protonation and loss of water to give pyrazine, which was detected as its radical-anion by e.s.r. spectroscopy.[115] In aqueous solution, pyrazine mono-oxide is reduced irreversibly to pyrazine, which can be reduced further at more negative potentials.

Pteridine (11) undergoes reversible reduction to the 5,8-dihydro-compound in aqueous buffer solutions. These two compounds condense at a measurable rate to give the dimer (12), which can be reduced in an irreversible step at more negative potentials, with carbon–carbon bond cleavage, to form 7,8-dihydropteridine (Scheme 19).[116]

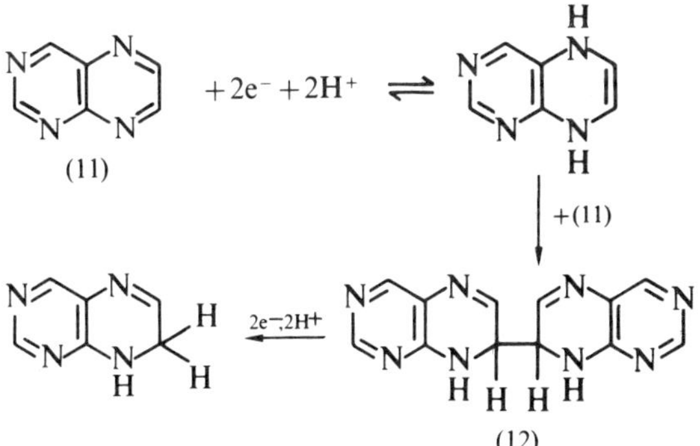

Scheme 19

**Oxygen (except Carbonyl), Sulphur, and Phosphorus Compounds.**—Alkyl benzyl ethers are cleaved to give the alkanol and toluene at a mercury cathode in DMF with tetrabutylammonium perchlorate as electrolyte (Scheme 20). The reaction is easier in the presence of a five-fold excess of biphenyl as the electron carrier. 4-Nitrobenzyl ethers are similarly cleaved to the alkanol, and this process occurs at

$$PhCH_2OBu \xrightarrow{e^-} PhCH_2\cdot + BuO^- \xrightarrow[2H^+]{e^-} PhCH_3 + BuOH$$

Scheme 20

---

[112] S. Hünig, F. Linhart, and D. Scheutzow, *Justus Liebigs Ann. Chem.*, 1975, 2102.
[113] V. Kadis, J. Stradins, E. Lavrinovics, and P. Zarins, *Khim. Geterotsikl. Soedin.*, 1975, 675.
[114] A. V. Il'yasov, Yu. M. Kargin, Ya. A. Levin, I. D. Morozova, A. A. Vafina, B. V. Mel'nikov, A. Sh. Mukhtarov, M. S. Skorobogatova, and E. I. Zoroatskaya, *Izv. Akad. Nauk, SSSR, Ser. Khim.*, 1975, 2194.
[115] J. Volke and S. Beran, *Coll. Czech. Chem. Commun.*, 1975, **40**, 2232.
[116] D. L. McAllister and G. Dryhurst, *J. Electroanal. Chem. Interfacial Electrochem.*, 1975, **59**, 75.

even lower potentials.[117] These processes are suggested as routes for the removal of benzyl ethers used as alcohol-protecting groups. The reaction involves addition of an electron to the benzene ring followed by cleavage of the benzyl carbon–oxygen bond. Pinacols with reducible aromatic hydrocarbon substituents undergo a related carbon–oxygen and carbon–carbon bond cleavage in DMF or in DMF plus a phenol, at a mercury cathode, around the potential for reduction of the hydrocarbon residue.[118] Thus fluorenone pinacol gives fluorene and fluoren-9-ol (Scheme 21). It is not possible to decide which bond, *i.e.* carbon–carbon or carbon–oxygen, is cleaved first.

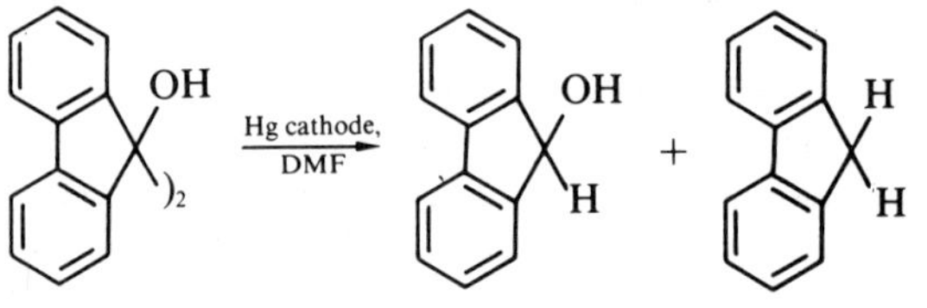

**Scheme 21**

Triethyloxonium and ethyltetramethyleneoxonium ions show irreversible one-electron waves on polarography in dichloromethane.[119] The reduction of some ten peroxides has been studied by polarography in methanol.[120]

The cathodic cleavage of diarylsulphones (13) has been extensively studied.[121–123] Polarography and cyclic voltammetry show that in aprotic media the

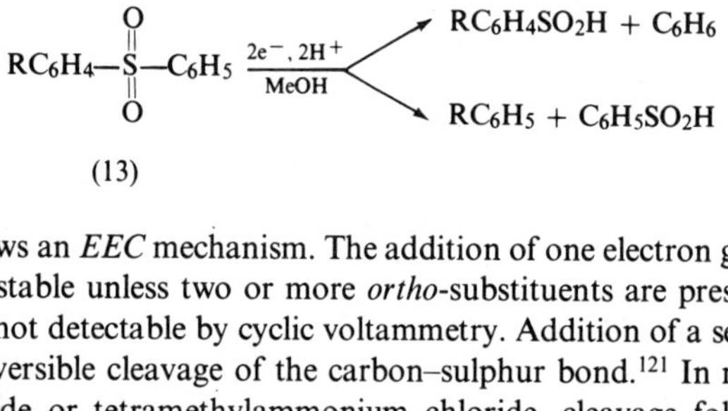

$$RC_6H_4-\overset{O}{\underset{O}{\overset{\|}{S}}}-C_6H_5 \xrightarrow[\text{MeOH}]{2e^-, 2H^+} \begin{array}{l} RC_6H_4SO_2H + C_6H_6 \\ \\ RC_6H_5 + C_6H_5SO_2H \end{array}$$

(13)

cleavage follows an *EEC* mechanism. The addition of one electron gives a radical-anion that is stable unless two or more *ortho*-substituents are present, when the radical ion is not detectable by cyclic voltammetry. Addition of a second electron results in irreversible cleavage of the carbon–sulphur bond.[121] In methanol with lithium chloride or tetramethylammonium chloride, cleavage follows an *ECE* mechanism, and only one polarographic wave is seen.[121] Phenyl groups with electronegative substituents are cleaved quantitatively from the sulphonyl group, and those with *ortho*-substituents are cleaved preferentially.[122] The yields of products that do not contain sulphur from the reductive cleavage of some sulphones in methanol are given in Table 1.[123]

---

[117] V. G. Mairanovskii and N. F. Loginova, *Zh. Obshch. Khim.*, 1975, **45**, 2112.
[118] M. A. Michel, G. Mousset, J. Simonet, and H. Lund, *Electrochim. Acta*, 1975, **20**, 143.
[119] P. H. Plesch and F. G. Thomas, *J. Chem. Soc., Perkin Trans. 2.*, 1975, 1532.
[120] N. V. Dzumedzei, A. A. Turovskii, and N. A. Turovskii, *Dopov. Akad. Nauk Ukr. RSR, Ser. B*, 1975, 810 (*Chem. Abs.*, 1976, **84**, 23 711).
[121] L. Horner and E. Meyer, *Ber. Bunsenges. Phys. Chem.*, 1975, **79**, 136.
[122] L. Horner and E. Meyer, *Justus Liebigs Ann. Chem.*, 1975, 2053.
[123] L. Horner and E. Meyer, *Ber. Bunsenges. Phys. Chem.*, 1975, **79**, 143.

**Table 1** Products from reductive cleavage of the sulphones (13) in methanol

| Sulphone (13) | % Yield | |
|---|---|---|
| R | RPh | $C_6H_6$ |
| 4-$NH_2$ | 11 | 83 |
| 2-$NH_2$ | 70 | 30 |
| 4-OMe | 75 | 25 |
| 2-OMe | 18 | 82 |
| 4-$MeO_2C$ | 99 | 0 |
| 4-CN | 98 | 2 |

Triphenylphosphine phenylimide (14) forms a radical-anion that is stable in aprotic solvents. A second polarographic wave at more negative potentials is due to addition of a second electron and cleavage of the phosphorus–nitrogen bond to produce triphenylphosphine and aniline.[124] Reduction of the tetraphenylphosphonium cation (15) in DMF or AN proceeds by addition of two electrons and

$Ph_3P=NPh$         $Ph_4P^+$

(14)              (15)

cleavage of a phosphorus–carbon bond to form triphenylphosphine and benzene.[125] Hydroxide ions liberated in this step attack tetraphenylphosphonium ion to form triphenylphosphine oxide. For cyclopolyphosphines of the general formula $(RP)_n$, a linear relationship which is independent of the ring size, $n$, exists between the reduction potential in THF and the inductive effect of the R group, indicating that $p\pi$–$d\pi$ bonding between phosphorus atoms is relatively unimportant.[126]

**Halides.**—Carbon tetrachloride shows four reduction waves at a glassy carbon electrode and is finally converted into methane.[127] The facile controlled-potential reduction of the trichloromethyl group with loss of one chlorine atom has been utilized in the final step of a synthesis of armentomycin (16). Reduction of the chloro-compound (17) afforded a synthesis of dehydroarmentraycin (18).[128] Alkyl

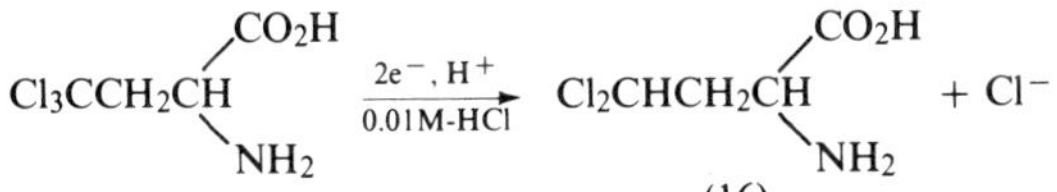

$Cl_3CCH_2\underset{NH_2}{\overset{CO_2H}{\underset{|}{CH}}} \xrightarrow[0.01M\text{-}HCl]{2e^-, H^+} Cl_2CHCH_2\underset{NH_2}{\overset{CO_2H}{\underset{|}{CH}}} + Cl^-$

(16)

$Cl_3CCHCl\underset{NHCOCH_2Ph}{\overset{CO_2H}{\underset{\|\ O}{CH}}} \xrightarrow[\text{hydrolyse}]{2e^-} Cl_2C=CH\underset{NH_2}{\overset{CO_2H}{\underset{|}{CH}}}$

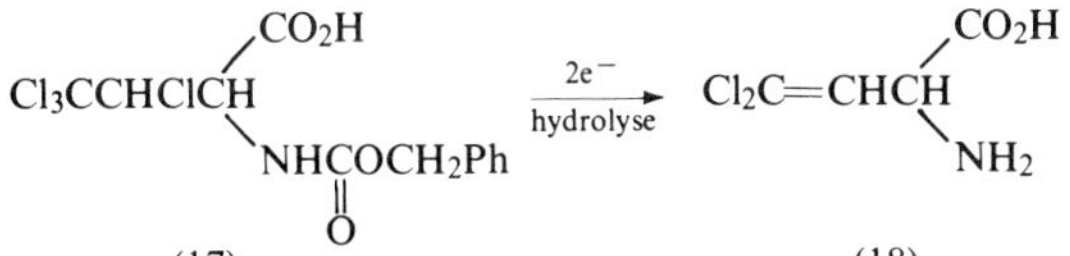

(17)                                    (18)

---

[124] C. M. Pak and W. M. Gulick, *Taehan Hwahak Hoechi*, 1974, **18**, 341 (*Chem. Abs.*, 1975, **82**, 49 224).
[125] J. M. Savéant and Su Khac Bink, *Electrochim. Acta*, 1975, **20**, 21.
[126] J. T. DuPont, L. R. Smith, and J. L. Mills, *J. Chem. Soc., Chem., Commun.*, 1974, 1001.
[127] F. L. Lambert, B. L. Hasslinger, and R. L. Franz, *J. Electrochem. Soc.*, 1975, **122**, 737.
[128] Y. Urabe, T. Iwasaki, K. Matsumoto, and M. Miyoshi, *Tetrahedron Lett.*, 1975, 997.

radicals formed by cathodic reduction of allyl halides (X=Cl or Br) have a long enough lifetime to undergo dimerization to hexa-1,5-diene:[129]

$$2H_2C{=}CHCH_2X + 2e^- \longrightarrow H_2C{=}CHCH_2CH_2CH{=}CH_2 + 2X^-$$

*gem*-Dihalogeno-disulphonylmethanes are reduced to a stable halogeno-disulphonylmethane carbanion[130] at a platinum cathode in DMF:

$$(RSO_2)_2CCl_2 + 2e^- \longrightarrow (RSO_2)_2\bar{C}Cl + Cl^-$$

Organomercury intermediates are involved in the reaction at a mercury cathode.

α-Bromo-ketones are known to be reduced at less negative potentials than bromo-alkanes, and this has been demonstrated again in a polarographic study of bromopyruvic acid and its ethyl ester in aqueous buffers.[131] The electrochemical reduction of highly branched α,α′-dibromo-ketones in acetic acid is a convenient synthetic route to highly branched α-acetoxy-ketones (Scheme 22).[132] This reaction involves reductive cleavage of one halogen followed by rapid solvolysis of the second halogen and reaction of the carbonium ion intermediate with the solvent. Solvolysis of the monobromo intermediate from non-branched dibromo-ketones is relatively slow, and the intermediate undergoes preferential reductive removal of bromine in place of the reaction shown in Scheme 22.

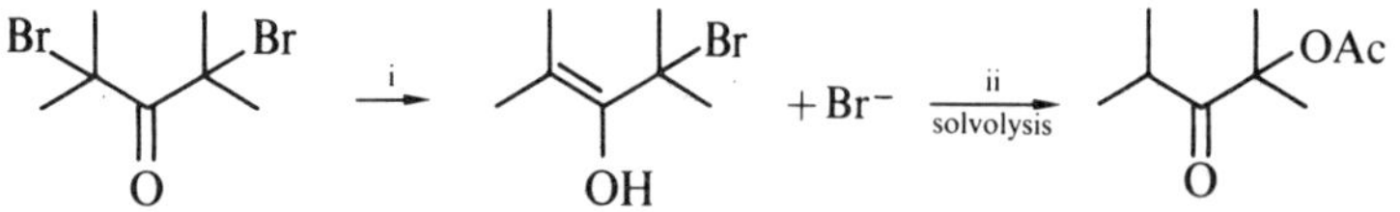

Reagents: i, H⁺, 2e⁻, HOAc, NaOAc; ii, HOAc

**Scheme 22**

A series of experiments have demonstrated that a nitrogen radical is formed in the first step of the reduction of *N*-halogeno-amides.[133] Radicals formed in the reduction:

$$ClNHCO_2Et + e^- \longrightarrow Cl^- + \cdot NHCO_2Et$$

are trapped by intermolecular addition to olefins that have been made into a paste with carbon and used as the cathode. A chain reaction is initiated (Scheme 23). Nitrogen radicals (19) have also been detected because they undergo intramolecular hydrogen abstraction (Scheme 24).

Further examples (*e.g.*, see Scheme 25) have been given of the formation of cyclopropanes by reduction of 1,3-dibromo-alkanes in DMF.[134] Reduction of the 1,5-dibromo-compound (21) at lead or carbon cathodes in acetone and lithium perchlorate also causes cyclizations[135] in 75% yield. The reduction of 1,2-dibromo-compounds (22) has been used in a synthesis of acetylenes.[136]

[129] M. M. Baizer, U.S. P. 3 876 514 (*Chem. Abs.*, 1975, **83**, 9162).
[130] J. G. Gourey, G. Jeminet, and J. Simonet, *Bull. Soc. Chim. Fr.*, 1975, 1713.
[131] D. Fleury and J. Moiroux, *Electrochim. Acta*, 1975, **20**, 369.
[132] A. J. Fry and J. J. O'Dea, *J. Org. Chem.*, 1975, **40**, 3625.
[133] D. Berube, J. Caza, F. M. Kimmerle, and J. Lessard, *Can. J. Chem.*, 1975, **53**, 3060.
[134] M. M. Gol'din, A. I. D'yachenko, and L. G. Foektistov, *Izv. Akad. Nauk. SSSR, Ser. Khim.*, 1975, 2605.
[135] V. N. Leibzon, A. S. Mendkovich, T. A. Klimova, M. M. Krayushkin, S. G. Mairanovskii, S. S. Novikov, and V. V. Sevast'yanova, *Elektrokhimiya*, 1975, **11**, 349.
[136] K. M. Smionov and A. P. Tomilov, *Elektrokhimiya*, 1975, **11**, 784.

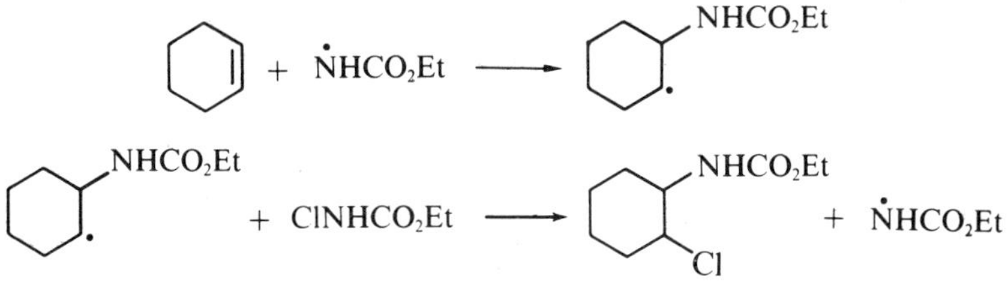

## Scheme 23

$$\text{C}_6\text{H}_{10} + \overset{\bullet}{\text{N}}\text{HCO}_2\text{Et} \longrightarrow \text{C}_6\text{H}_{10}(\text{NHCO}_2\text{Et})_2^{\bullet}$$

$$\text{C}_6\text{H}_{10}\overset{\bullet}{-}\text{NHCO}_2\text{Et} + \text{ClNHCO}_2\text{Et} \longrightarrow \text{C}_6\text{H}_{10}(\text{NHCO}_2\text{Et})(\text{Cl}) + \overset{\bullet}{\text{N}}\text{HCO}_2\text{Et}$$

## Scheme 24

$$\text{MeCH}_2(\text{CH}_2)_2\text{CO}\overset{\bullet}{\text{N}}\text{Bu}^t \longrightarrow \text{Me}\overset{\bullet}{\text{C}}\text{H}(\text{CH}_2)_2\text{CONHBu}^t$$
$$(19) \qquad\qquad (20)$$

$$(20) + \text{MeCH}_2(\text{CH}_2)_2\text{CONClBu}^t \rightarrow \text{Me}\underset{\underset{\text{Cl}}{|}}{\text{CH}}(\text{CH}_2)_2\text{CONHBu}^t + (19)$$

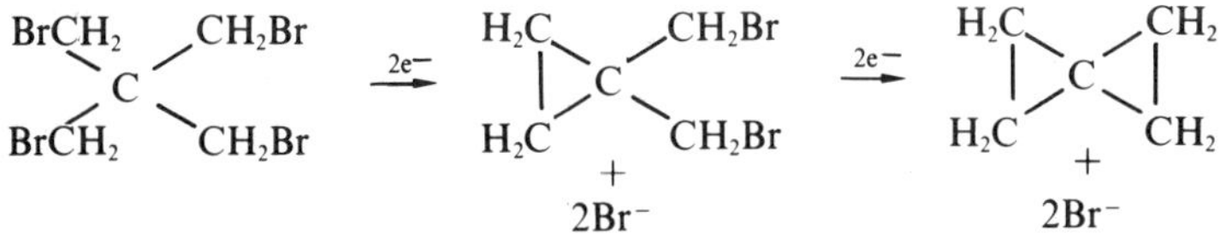

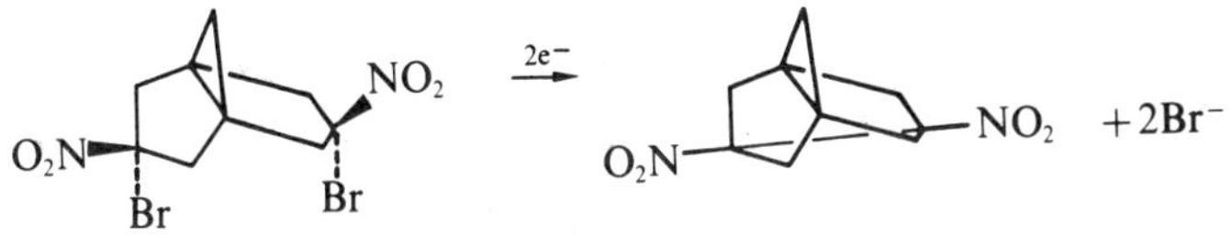

## Scheme 25

[Scheme showing reduction of BrCH₂–C(CH₂Br)–CH₂Br/BrCH₂ with 2e⁻ steps to form cyclic products with loss of 2Br⁻]

[Scheme showing 2e⁻ reduction of dibromo-dinitro bicyclic compound (21) to dinitro bicyclic compound + 2Br⁻]

$$\text{Br}_2\text{FC}-\text{CHBr}_2 \xrightarrow{\text{Hg cathode}} \text{FC}\equiv\text{CH} + \text{HC}\equiv\text{CH} + \text{FBrC}=\text{CHBr}$$
(22)

Alkyl chlorides are not normally reducible; however, two groups of workers have found that electrolytically generated radical-anions from biphenyl, naphthalene, and other aromatic hydrocarbons in DMF will transfer an electron to alkyl chlorides, causing fission of the carbon–chlorine bond.[137, 138] The alkane[137, 138] is isolated along with products of coupling[137] of alkyl radicals and hydrocarbon radical-anions. So far, the generality and usefulness of this reaction have not been tested with a wide range of alkyl halides. The electroreductive cyclization of

---

[137] J. Simonet, M. A. Michel, and H. Lund, *Acta Chem Scand., Ser. B*, 1975, **29**, 489.
[138] J. W. Sease and R. C. Reed, *Tetrahedron Lett.*, 1975, 393.

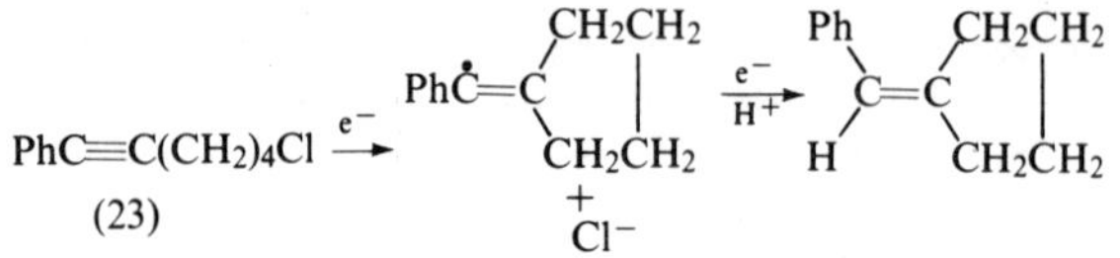

**Scheme 26**

PhC≡C(CH₂)₄Cl (23) $\xrightarrow{e^-}$ PhĊ=C(CH₂CH₂–CH₂CH₂)·Cl⁻ $\xrightarrow[H^+]{e^-}$ PhC(H)=C(CH₂CH₂–CH₂CH₂)

6-chloro-1-phenylhex-1-yne (23) in DMF (see Scheme 26) could be formulated as a related intramolecular electron transfer followed by reaction between the alkyl radical that is generated and the ethyne bond. The authors regard the reaction as an intramolecular nucleophilic attack of the ethyne radical-anion on the alkyl halide centre.[139] 6-Chloro-1-phenylhexa-1,2-diene undergoes a related electroreductive cyclization.[139]

Electrochemical dechlorination of polychloro-aromatics has received detailed attention. All the possible chlorinated benzenes[140] and biphenyls with all chlorine substituents in one ring[141] have been examined by polarography and preparative electrolysis in DMSO, and the product distribution has been determined after reduction at the foot of the first polarographic wave. Polychlorobenzene-1,4-disulphonic acids have been similarly examined in a phosphate buffer at pH 8.[142] The dechlorination reaction has been extended to chloro-pyridines in DMF.[143] A patent has been issued for the selective debromination of bromodichlorobenzene derivatives (24) (Scheme 27).[144]

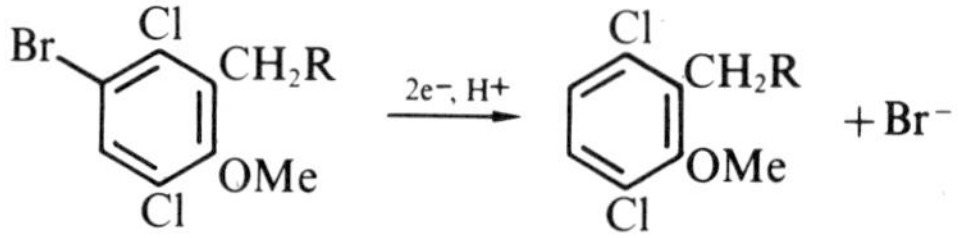

[Br,Cl-substituted benzene with CH₂R and OMe] $\xrightarrow{2e^-, H^+}$ [Cl-substituted benzene with CH₂R and OMe] + Br⁻

(24) R = OH, SMe, alkyl, or Ph

**Scheme 27**

The radical-anions of the halogenophenyl azides (25) and (26) are a new class of radical-anion which undergo cleavage of the carbon–halogen bond.[145]

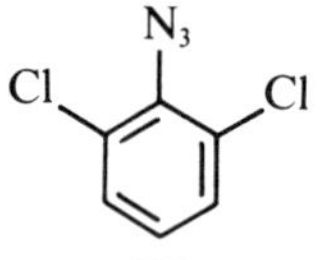

(25) 2,6-dichlorophenyl azide

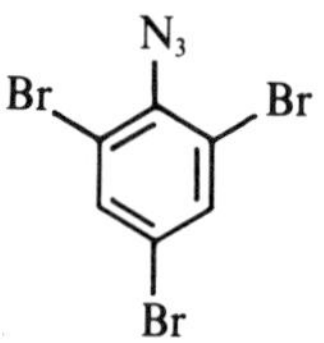

(26) 2,4,6-tribromophenyl azide

---

[139] W. M. Moore, A. Salajegheh, and D. G. Peters, *J. Am. Chem. Soc.*, 1975, **97**, 4954.
[140] S. O. Farwell, F. A. Beland, and R. D. Geer, *J. Electroanal. Chem. Interfacial Electrochem.*, 1975, **61**, 303.
[141] S. O. Farwell, F. A. Beland, and R. D. Geer, *J. Electroanal. Chem. Interfacial Electrochem.*, 1975, **61**, 315.
[142] E. A. Zalogina, E. Yu. Khmel'nitskaya, G. A. Mazentseva, and N. S. Dokunikhin, *Zh. Obshch. Khim.*, 1975, **45**, 611.
[143] M. Maruyama and K. Murakami, *Nippon Kagaku Kaishi*, 1975, 2119 (*Chem. Abs.*, 1976, **84**, 66 877).
[144] H. M. Becher and R. Sehring, Ger. P. 2 331 711 (*Chem. Abs.*, 1975, **82**, 139 622).
[145] N. I. Malyugina, N. L. Gusarskaya, and A. V. Oleinik, *Zh. Obshch. Khim.*, 1975, **45**, 1837.

## Organic Electrochemistry – Synthetic Aspects

An aqueous alcoholic solution of iodobenzene undergoes reduction at a mercury pool to benzene and varying amounts (up to 9%) of diphenylmercury, depending on the concentration of the electrolyte, which is tetraethylammonium perchlorate.[146] The radical-anion from 4-fluorobenzonitrile decomposes by two pathways in DMF, depending on its concentration.[147] In dilute solution ($<2\cdot5\times10^{-4}$ mol l$^{-1}$) decomposition follows unimolecular kinetics by cleavage of the carbon–fluorine bond, the ultimate products being benzonitrile and fluoride ion. In more concentrated solution a bimolecular process occurs, and this has been identified as dimerization of the radical-anion (Scheme 28): Loss of fluoride ion then leads to 4,4'-dicyanobiphenyl.

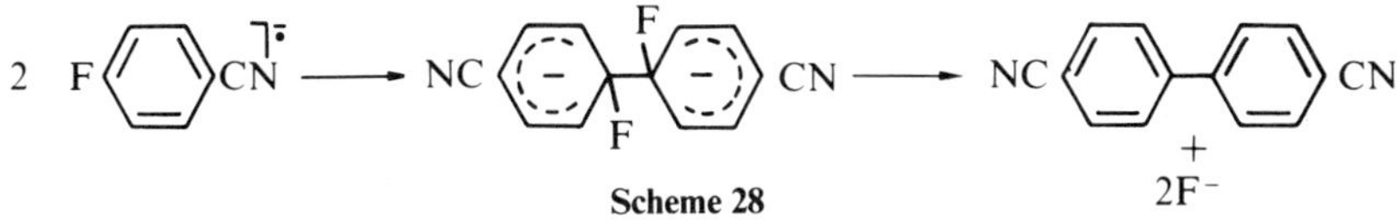

$$2 \;\; F\text{–}\langle\rangle\text{–}CN^{\overline{\cdot}} \longrightarrow NC\text{–}\langle\rangle\text{–}\langle\rangle\text{–}CN \longrightarrow NC\text{–}\langle\rangle\text{–}\langle\rangle\text{–}CN$$
$$\hspace{7cm} F \hspace{2cm} F \hspace{5cm} + \; 2F^-$$

**Scheme 28**

Cleavage of the radical anions from aryl halides gives an aryl radical which can either abstract a hydrogen atom from the solvent or undergo further electron transfer to form a carbanion which abstracts a proton from some source, probably extraneous water. These two processes can be distinguished by carrying out the reaction in DMF + 1% D$_2$O.[148] Radical-anions such as those from 1-bromopyrene and from 4-chloro- and 4-bromo-benzophenone, which decompose relatively slowly and so have time to diffuse from the electrode surface, do not incorporate deuterium in the product; the products have acquired their hydrogen by radical abstraction from DMF. Radical-anions from 1-chloropyrene and 2-bromo-4-methoxybenzanilide, which decompose very rapidly at the electrode surface, generate aryl radicals that can be reduced, and so deuterium is introduced into the product since the so-formed carbanion is quenched by D$_2$O.

**Simple Inorganic Materials.**—The reduction of carbon disulphide at a platinum cathode in DMF has again received attention.[149] The irreversible one-electron reaction leads to CS and CS$_3^-$. Carbon sulphide is detected by its evaporation in a stream of nitrogen and then trapping it as a rhodium complex (27). The two species that are formed first, i.e. CS and CS$_3^-$, react slowly in solution to give the cyclic sulphur compound (28), previously described.[150]

$$CS \; + \; RhCl(PPh_3)_3 \longrightarrow RhCl(CS)(PPh_3)_2 \; + \; Ph_3P$$
$$\hspace{5cm} (27)$$

$$2CS_2 \; + \; 2e^- \longrightarrow CS \; + \; CS_3^{2-} \xrightarrow{\text{slow}}$$

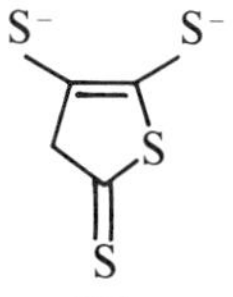

(28)

---

[146] S. G. Mairanovskii, T. Ya. Rubinskaya, and C. V. Proskurovskaya, *Elektrokhimiya*, 1975, **11**, 1386.
[147] M. R. Asirvatham and M. D. Hawley, *J. Am. Chem. Soc.*, 1975, **97**, 5024.
[148] J. Grimshaw and J. Trocha-Grimshaw, *J. Chem. Soc., Perkin Trans. 2*, 1975, 215.
[149] G. Bontempelli, F. Magno, and G. A. Mazzocchin, *J. Electroanal. Chem. Interfacial Electrochem.*, 1975, **63**, 231.
[150] S. Wawzonek and S. M. Heilmann, *J. Org. Chem.*, 1974, **39**, 511.

## 3 Oxidation

**General.**—3,7-Dimethoxyphenothiazine (29) shows two reversible oxidation waves in AN.[151] The four species which are interconvertible by electron transfer and protonation–deprotonation are stable in AN and have distinctive u.v. spectra. Hence the system can be used to determine hydrogen ion concentration in AN. The relative basicity of a series of pyridines was determined.

$$\text{(29) (PH)} \xrightleftharpoons{-e^-} PH^{\cdot+} \xrightleftharpoons{-H^+} P^{\cdot} \xrightleftharpoons{-e^-} P^+$$

**Aliphatic Hydrocarbons and Ethers.**—Oxidation of alkyl-adamantanes (30) in AN leads to the formation of tertiary carbonium ions either by the cleavage of carbon–hydrogen or carbon–alkyl bonds, depending on the nature of the alkyl group.[152] Tertiary alkyl groups are lost preferentially (see Table 2), and the carbonium ions react with the solvent to form an acetamido product (31) or (32) after the addition of water. The halogen group in 2-iodo-adamantanes (33) is preferentially oxidized to give the 2-adamantyl carbonium ion, which subsequently reacts with the AN solvent. Some rearrangement of the secondary carbonium ion from 2-iodotetramethyladamantane (34) to the tertiary ion occurs before capture by the solvent. 2-Bromoadamantane (35) undergoes preferential oxidative cleavage of a carbon–hydrogen bond to form the 1-adamantyl carbonium ion.[153]

**Table 2** *Products from the oxidation of 1-alkyl-adamantanes (30) in acetonitrile*

| Alkyl substituent, R | | H | Et | Pr$^i$ | Bu$^t$ |
| --- | --- | --- | --- | --- | --- |
| % Yield | (31) | 74 | 77 | 75 | 7 |
|  | (32) | — | — | 9 | 62 |

---

[151] G. Cauquis, A. Deronzier, D. Serve, and E. Vieil, *J. Electroanal. Chem. Interfacial Electrochem.*, 1975, **60**, 205.
[152] G. J. Edwards, S. R. Jones, and J. M. Mellor, *J. Chem. Soc. Chem. Commun.*, 1975, 816.
[153] F. Vincent, R. Tardivel, and P. Mison, *Tetrahedron Lett.*, 1975, 603.

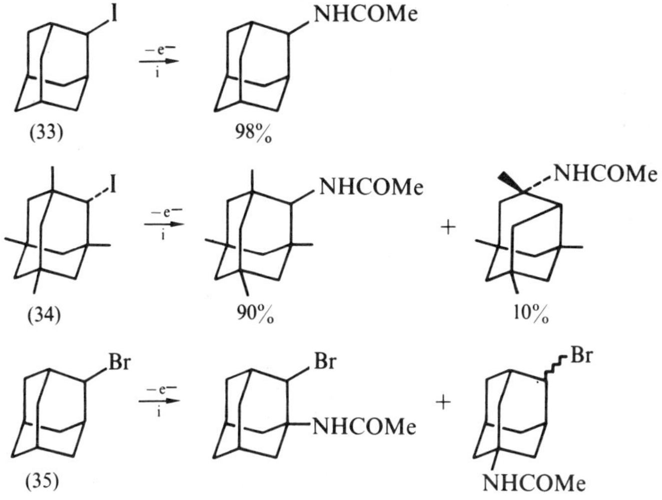

# Organic Electrochemistry – Synthetic Aspects

Reagents: i, H₂O, AN

**Scheme 29**

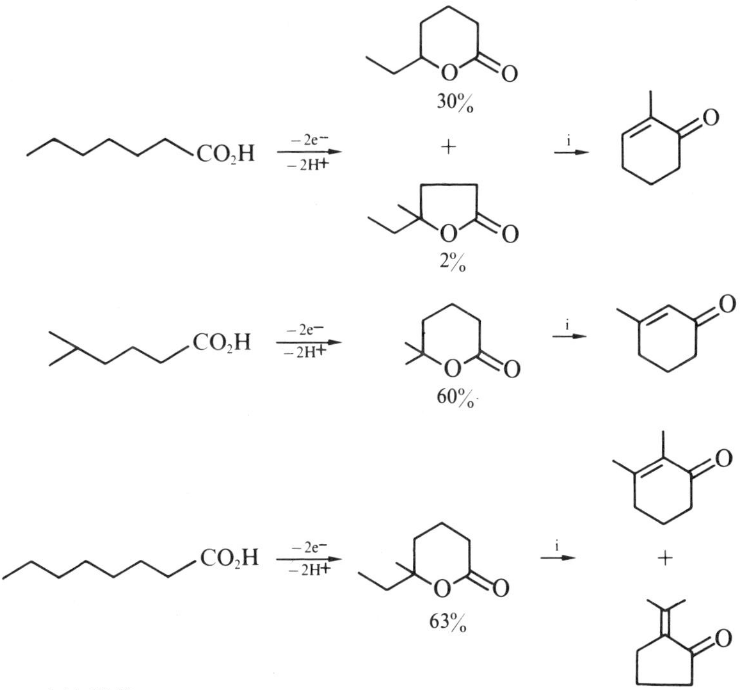

Reagents: i, MeSO₃H

**Scheme 30**

The oxidation of alkanoic acids at a platinum anode in methanesulphonic acid containing potassium methanesulphonate leads to cleavage of a carbon–hydrogen bond and the formation of a carbonium ion remote from the functional group. Isomerization of the carbon chain can then occur, to form a more stable carbonium ion, and the products that can be isolated are lactones formed by the reaction of the carboxylic acid function with the carbonium ion centre (Scheme 30).[154] These lactones undergo further rearrangement in methanesulphonic acid, and the ratio of products isolated depends upon the time for which the products have been exposed to this strongly acid medium. Other workers[155] have challenged the usually accepted view that the reactive species which undergo oxidation in these strongly acid media are the protonated paraffins. Results from the oxidation of aliphatic ketones[215, 216] should also be read in relation to the oxidation of carboxylic acid side-chains, since the mechanism proposed there could also be applied to the carboxylic acids.

Cyclopropanes with a phenyl or olefin substituent undergo ring opening during oxidation at a platinum anode in methanol,[156] and 1,1-dichloro-2-phenylcyclopropane (36) behaves in an analogous manner (Scheme 31).[157] These reactions involve

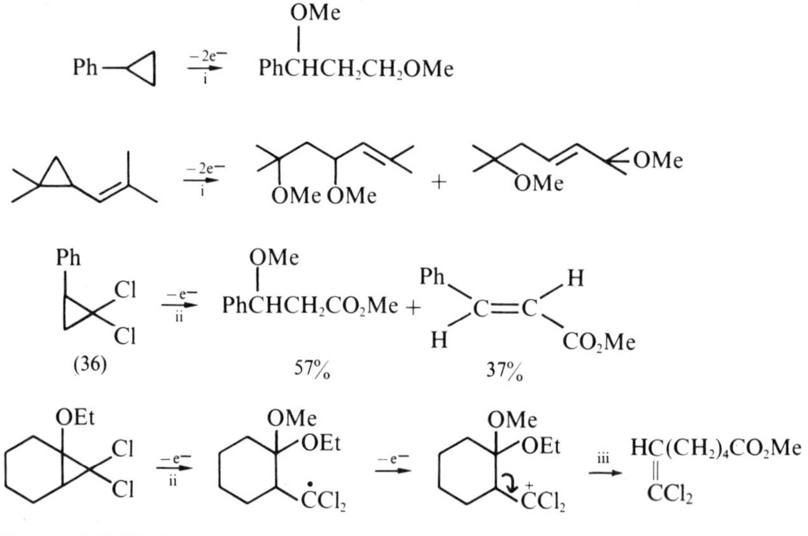

$$Ph-\triangleleft \xrightarrow[i]{-2e^-} PhCHCH_2CH_2OMe$$
$$\quad\quad\quad\quad\quad\quad\quad OMe$$

Reagents: i, MeOH, NaOMe; ii, MeOH, lutidine, NaClO$_4$; iii, MeOH

**Scheme 31**

carbon–carbon bond cleavage in the radical cation, assisted by relief of ring strain, to give a carbonium ion and a radical centre. The products can be rationalized on the basis of the formation of the most stable carbonium ion intermediates.[157]

β-Carotene gives a reversible two-electron oxidation wave in AN–dichloro-

---

[154] D. Pletcher and C. Z. Smith, *J. Chem. Soc., Perkin Trans. 1*, 1975, 948.
[155] C. Pitti, F. Bobilliart, A. Thiebault, and M. Herlem, *Anal. Lett.*, 1975, **8**, 241.
[156] R. Brettle and J. R. Sutton, *J. Chem. Soc., Perkin Trans. 1*, 1975, 1955.
[157] M. Klehr and H. J. Schäfer, *Angew. Chem. Int. Ed. Engl.*, 1975, **14**, 247.

ethane at +0.51 V vs. S.C.E.[158] The stable cyclobutadiene derivative (37) shows an irreversible oxidation wave at +0.6 V vs. S.C.E. in DME.[159]

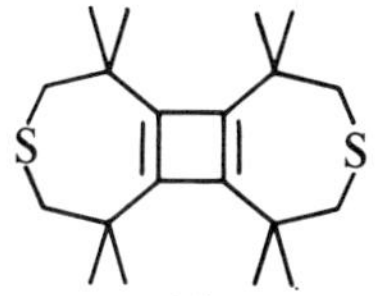

(37)

Improved yields from the electrochemical functionalization of olefins in methanol can sometimes be obtained by using sodium methylcarbonate as electrolyte, prepared *in situ* from sodium methoxide and carbon dioxide.[160] Functionalization of a number of alkenes,[160] norbornene,[161] norbornadiene,[161, 162] and bicyclo-[2.2.2]octadiene[162] follows the expected course where the intermediate carbonium ions can undergo rearrangement before being quenched by a nucleophile (Scheme 32). The intra-ring olefinic group in limonene and in 4-vinylcyclohexene is preferentially oxidized.[162] Electrochemical functionalization of enol acetates in acetic acid is an excellent method for the selective formation of α-acetoxy-ketones.[163] One

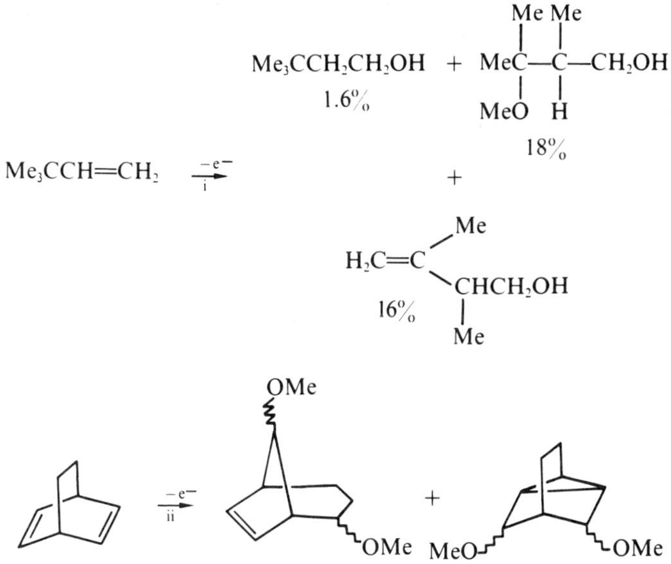

$Me_3CCH=CH_2 \xrightarrow[i]{-e^-}$

$Me_3CCH_2CH_2OH$
1.6%

+

$\begin{matrix} Me & Me \\ | & | \\ MeC—C—CH_2OH \\ | & | \\ MeO & H \end{matrix}$
18%

+

$H_2C=C\begin{matrix}Me\\CHCH_2OH\\|\\Me\end{matrix}$
16%

Reagents: i, MeOH, MeOCO$_2^-$ Na$^+$; ii, MeOH, Et$_4$N Tos

**Scheme 32**

---

[158] V. G. Mairanovskii, A. A. Engovatov, N. T. Ioffe, and G. I. Samokhvalov, *Elektrokhimiya*, 1975, **11**, 184.
[159] R. Breslow, R. W. Johnson, and A. Krebs, *Tetrahedron Lett.*, 1975, 3443.
[160] R. Brettle and J. R. Sutton, *J. Chem. Soc., Perkin Trans. 1*, 1975, 1947.
[161] A. J. Baggaley, R. Brettle, and J. R. Sutton, *J. Chem. Soc., Perkin Trans. 1*, 1975, 1055.
[162] T. Shono, A. Ikeda, J. Hayashi, and S. Hakozaki, *J. Am. Chem. Soc.*, 1975, **97**, 4261.
[163] T. Shono, M. Okawa, and I. Nishiguchi, *J. Am. Chem. Soc.*, 1975, **97**, 6144.

example of this process is given in Scheme 33 to show that the product ratio depends on the nucleophilicity of the solvent and supporting electrolyte; in all, nine enol acetates were examined.

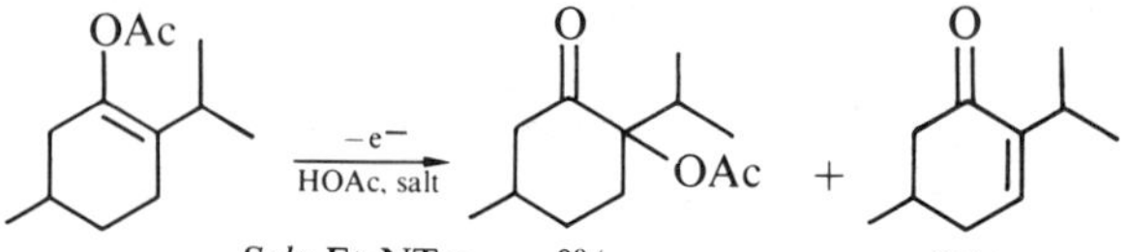

*Salt*: Et$_4$NTos  0%  90%
KOAc  60%  25%

**Scheme 33**

A patent has been issued for the electrochemical epoxidation of propylene at a graphite anode in an electrolyte containing alkali-metal bromide.[164] Electrochemically generated bromide reacts with the olefin to form the bromohydrin. The latter then forms the epoxide by reaction with base.

**Aromatic Compounds.**—Anodic oxidation of benzene in aqueous suspension in a divided cell gives quinone. When this solution is pumped into the cathode compartment of the cell the quinone is reduced to hydroquinone, which can be recovered.[165] Anodic oxidation of mesitylene in acetic acid and sodium acetate gives 2,4,6,-trimethylphenyl acetate and 3,5-dimethylbenzyl acetate from nuclear and side-chain oxidation respectively. A study of the influence of anode material on the yields of reaction products led to the choice of graphite for the production of 2,4,6-trimethylphenyl acetate in 40% yield from mesitylene.[166] Bis(methanesulphonyl) peroxide, prepared *in situ* by the electrochemical oxidation of sodium methanesulphonate:

$$2\text{MeSO}_3^- \xrightarrow{-e^-} \text{MeSO}_2\text{O}-\text{OSO}_2\text{Me}$$

will oxidize benzene to phenyl methanesulphonate and naphthalene to α-naphthyl methanesulphonate.[167]

Oxidation of hexamethylbenzene in dichloromethane containing trifluoroacetic acid generates the pentamethylbenzyl cation, which will react with added benzene, toluene, *p*-xylene, or mesitylene with substitution to form diphenylmethanes.[168]

A number of investigations have been made on the electrochemical hydroxylation of a benzene ring bearing deactivating substituents.[169—171] The proportion of isomers formed correlates well with the calculated positive charge distribution in the benzene radical-cation.[169] The anodic reactions give a very different product distribution to that found in electrophilic aromatic substitution, where the *meta*-product predominates for these deactivated benzenes.

Hydroxylation of trifluoromethylbenzene in trifluoroacetic acid[170] gives *ortho*- and *para*-substitution in the ratio 1:3, while hydroxylation of benzoic acid[171] gives a

---

[164] F. Beck and J. Heiss, Ger. P. 2 336 288 (*Chem. Abs.*, 1975, **83**, 87 412).
[165] F. A. Keidel, U.S. P. 3 884 776 (*Chem. Abs.*, 1975, **83**, 105 452).
[166] L. Eberson and K. Nyberg, *Acta Chem. Scand., Ser. B*, 1975, **29**, 168.
[167] C. J. Myall and D. Pletcher, *J. Chem. Soc., Perkin Trans 1*, 1975, 953.
[168] K. Nyberg and A. Trojanek, *Coll. Czech. Chem. Commun.*, 1975, **40**, 526.
[169] Z. Blum, L. Gedheim, and K. Nyberg, *Acta Chem. Scand., Ser. B.*, 1975, **29**, 715.
[170] N. L. Weinberg and C. N. Wu, *Tetrahedron Lett.*, 1975, 3367.
[171] Y. H. So, J. V. Becker, and L. L. Miller, *J. Chem. Soc., Chem. Commun.*, 1975, 262.

# Organic Electrochemistry – Synthetic Aspects 25

ratio of ortho-:para-substitution of 2.7:1. The acetamidation of benzoic acid in acetonitrile[171] gives ortho- and para-substitution in the ratio 5:1.

In the anodic cyanation of 1,4-dimethoxybenzene with methanol as solvent, (39) and (40) are formed together with 4-methoxybenzonitrile.[172] The latter arises by attack of cyanide ion on the radical-cation of (38) followed by oxidation and

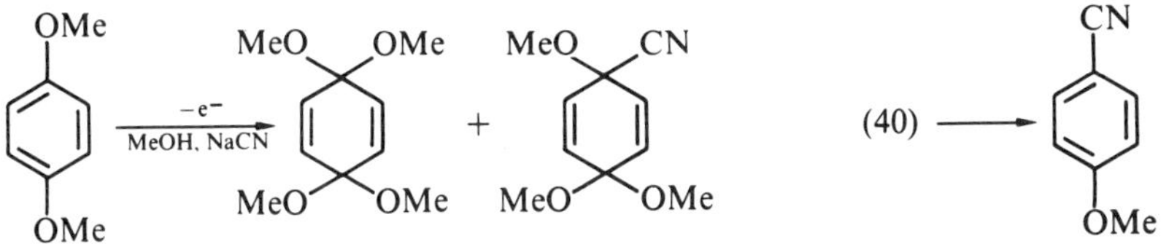

$$\underset{(38)}{\text{OMe-C}_6\text{H}_4\text{-OMe}} \xrightarrow[\text{MeOH, NaCN}]{-e^-} \underset{(39)}{\text{(MeO)}_2\text{C}_6\text{H}_4\text{(OMe)}_2} + \underset{(40)}{\text{(MeO)}_2\text{C}_6\text{H}_3\text{(CN)(OMe)}} \quad (40) \longrightarrow \text{NC-C}_6\text{H}_4\text{-OMe}$$

subsequent loss of a proton and formaldehyde. The ratio of products is dependent on anode potential; high potentials favour 4-methoxybenzonitrile. A comparison between the anodic cyanation of aromatic ethers in dichloromethane–water emulsions and the cyanation in AN containing Et$_4$NCN shows that the electrical resistance of the emulsion system is less, which is an advantage, but the product distribution is less selective than is the case in AN.[173] The product distribution from the anodic cyanation of the three methyl-anisoles in methanol has been determined.[174] 2-Methoxytoluene gives principally one product by cyanation *para* to the methoxy-group, 3-methoxytoluene gives three products by substitution *ortho* and *para* to the methoxy-group, while 4-methoxytoluene undergoes methoxylation of the side-chain.

The anodic acetoxylation of *p*-dimethoxybenzene leads to one product in good yield (Scheme 34); acetoxylation of the other dimethoxybenzenes leads to mixture of products in poor yield.[175]

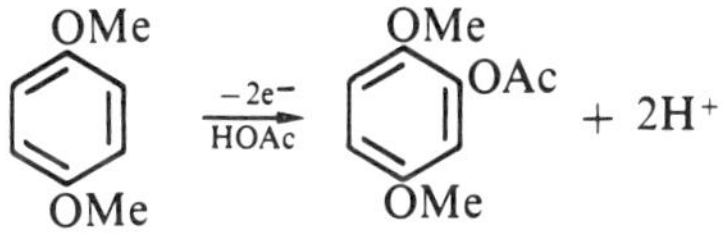

$$\underset{\text{OMe}}{\overset{\text{OMe}}{\text{C}_6\text{H}_4}} \xrightarrow[\text{HOAc}]{-2e^-} \underset{\text{OMe}}{\overset{\text{OMe}}{\text{C}_6\text{H}_3\text{OAc}}} + 2\text{H}^+$$

**Scheme 34**

The formation of (41) from hydroquinone bis-(2'-hydroxyethyl) ether by oxidation in methanol has been shown to proceed *via* a dimethoxy compound which is transformed into (41) by mild acid treatment (Scheme 35).[176]

Oxidation of 4-methoxyphenol in methanol[177] gives a good yield of quinone dimethylketal, and a good process for the oxidation of alkyl-phenols to quinones in aqueous acetone containing sulphuric acid has been described.[178] 4-Methoxyphenol

---

[172] N. L. Weinberg, D. H. Marr, and C. N. Wu, *J. Am. Chem. Soc.*, 1975, **97**, 1499.
[173] L. Eberson and B. Helgee, *Acta Chem. Scand., Ser. B.*, 1975, **29**, 451.
[174] K. Yoshida, M. Shigi, and T. Fueno, *J. Org. Chem.*, 1975, **40**, 63.
[175] K. Yoshida, M. Shigi, T. Kanbe, and T. Fueno, *J. Org. Chem.*, 1975, **40**, 3805.
[176] P. Margaretha and P. Tissot, *Helv. Chim. Acta*, 1975, **58**, 933.
[177] A. Nilsson, A. Ronlan, and V. D. Parker, *Tetrahedron Lett.*, 1975, 1107.
[178] D. Degner, Ger. P. 2 360 494 (*Chem. Abs.*, 1975, **83**, 123 305).

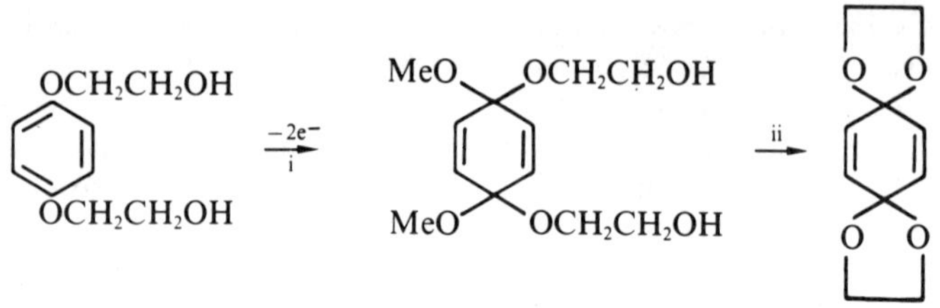

**Scheme 35**

Reagents: i, MeOH, KOH; ii, H⁺, Et₂O    (41)

has been described as a product of the oxidation of phenol in methanol, but its further oxidation was not noted.[179]

Anodic fluorination of a series of *para*-disubstituted benzenes in hydrogen fluoride involves the addition of fluoride ion to the radical-cation and then oxidation of the intermediate radical to a cation which may either undergo elimination of a suitable substituent or the further addition of fluoride ion (Scheme 36).[180]

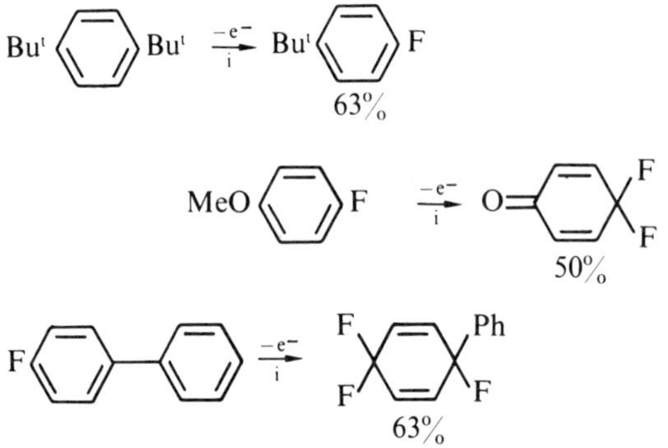

Reagents: i, HF, Et₄NI

**Scheme 36**

The anodic oxidation of a series of α,ω-diaryl-alkanes in dichloromethane–trifluoroacetic acid gives either cyclized or dimerized products, depending on the length of the alkane chain.[181] Two types of reaction, taken from a lengthy series of examples, are given in Scheme 37.

Trialkylamines show an oxidation wave at +0.55 V in AN with lithium perchlorate. Oxidation of laudanosine (42) at this potential causes cyclization, just as does oxidation at +1.1 V, when an electron is removed from a dimethoxybenzene ring. Intramolecular electron transfer from a benzene ring to the first formed amine radical-cation has been proposed as a step in the mechanism of the oxidation at this low potential of +0.55 V.[182]

[179] K. Sasaki and A. Takeda, *Jpn. Kokai.* 75 47 935 (*Chem. Abs.*, 1975, **83**, 78 850).
[180] I. N. Rozkhov and I. Ya. Aliev, *Tetrahedron*, 1975, **31**, 977.
[181] V. D. Parker and A. Ronlan, *J. Am. Chem. Soc.*, 1975, **97**, 4714.
[182] L. L. Miller, F. R. Stermitz, J. Y. Becker, and V. Ramachandran, *J. Am. Chem. Soc.*, 1975, **97**, 2922.

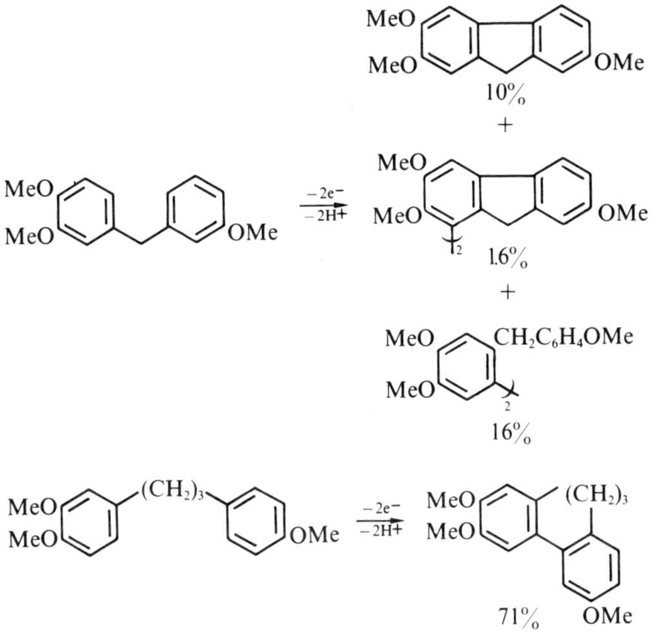

## Scheme 37

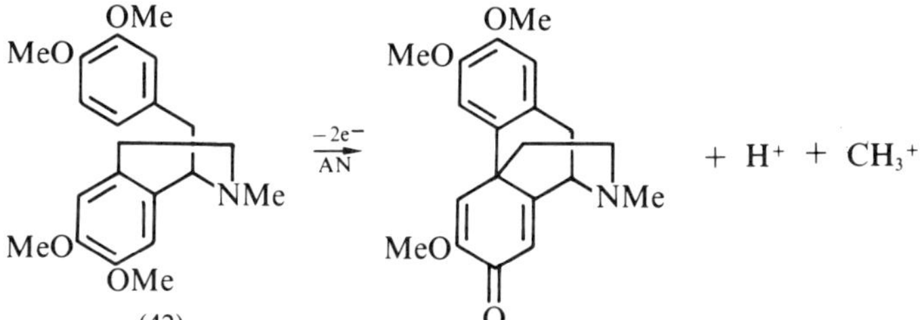

[Scheme 37: electrochemical oxidation (−2e−, −2H+) of a bis(3,4-dimethoxyphenyl)methane giving 2,3,6,7-tetramethoxyfluorene (10%), a 2,3,7-trimethoxyfluorene isomer (1.6%), a (MeO)$_2$C$_6$H$_3$-CH$_2$-C$_6$H$_4$OMe product (16%); and a (CH$_2$)$_3$-linked bis(dimethoxyphenyl) substrate cyclizing to the corresponding biphenyl (71%).]

[Structure (42): a methoxy-substituted benzyl-NMe tetrahydroisoquinoline-type substrate undergoes −2e−/AN oxidation to a dienone product + H$^+$ + CH$_3^+$.]

4-Methoxybenzyl ethers are oxidatively cleaved at $+1.65$ V vs. S.C.E. in aqueous AN (e.g., see Scheme 38), and this reaction has been proposed as part of a procedure for protecting alcohol functions during a synthetic sequence. Examples are given of the deprotection of primary, secondary, and tertiary alcohols in 74—98% yield.[183]

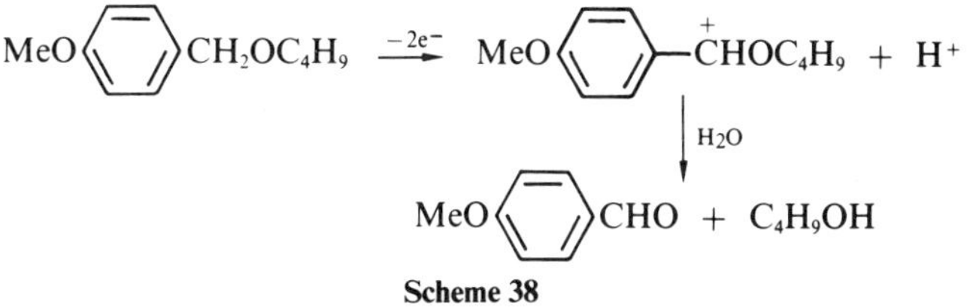

$$\text{MeO-C}_6\text{H}_4\text{-CH}_2\text{OC}_4\text{H}_9 \xrightarrow{-2e^-} \text{MeO-C}_6\text{H}_4\text{-}\overset{+}{\text{C}}\text{HOC}_4\text{H}_9 + \text{H}^+$$

$$\downarrow \text{H}_2\text{O}$$

$$\text{MeO-C}_6\text{H}_4\text{-CHO} + \text{C}_4\text{H}_9\text{OH}$$

**Scheme 38**

---

[183] S. M. Weinreb, G. A. Epling, R. Comi, and M. Reitano, *J. Org. Chem.*, 1975, **40**, 1356.

Secondary alkylphenylcarbinols[184] are oxidized in AN to the corresponding ketone when the alkyl group is primary. When the alkyl group is secondary or tertiary, preferential cleavage of the carbon–alkyl bond occurs to give benzaldehyde (Scheme 39). Tertiary dialkylphenylcarbinols[184] undergo oxidative cleavage of a carbon–alkyl bond.

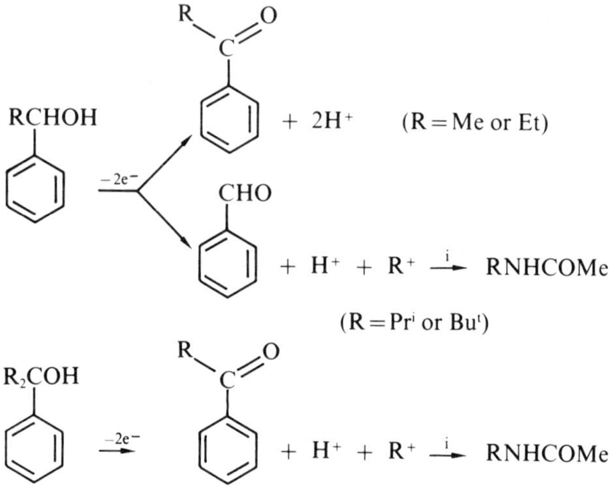

Scheme 39

Reagents: i, AN, $H_2O$

Electrochemical oxidation of benzofurans (43)[185] in methanol containing sulphuric acid, sodium perchlorate, or potassium hydroxide as electrolyte gave the 2,3-dialkoxy product. The anodic oxidation of furans in methanol has been adapted

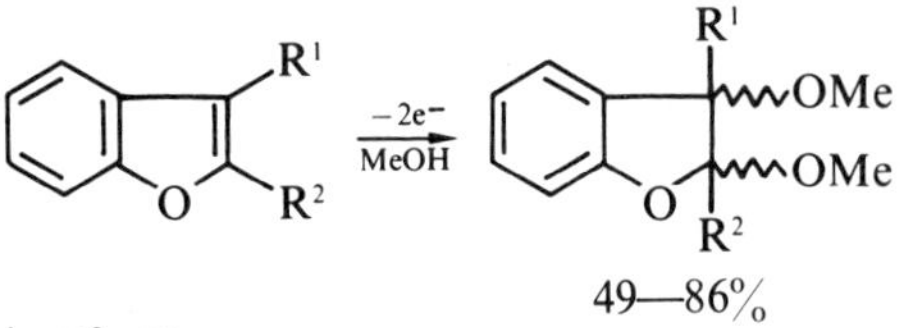

(43) $R^1 = R^2 = H$
or $R^1 = $ Me, $R^2 = CO_2Et$

49—86%

to a process for the manufacture of α-ketoglutaric acid from furoic acid,[186] and the anodic oxidation[187] of furans with aromatic or potentially conjugating 2-substituents gives the corresponding 5-methoxyfurans (44) in 42—76% yield.

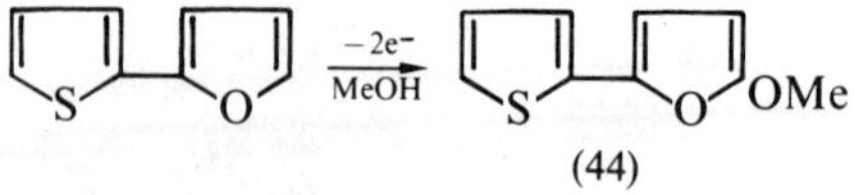

(44)

[184] E. A. Mayeda, *J. Am. Chem. Soc.*, 1975, **97**, 4012.
[185] J. Srogl, M. Janda, I. Stibor, and R. Rozinek, *Synthesis*, 1975, 717.
[186] C. J. V. Scanio, U.S. P. 3 871 977 (*Chem. Abs.*, 1975, **83**, 68 078).
[187] I. Stibor, J. Srogl, and M. Janda, *J. Chem. Soc., Chem. Commun.*, 1975, 397.

α,α′-Dimethoxystilbene affords benzil and benzil monoketal on oxidation in AN, in a total yield of 80%.[188]

**Carboxylic Acids.**—Electrochemical oxidation of triphenylacetic acid in AN in the cavity of an e.s.r. spectrometer does not produce $Ph_3CCO_2^{\cdot}$ radicals in detectable concentration, as previously stated. The first detectable intermediate is the triphenylcarbonium ion, which is reduced, either by switching the electrode potential to −0.35 V or by making the anode open circuit, to the triphenylmethyl radical.[189] The observed e.s.r. signal is due to the latter radical.

A synthesis of the insect sex attractant brevicomin (45) has been achieved, using the Kolbe electrolysis route to elaborate the carbon chain.[190]

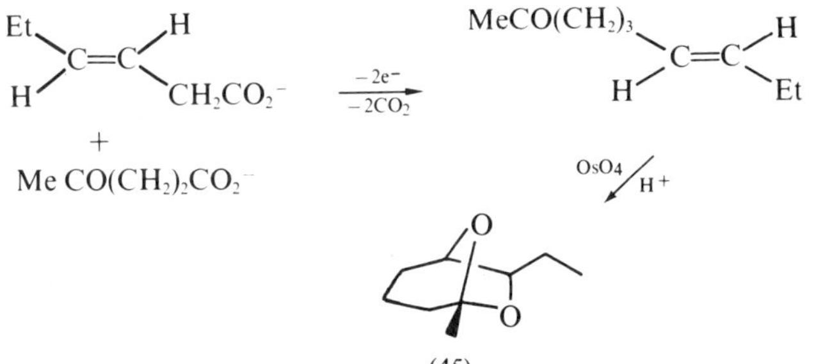
Et
H
C=C
H
CH2CO2−
+
Me CO(CH2)2CO2−
−2e−
−2CO2
MeCO(CH2)3
H
C=C
H
Et
OsO4
H+
O
O

(45)

The two-electron oxidation of carboxylates to carbonium ions has been utilized at one stage in a synthesis (Scheme 40) of methyl jasmonate (46).[191] The *N*-acyl

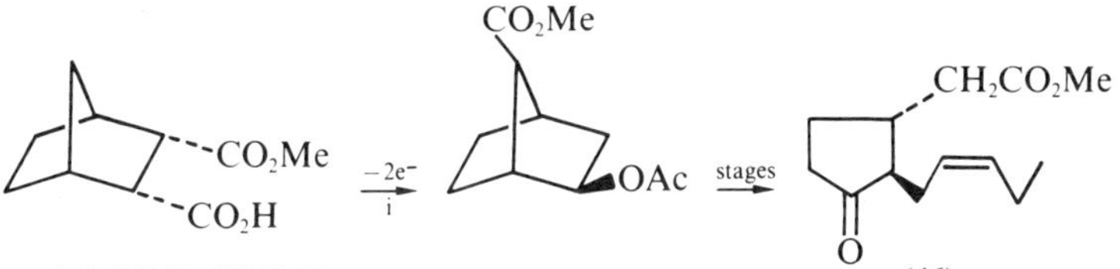
CO2Me
CO2H
−2e−
i
CO2Me
OAc
stages
CH2CO2Me
O

Reagents: i, AcOH, Bu$^t$OH, $Et_3N$ (46)

**Scheme 40**

derivatives of α-alanine are oxidized in methanol containing sodium hydroxide to *N*-(α-methoxyethyl)-carboxamides in 85% yield.[192] This two-electron oxidation of carboxylates has also been applied to β-keto-carboxylates, which are converted in good yield into αβ-ethyleneketones (Scheme 41).[193]

Deuteriated butyric acid has been used as a substrate to investigate the rearrangement of carbonium ions formed in the two-electron oxidation of carboxylates.[194]

188 M. A. Michel, P. Martigny, and J. Simonet, *Tetrahedron Lett.*, 1975, 3143.
189 R. D. Goodin, J. C. Gilbert, and A. J. Bard, *J. Electroanal. Chem. Interfacial Electrochem.*, 1975, **59,** 163.
190 J. Knolle and H. J. Schäfer, *Angew. Chem. Int. Ed. Engl.*, 1975, **14,** 758.
191 S. Torii, H. Tanaka, and T. Mandai, *J. Org. Chem.*, 1975, **40,** 2221.
192 M. Mitzlaff and H. Schnabel, Ger. P. 2 336 976 (*Chem. Abs.*, 1975, **83,** 9239).
193 D. Lelandais and M. Chkir, *C.R. Hebd. Seances Acad. Sci., Ser. C*, 1975, **281,** 731.
194 E. Laurent and M. Thornalla, *Tetrahedron Lett.*, 1975, 4411.

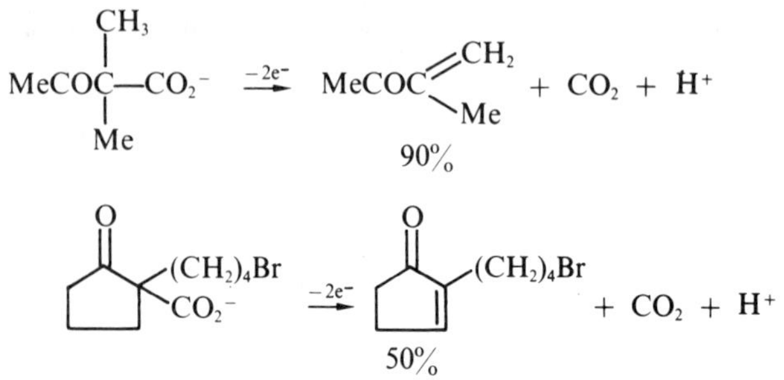

$$\text{MeCOC(CH}_3\text{)(Me)CO}_2^- \xrightarrow{-2e^-} \text{MeCOC(=CH}_2\text{)Me} + CO_2 + H^+$$

90%

[cyclopentanone-2-(CH$_2$)$_4$Br-2-CO$_2^-$] $\xrightarrow{-2e^-}$ [2-(CH$_2$)$_4$Br-cyclopentenone] + $CO_2$ + $H^+$

50%

**Scheme 41**

The reaction was carried out in AN, $N$-alkyl-acetamides were isolated after addition of water, and the distribution of the deuterium label was determined by n.m.r. spectroscopy. The results indicate that one group of rearrangements proceeds *via* a protonated cyclopropane intermediate (Scheme 42).

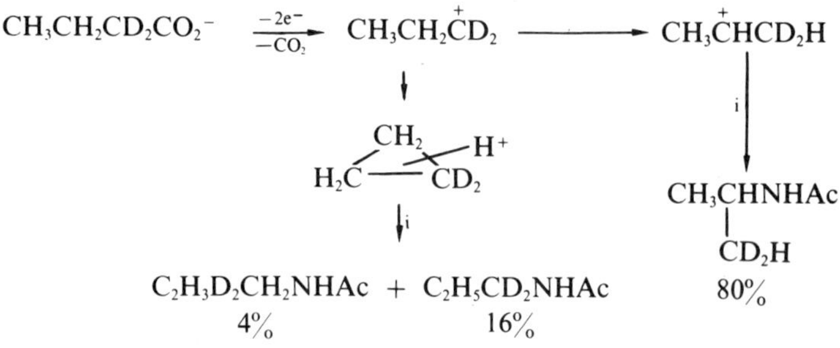

$$CH_3CH_2CD_2CO_2^- \xrightarrow[-CO_2]{-2e^-} CH_3CH_2\overset{+}{C}D_2 \longrightarrow CH_3\overset{+}{C}HCD_2H$$

$$\downarrow \qquad\qquad\qquad\qquad\qquad\qquad \downarrow i$$

$$H_2C\overset{CH_2}{\underset{CD_2}{\diagup\!\!\!\diagdown}}H^+ \qquad\qquad CH_3CHNHAc$$

$$\downarrow i \qquad\qquad\qquad\qquad\qquad CD_2H$$

$$C_2H_3D_2CH_2NHAc + C_2H_5CD_2NHAc \qquad 80\%$$
$$4\% \qquad\qquad 16\%$$

Reagents: i, AN, H$_2$O

**Scheme 42**

Oxidation of a 1,2-dicarboxylic acid to give an olefin with loss of carbon dioxide has been applied to the synthesis of bicyclo[2.2.2]octadienes (Scheme 43).[195]

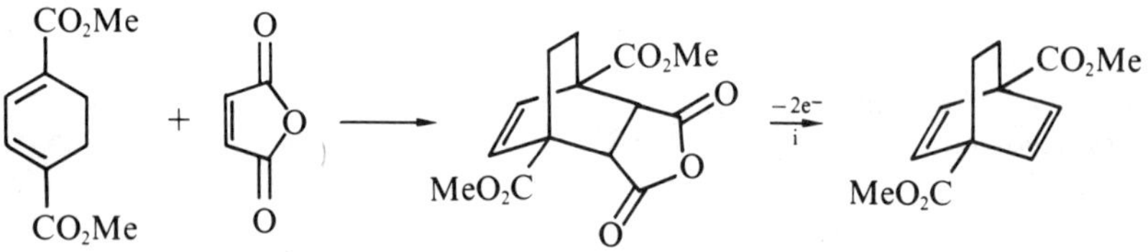

[cyclohexadiene-1,4-di-CO$_2$Me] + [maleic anhydride] $\longrightarrow$ [bicyclic adduct with CO$_2$Me, MeO$_2$C and anhydride] $\xrightarrow[i]{-2e^-}$ [bicyclo[2.2.2]octadiene with CO$_2$Me, MeO$_2$C]

Reagents: i, AN, H$_2$O, pyridine, Et$_3$N

**Scheme 43**

**Boron Compounds.**—The electrochemical reaction of trialkylboranes in AN containing tetra-alkylammonium iodides as supporting electrolyte in an undivided cell between platinum electrodes leads to aliphatic nitriles. This synthesis is the result of coupled anodic and cathodic reactions (Scheme 44).[196]

[195] G. B. Warren and J. J. Bloomfield, *J. Org. Chem.*, 1973, **38**, 4011.

Organic Electrochemistry – Synthetic Aspects                                31

$$\text{Cathode reaction: } CH_3CN \xrightarrow{+e^-} \overline{C}H_2CN + \tfrac{1}{2}H_2$$

$$\text{Anode reaction: } 3I^- \xrightarrow{-2e^-} I_3^-$$

$$R_3B + I_3^- \longrightarrow RI$$

$$RI + \overline{C}H_2CN \longrightarrow RCH_2CN + I^-$$

**Scheme 44**

**Nitrogen, Phosphorus, and Arsenic Compounds.**—Aliphatic nitroxyl radicals $R_2\dot{N}O$ undergo reversible one-electron oxidation in AN containing lithium perchlorate to form the $R_2\overset{+}{N}O$ ion.[197] The oxidation of t-butylamine, acting as the solvent for sodium perchlorate, at a carbon anode gives $Bu^tN{=}NBu^t$.[198]

The electrochemical oxidation of $N$-methyl-amides leads to stabilized carbonium ions of the type $R^1CO\overset{+}{N}(R^2){=}CH_2$. These are new electrophilic species that are useful in organic synthesis (Scheme 45). Electrolysis of aromatic carboxylic acids in

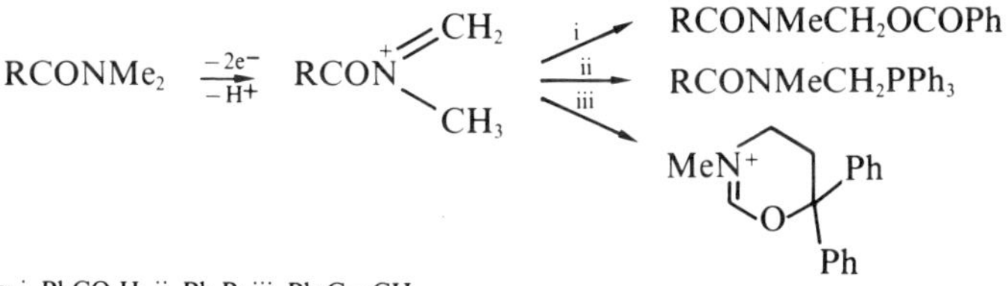

$$RCONMe_2 \xrightarrow[-H^+]{-2e^-} RCO\overset{+}{N}{\diagup}^{CH_2}_{\diagdown CH_3} \xrightarrow{\begin{array}{c}i\\ \underset{iii}{\longrightarrow}\\ ii\end{array}} \begin{array}{l} RCONMeCH_2OCOPh\\ RCONMeCH_2PPh_3\\ Me\overset{+}{N}\!\!\diagup\!\!\diagdown Ph\\ \phantom{xxx}O\diagdown\!\!\diagup\\ \phantom{xxxxxx}Ph \end{array}$$

Reagents: i, $PhCO_2H$; ii, $Ph_3P$; iii, $Ph_2C{=}CH_2$

**Scheme 45**

amide solvents leads to the formation of such intermediates, which are trapped by the added acid.[199] Solutions of these intermediates can be prepared by oxidation of an amide in AN containing lithium perchlorate, and examples are available to illustrate their reactivity towards triphenylphosphine as a nucleophile and to olefins in a $(4\pi + 2\pi)$ electrocyclic reaction.[200]

The related oxidation of $NN$-dialkyl-carbamates (47) in methanol has been studied mechanistically. Evidence from the isotope effect of deuterium substitution α- to the nitrogen atom confirms the view that these reactions in methanol are initiated by electron abstraction from the amide function and not by attack from electrochemically generated methoxyl radicals.[201]

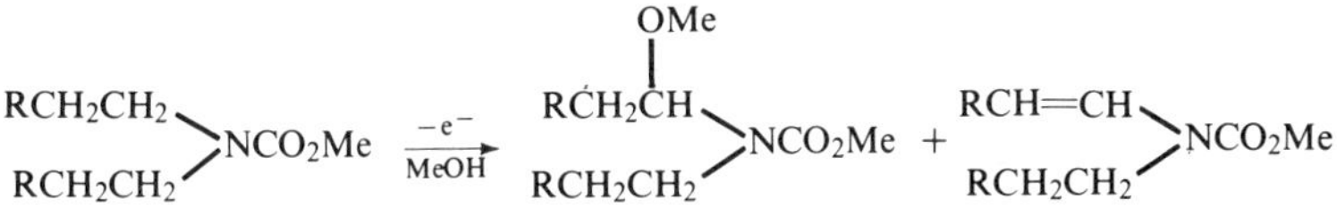

$$\begin{array}{c}RCH_2CH_2\\ RCH_2CH_2\end{array}\!\!\!\!\diagup NCO_2Me \xrightarrow[MeOH]{-e^-} \begin{array}{c}R\dot{C}H_2CH\\ RCH_2CH_2\end{array}\!\!\!\!\diagup\overset{\displaystyle OMe}{\underset{\phantom{x}}{N}}CO_2Me + \begin{array}{c}RCH{=}CH\\ RCH_2CH_2\end{array}\!\!\!\!\diagup NCO_2Me$$

(47)

[196] Y. Takahashi, M. Tokuda, M. Itoh, and A. Suzuki, *Chem. Lett.*, 1975, 523.
[197] W. Suemmermann and U. Deffner, *Tetrahedron*, 1975, **31**, 593.
[198] A. U. Blackham, S. Kwak, and J. L. Palmer, *J. Electrochem. Soc.*, 1975, **122**, 1081.
[199] N. Hirai, Y. Nishimura, K. Takeshita, and S. Arita, *Asahi Garasu Kogyo Gijutsu Shoreikai Kenkyu Hokoku*, 1974, **24**, 79 (*Chem. Abs.*, 1975, **83**, 138 930).
[200] M. Genies, *Bull. Soc. Chim. Fr.*, 1975, 389.
[201] T. Shono, H. Hamaguchi, and Y. Matsumura, *J. Am. Chem. Soc.*, 1975, **97**, 4264.

Electrochemical oxidation of laurohydroxamic acid (48) in AN leads to cleavage of the carbon–nitrogen bond and formation of hyponitrous acid.[202]

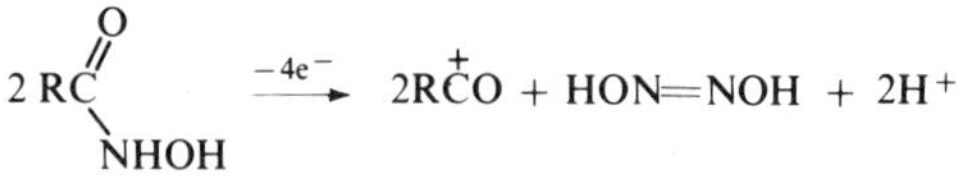

$$2 \text{ RC(O)NHOH} \xrightarrow{-4e^-} 2\text{RCO}^+ + \text{HON=NOH} + 2\text{H}^+$$

(48)

Anodic oxidation of aziridines[203] generates the aza-alkyl cation, which reacts with the solvent methanol (Scheme 46). A stable, identifiable reaction product was isolated from the reaction mixture by reduction with lithium aluminium hydride.

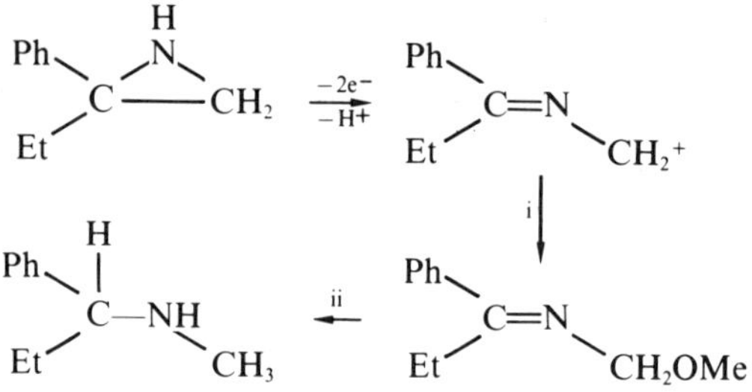

Reagents: i, MeOH; ii, LiAlH₄

**Scheme 46**

Oxidation of aniline in an aqueous buffer solution gives aniline black. The product is inhomogeneous, but with a high density of unpaired electrons, and is a good conductor of electricity.[204]

Methoxy substituents stabilize the reactive intermediates generated by the electro-oxidation of diphenylamines in AN. The hexamethoxy-compound (49)[205] can be shown to undergo a series of reversible electron- and proton-transfer reactions (Scheme 47). The 4,4′-dimethoxy-compound gives more reactive intermediates, which rapidly self-condense to form 2,6-dimethoxy-9,10-di-(4-methoxyphenyl)dihydrophenazine.[205]

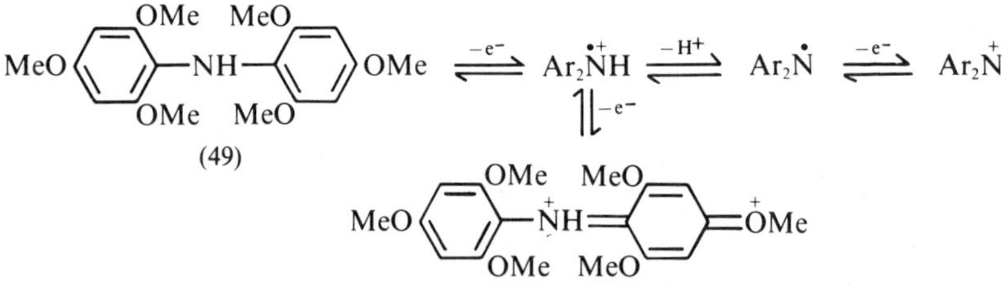

**Scheme 47**

[202] F. Barba, A. Soler, and F. A. Martinez, *Ion (Madrid)*, 1975, **35**, 13.
[203] P. G. Gassman, I. Nishiguchi, and H. Yamamoto, *J. Am. Chem. Soc.*, 1975, **97**, 1600.
[204] F. Bergakad, E. Freiberg, and L. Dunsch, *J. Prakt. Chem.*, 1975, **317**, 409; L. Dunsch, *J. Electroanal. Chem. Interfacial. Electrochem.*, 1975, **61**, 61.
[205] D. Serve, *J. Am. Chem. Soc.*, 1975, **97**, 432.

# Organic Electrochemistry – Synthetic Aspects

In an elegant synthesis of indazoles, $o$-nitrobenzylamines are first reduced at a mercury cathode in an ammonium acetate buffer and the so-formed hydroxylamino-compound is oxidized *in situ* to the nitroso-compound, which cyclizes spontaneously to the indazole (Scheme 48).[206]

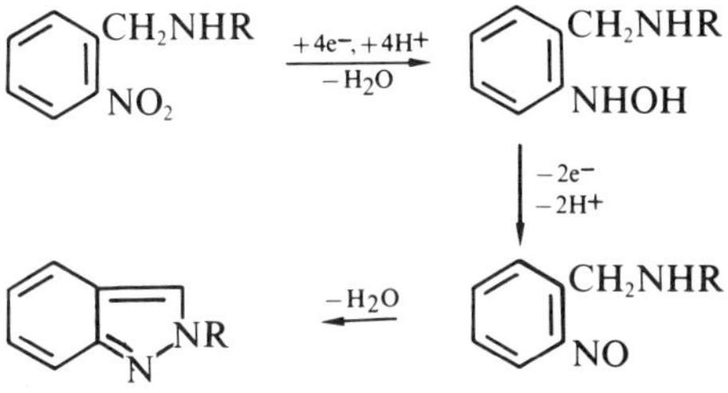

**Scheme 48**

Tetrahydrocinnolines (50) show two oxidation waves in AN containing lithium perchlorate, with sodium carbonate present as a buffer. Oxidation at the potential of the first wave leads to the 1,4-dihydrocinnoline and oxidation at the potential of the second wave completes the dehydrogenation to form cinnoline.[207] Similarly, the cyclic aldehyde-ammonia compounds formed *in situ* in AN undergo oxidative dehydrogenation to *sym*-triazines (51).[208] Steric and electronic effects on the half-wave potentials for the oxidation of 1,4-dihydropyridines have been examined in some detail.[209]

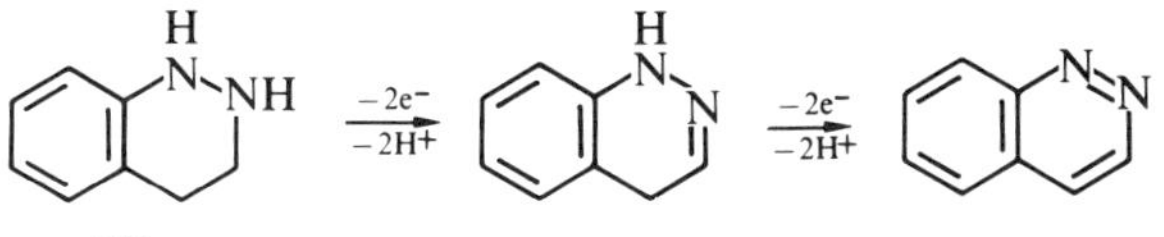

(50)

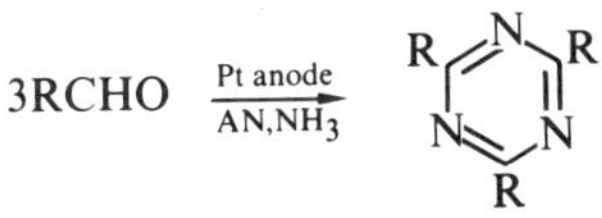

$$3RCHO \xrightarrow{\text{Pt anode}}_{AN, NH_3}$$

(51)

Papers on the electrochemical oxidation of pyrazolines have continued to appear during this year (Scheme 49). Oxidation of pyrazolines in AN containing a pyridine base results in rapid deprotonation of the first-formed radical-anion. The radical that is formed is then oxidized to a carbonium ion located on the 4-position of the pyrazoline ring. A hydrogen on the 5-position is eliminated as a proton to give a pyrazole.[210] If the 5-position bears two substituents, one of these migrates to the

---

[206] R. Hazard and A. Tallec, *Bull. Soc. Chim. Fr.*, 1975, 679.
[207] G. Cauquis, B. Chabaud, and M. Genies, *Bull. Soc. Chim. Fr.*, 1975, 583.
[208] B. F. Becker and H. P. Fritz, *Chem. Ber.*, 1975, **108**, 3292.
[209] J. Stradins, J. Beilis, J. Uldrikjis, G. Duburs, A. E. Sausins, and B. Cekavicius, *Khim. Geterotsikl. Soedin.*, 1975, 1525; J. Stradins, G. Duburs, J. Beilis, J. Uldrikjis, A. Sausins, and B. Cekavicius, *ibid.*, p. 1530.
[210] F. Pragst, H. Köppel, W. Jugel, and F. G. Weber, *J. Electroanal. Chem. Interfacial Electrochem*, 1975, **60**, 323.

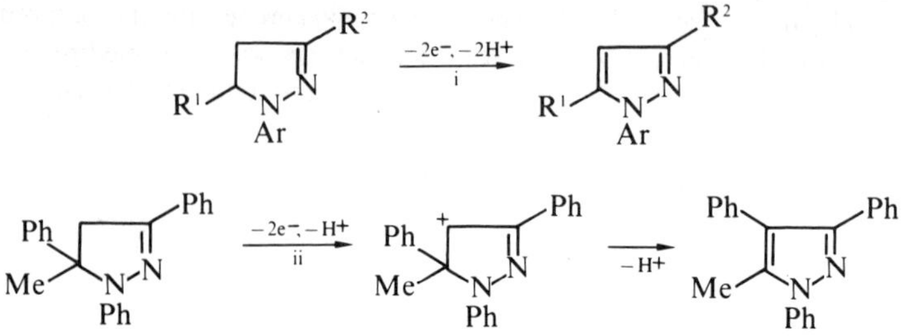

Reagents: i, AN, pyridine; ii, AN, collidine

**Scheme 49**

4-position and then elimination of a proton forms a pyrazole.[211] In some cases, pyridine will trap the intermediate carbonium ion by forming a quaternary ammonium salt, and this reaction is suppressed by using the more hindered collidine as the base.[211]

A number of radical-anions of phosphorus compounds such as (52) have been characterized by e.s.r. spectroscopy.[212] Anodic oxidation of triphenylstibine in AN gives (53), which may be isolated as its perchlorate.[213]

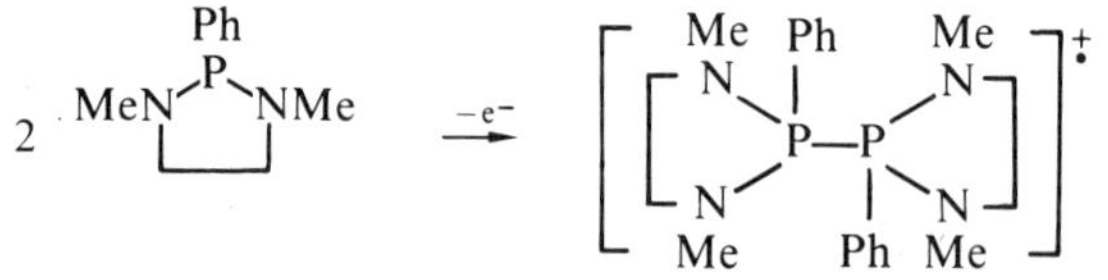

(52)

$$Ph_3Sb + H_2O \xrightarrow[AN]{-2e^-, -2H^+} Ph_3SbOSbPh_3$$

(53)

**Oxygen and Sulphur Compounds.**—Both *cis*- and *trans*-1,2-diols are cleaved in good yields by anodic oxidation in methanol to give the acetal of a dialdehyde (Scheme 50).[214] The reaction is an excellent alternative to the cleavage by lead(IV) acetate and other chemical reagents.

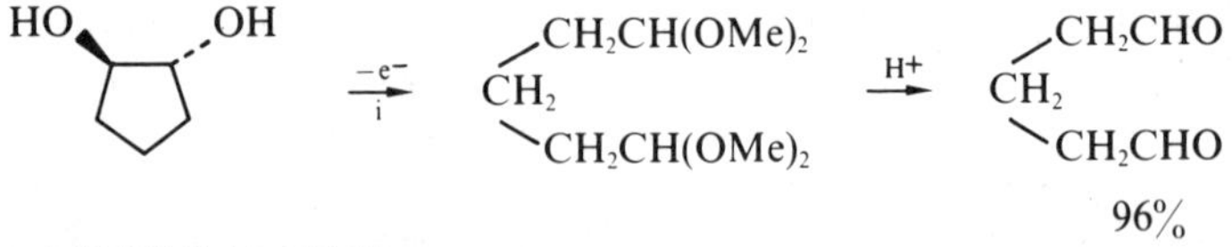

Reagents: i, MeOH, $Et_4N^+$ $C_7H_7SO_3^-$

**Scheme 50**

[211] F. Pragst and C. Boeck, *J. Electroanal. Chem. Interfacial Electrochem.*, 1975, **61**, 47.
[212] W. B. Gara and B. P. Roberts, *J. Chem. Soc., Chem. Commun.*, 1975, 949.
[213] G. Schiavon, S. Zecchin, G. Cogoni, and G. Bontempelli, *J. Electroanal. Chem. Interfacial Electrochem.*, 1975, **59**, 195.
[214] T. Shono, Y. Matsumura, T. Hashimoto, K. Hibino, H. Hamaguchi, and T. Aoki, *J. Am. Chem. Soc.*, 1975, **97**, 2546.

Electrochemical oxidation of aliphatic ketones in AN generates the ketone radical-cation as a first intermediate. α-Branched ketones then undergo α-cleavage and further abstraction of one electron to form two relatively stable carbo-cations, which react with the solvent (Scheme 51).[215] The radical-cation from ketones which lack this α-branch abstracts a proton from the alkyl chain to generate a carbocation centre which can undergo rearrangement before being quenched by the solvent.[216] Proton abstraction from γ-, δ-, and ε-positions is observed for straight-chain ketones. Two examples of proton abstraction followed by rearrangement are given in Scheme 51.

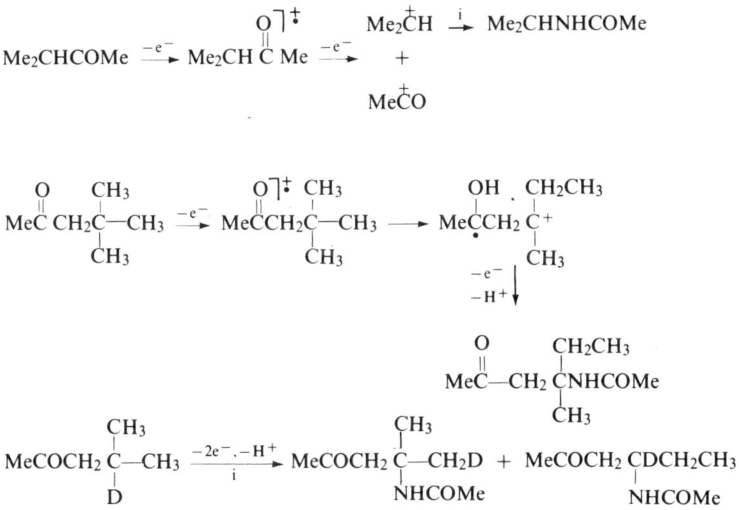

$$Me_2CHCOMe \xrightarrow{-e^-} Me_2CH \overset{O}{\underset{\|}{C}} Me \overset{+\cdot}{\phantom{]}} \xrightarrow{-e^-} \begin{array}{c} Me_2\overset{+}{C}H \xrightarrow{i} Me_2CHNHCOMe \\ + \\ Me\overset{+}{C}O \end{array}$$

$$\underset{\underset{CH_3}{|}}{MeC} \overset{O}{\underset{\|}{C}} CH_2 \overset{|}{\underset{|}{C}} - CH_3 \xrightarrow{-e^-} \underset{\underset{CH_3}{|}}{Me\overset{O}{\underset{\|}{C}}CH_2 \overset{|}{\underset{|}{C}} - CH_3} \overset{+\cdot}{\phantom{]}} \longrightarrow \underset{\underset{CH_3}{|}}{Me\overset{OH}{\underset{\cdot}{C}}CH_2 \overset{CH_2CH_3}{\underset{|}{C}^+}} \xrightarrow[-H^+]{-e^-}$$

$$\underset{D}{MeCOCH_2 \overset{CH_3}{\underset{|}{C}} - CH_3} \xrightarrow[i]{-2e^-, -H^+} MeCOCH_2 \overset{CH_3}{\underset{NHCOMe}{\overset{|}{C}} - CH_2D} + MeCOCH_2 \underset{NHCOMe}{\overset{CH_3}{\underset{|}{C}}DCH_2CH_3}$$

Also: $Me\overset{O}{\underset{\|}{C}} - CH_2 \overset{CH_2CH_3}{\underset{CH_3}{\overset{|}{C}}NHCOMe}$

Reagents: i, AN, H₂O

**Scheme 51**

Acetophenone, on oxidation at a platinum anode in dilute sulphuric acid, gives phenol (2—10%), formaldehyde, and resin.[217] Oxidation at low temperatures of the lithium enolates derived from esters causes dimerization to form succinate esters (Scheme 52).[218]

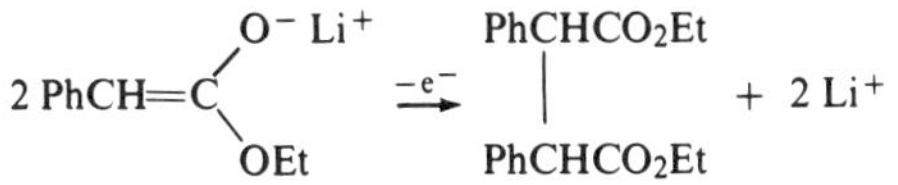

$$2\ PhCH{=}C\underset{OEt}{\overset{O^-Li^+}{\diagup}} \xrightarrow{-e^-} \begin{array}{c} PhCHCO_2Et \\ | \\ PhCHCO_2Et \end{array} + 2\ Li^+$$

**Scheme 52**

U.v. spectroscopy combined with electrochemistry has been used to find evidence for the existence of the pseudohalogen (OCN)₂. The substance may have been

---
[215] J. Y. Becker, L. L. Miller, and T. M. Siegel, *J. Am. Chem. Soc.*, 1975, **97**, 849.
[216] J. Y. Becker, L. R. Byrd, L. L. Miller, and Y. H. So, *J. Am. Chem. Soc.*, 1975, **97**, 853.
[217] A. V. Solomin and E. I. Kryuchkova, *Elektrokhimiya*, 1975, **11**, 1053.
[218] M. Tokuda, T. Shigei, and M. Itoh, *Chem. Lett.*, 1975, 621.

observed as an unstable species, with absorption at 394 nm, by oxidation of tetra-alkylammonium isocyanates in AN.[219] Some further examples have been given of the electrochemical oxidation of thiols to disulphides in AN containing sodium perchlorate.[220] Cyclic trithiocarbonates undergo irreversible two-electron oxidation in AN followed by reaction with water in the solvent to give the cyclic dithiocarbonate and elemental sulphur.[221]

## 4 Electrochemical Halogenation

**Fluorination.**—A study, aimed at increasing product yields, has been made of the electrochemical fluorination of propene.[222] Controlled conditioning of the anode was critical to the reproducibility of reactions. Control of anode potential within specified limits minimized the breakdown of reaction products. Patents have been issued for the electrochemical fluorination of hydrocarbons[223] at a graphite anode in a melt of $KF \cdot 2HF$ at about 93 °C and for the fluorination of alkanoic acids.[224] The fluorination of ethylene glycol and its ethers and acetates[225] and of $N$-alkyl-morpholines and $N$-alkyl-piperidines[226] has been examined. $NN$-Dimethylaniline yields perfluoro-$NN$-dimethylcyclohexylamine; other $NN$-dialkyl-anilines behave similarly.[227] Fluorination of alkyl-pyridines gives a low yield of the perfluoroalkyl-piperidine together with fluorocarbons that have the same carbon content as the original alkyl-pyridine.[228]

**Chlorination and Bromination.**—The electrosynthesis of chloroform by electrolysis of sodium chloride at a lead dioxide anode in aqueous ethanol has been described.[229] Kinetics and mechanism of the anodic bromination of toluene and $p$-xylene in acetic acid have been examined, and the reaction has been shown to have a photochemical component.[230]

## 5 Organometallic Compounds

**General.**—Only the synthetic aspects of the electrochemistry of organometallic compounds are reviewed here. In addition, the literature contains a large body of data on the polarography and cyclic voltammetry of organometallic compounds. Complexes of the type $[ML_3]^{2+}$, where the ligand is 2,2′-bipyridyl, show a series of polarographic waves where electrons are added either to a metal orbital or

---

[219] G. Cauquis and G. Pierre, *Bull. Soc. Chim. Fr.*, 1975, 997.
[220] H. Berge and H. Millat, *Z. Chem.*, 1975, **15**, 37 (*Chem. Abs.*, 1975, **82**, 146 950).
[221] P. R. Moses, J. Q. Chambers, J. O. Sutherland, and D. R. Williams, *J. Electrochem. Soc.*, 1975, **122**, 608.
[222] F. G. Drakesmith and D. A. Hughes, *J. Appl. Electrochem.*, 1976, **6**, 23.
[223] K. L. Mills, U.S. P. 3 882 001 (*Chem. Abs.*, 1975, **83**, 87 414).
[224] T. Suzuki and S. Yahara, Jpn. Kokai. 75 30 827 (*Chem. Abs.*, 1975, **83**, 58 136).
[225] V. V. Berenblit, Yu. P. Dolnakov, G. A. Davydov, V. L. Grachev, and S. V. Sokolov, *Zh. Prikl. Khim. (Leningrad)*, 1975, **48**, 2206.
[226] V. S. Plashkin, *Zh. Prikl. Khim. (Leningrad)*, 1975, **48**, 706.
[227] V. S. Plashkin and Yu. P. Dolnakov, *Zh. Prikl. Khim. (Leningrad)*, 1975, **48**, 702.
[228] V. J. Davis and R. N. Haszeldine, *J. Chem. Soc., Perkin Trans. 1*, 1975, 1263.
[229] E. Dzhafarov, F. G. Bairamov, and V. A. Mukhtarov, *Dokl. Akad. Nauk Az. SSR*, 1975, **31**, 24.
[230] G. Gasalbore, M. Mastragostino, and S. Valcher, *J. Electroanal. Chem. Interfacial Electrochem.*, 1975, **61**, 33.

predominantly to a ligand $\pi^*$-orbital. A conclusion as to which orbital is involved can be drawn by comparing the half-wave potentials of analogues of 2,2'-bipyridyl.[231, 232] If an electron is added to a $\pi^*$-orbital there should be a correlation between the half-wave potential for the complex and that for the free ligand. Such a relationship does not hold if the electron is added to a metal orbital.[232] The electrochemical rate constant for electron transfer to the ligand $\pi$-system (0.1—0.3 cm s$^{-1}$) is significantly different to that for transfer to the metal orbital (0.8—1.3 cm s$^{-1}$).[233]

[Mesotetraphenylporphinato(2)]nickel undergoes reversible oxidation to form a $\pi$-radical-cation at 300 K which shows the expected e.s.r. spectrum. At 77 K the e.s.r. spectrum changes to that expected for the (TPP)Ni$^{III}$ species, and the change is reversible.[234]

Poly-1,1'-ferrocenes (54) show reversible stepwise oxidation, and all the ferrocene groups are oxidizable to ferricinium ions. The mixed-valence ions undergo intramolecular charge transfer when irradiated with near-i.r. light, absorption of light leading to a vibrationally excited oxidation state isomer.[235]

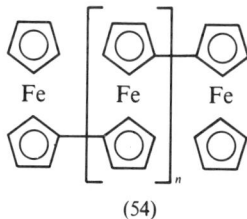

(54)

**Synthesis from Electrode Material.**—The electrosynthesis of metal alkoxides and metal acetonylacetonates by consumption of the metal used as anode material has been described. The solvent for the preparation of alkoxides is the alcohol itself, and some sodium bromide is added as the conducting salt. Phenoxides are obtained by using a solution of phenol in THF. Acetonylacetates are formed by using a solution of acetylacetone in aqueous alcohol. Derivatives of Fe$^{II}$, Co$^{II}$, and Ni$^{II}$ were prepared.[236]

**Induced Ligand Exchange.**—[Cr(phen)$_2$Cl$_2$]$^+$ is reduced with a half-wave potential of $-0.73$ V vs. S.C.E. to the Cr$^{II}$ state, and this rapidly undergoes ligand-exchange reactions. In the presence of excess 1,10-phenanthroline (phen), [Cr(phen)$_3$]$^{2+}$ is formed and can be oxidized to the Cr$^{III}$ state with a half-wave potential of $-0.49$ V.[237]

Cl$_2$Ni(Ph$_3$P)$_2$ is reduced in DME to the Ni$^0$ state and then undergoes ligand exchange with bromotetrafluorobenzene (Scheme 53).[238]

[231] T. Saji and S. Aoyagui, *J. Electroanal. Chem. Interfacial Electrochem.*, 1975, **63**, 405.
[232] T. Saji and S. Aoyagui, *J. Electroanal. Chem. Interfacial Electrochem.*, 1975, **58**, 401.
[233] T. Saji and S. Aoyagui, *J. Electroanal. Chem. Interfacial Electrochem.*, 1975, **63**, 31.
[234] D. Dolphin, T. Niem, R. H. Felton, and I. Fujita, *J. Am. Chem. Soc.*, 1975, **97**, 5288.
[235] G. M. Brown, T. J. Meyer, D. O. Cowan, C. Le Vanda, F. Kaufman, P. V. Roling, and M. D. Rausch, *Inorg. Chem.*, 1975, **14**, 506.
[236] H. Lehmkuhl and W. Eisenbach, *Justus Liebigs Ann. Chem.*, 1975, 672.
[237] D. M. Soignet and L. G. Hargis, *Inorg. Chem.*, 1975, **14**, 941.
[238] W. B. Hughes and D. R. Fahey, U.S. P. 3 887 441 (*Chem. Abs.*, 1975, **83**, 123 303).

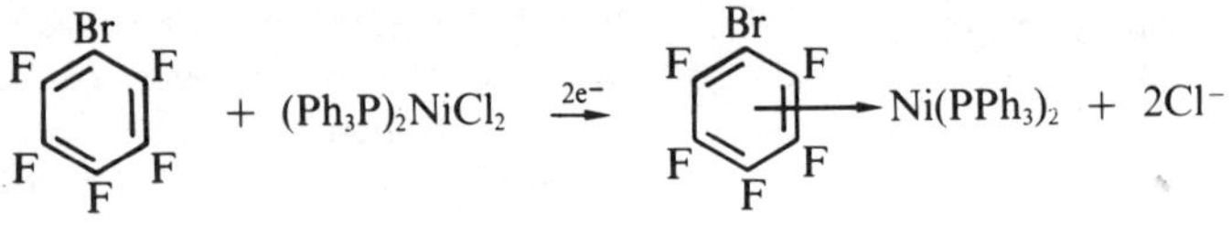

Scheme 53

Iridium and rhodium complexes of 1,2-di(diphenylphosphino)ethane (DPE) undergo a two-electron reduction under an atmosphere of carbon monoxide to form solutions of a metal($-1$) complex, which can be converted into a crystalline derivative by reaction with triphenyltin chloride (Scheme 54).[239]

$$[M(DPE)_2]^+ \xrightarrow[2CO]{2e^-} DPE + [M(DPE)(CO)_2]^-$$

$$\downarrow i$$

M = Ir or Rh  
DPE = Ph$_2$PCH$_2$CH$_2$PPh$_2$

$[M(DPE)(CO)_2(SnPh_3)]$

Reagent: i, Ph$_3$SnCl

Scheme 54

Transformations of some metal carbonyls during electrochemical oxidation or reduction (Scheme 55) have been recorded.[240, 241] Electrochemical reduction of $\mu$-oxo-bis(irontetraphenylporphyrin) in DMF leads to cleavage of the $\mu$-oxo bridge with uptake of two electrons.[242]

$$2\,[M(CO)_6] \xrightarrow[i]{2e^-} [M_2(CO)_{10}]^{2-} + 2CO \quad (M = Cr, Mo, or W)$$

$$[Fe(CO)_5] \xrightarrow{2e^-} [Fe_2(CO)_8]^{2-} + 2CO$$

$$[M_2(CO)_{10}] \xrightarrow[ii]{-2e^-} [M(CO)_5(MeCN)]^+ \quad (M = Mn or Re)$$

$$[Cr(CO)_6] \xrightarrow[iii]{e^-} [Cr_2(CO)_{10}]^{2-} + [HCr_2(CO)_{10}]^-$$

Reagents: i, AN, THF, or CH$_2$Cl$_2$; ii, AN; iii, DMF with a trace of water

Scheme 55

**Induced Ligand Reactivity.**—Cobalticinium salts are reduced in two one-electron stages in DMF. Solutions of the anion obtained by reduction at $-1.9$ V offer possibilities for synthesis. The ions show behaviour similar to that expected of a carbanion and will, for example, react with carbon dioxide by addition to one cyclopentadiene ring. This reaction has been made the basis for synthesis of substituted cobalticinium salts (Scheme 56).[243]
Electrochemical oxidation of alkyl-cobalt(III) chelates (56) leads to reactive

[239] G. Pilloni, G. Zotti, and M. Martelli, *Inorg. Chim. Acta*, 1975, **13**, 213.
[240] C. J. Pickett and D. Pletcher, *J. Chem. Soc., Dalton Trans.*, 1975, 879.
[241] P. Lemoine and M. Gross, *C.R. Hebd. Seances Acad. Sci., Ser. C*, 1975, **280**, 797.
[242] K. M. Kadish, G. Larson, D. Lexa, and M. Momenteau, *J. Am. Chem. Soc.*, 1975, **97**, 282.
[243] N. El Murr and E. Laviron, *Tetrahedron Lett.*, 1975, 875.

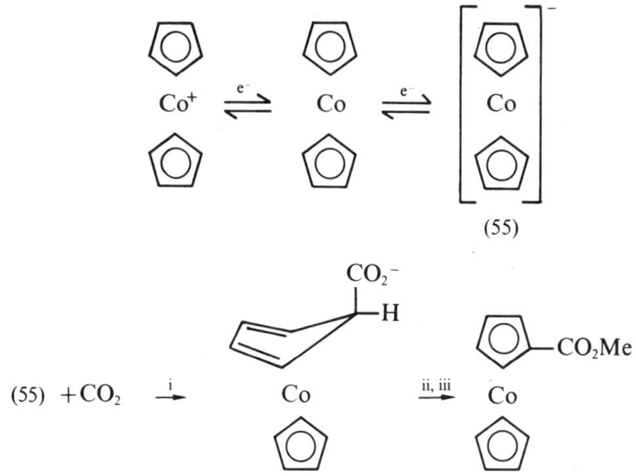


Reagents: i, DMF; ii, MeI; iii, $Ph_3C^+$ $BF_4^-$

**Scheme 56**

cobalt(IV) species in which the alkyl group is susceptible to nucleophilic substitution.[244] Nucleophiles will displace cobalt from the alkyl group, and the metal is recovered as a cobalt(III) complex (Scheme 57). The related rhodium(III) chelates behave in a similar manner.

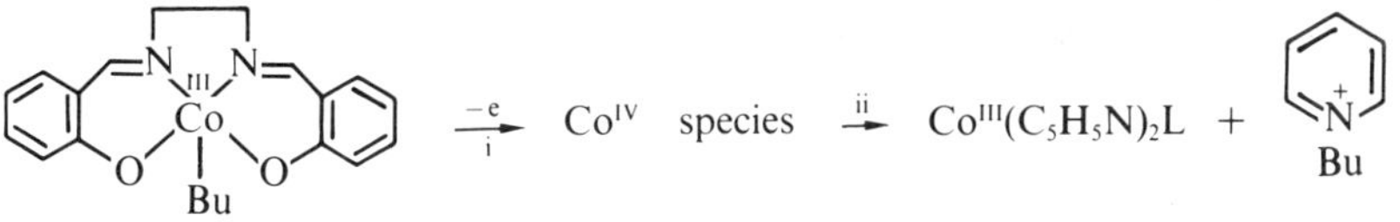


(56) $[Co^{III}(Bu)L]$

Reagents: i, AN; ii, $C_5H_5N$

**Scheme 57**

[244] I. Ya. Levitin, A. L. Sigan, and M. E. Vol'pin, *J. Chem. Soc., Chem. Commun.*, 1975, 469.

# 2
# Membrane Phenomena

BY N. LAKSHMINARAYANAIAH

## 1 Introduction

This Report covers, in the extensive membrane literature of 1974, 1975, 1976, and up to about September of 1977, the topics (except lipid bilayer membranes, excluded due to limitations of space) considered previously in this series.[1-3] During this period, various phenomena arising in lipid bilayer membranes, as opposed to those observed in thick polymeric membranes, have attracted the attention of several schools of researchers all over the world. This may be due to the fact that a lipid bilayer is considered to act as a better model for the biological membrane than a liquid or solid polymeric membrane, and so is used in several studies to understand the mechanisms of material transport and excitability in biological membranes. In addition, this is an area where funds for research are more readily available. Consequently, in the case of polymeric membranes, the major thrust has been in areas where funds for research are relatively easily available. These areas seem to be those where polymeric membranes are exploited as (i) useful separating devices in operations, experimental and otherwise, such as reverse osmosis, hyperfiltration, *etc.*, and (ii) sensing devices to detect and estimate ions, gases, enzymes, *etc*. A rough breakdown of papers appearing in the period under consideration is about 700 related to bilayer membranes and about 900 concerned with solid and liquid membranes, of which nearly 400 deal with several aspects of membranes used as sensing devices, *i.e.* as electrodes.

Membrane processes have acquired so much importance during the past decade that holding symposia in recent years has become almost an annual feature of certain sections of some professional associations and/or societies in different parts of the world. International symposia on fresh water from sea, dealing with several aspects of membranes, are held periodically in different countries of Europe, and the proceedings are published each time in about four volumes (editor: A. Delyaniss; publisher: Amaroussion, Greece). Mano[4] has edited the proceedings of the international symposium on macromolecules in which Meares,[5] in his presentation, has reviewed the separation processes of hyperfiltration and electrodialysis

---

[1] N. Lakshminarayanaiah, in 'Electrochemistry', ed. G. J. Hills (Specialist Periodical Reports), The Chemical Society, London, 1972, Vol. 2, Ch. 5.
[2] N. Lakshminarayanaiah, in 'Electrochemistry', ed. H. R. Thirsk (Specialist Periodical Reports), The Chemical Society, London, 1974, Vol. 4, Ch. 6.
[3] N. Lakshminarayanaiah, in 'Electrochemistry', ed. H. R. Thirsk (Specialist Periodical Reports), The Chemical Society, London, 1975, Vol. 5, Ch. 3.
[4] Proceedings of the International Symposium on Macromolecules, ed. E. B. Mano, Elsevier, Amsterdam, The Netherlands, 1975.
[5] P. Meares, in ref. 4, p. 131.

as applied to desalination of water using ion-selective polymer membranes. In addition, Meares[6] has presented a succinct review concerning molecular interpretation of material transport in polymers in which separation processes involving mixtures of gases and of organic liquids are considered. Proceedings of the study week held in April 1975 of the Pontifical Academy of Sciences are published in the form of a book[7] in which Sollner,[8] in his contribution, has discussed the use of models in the study of complex effects at mosaic membranes. The proceedings of the 1974 ICN–UCLA conference on membranes are published in the *Journal of Supramolecular Structure*,[9] a Journal started in 1973. Two other new Journals[10,11] devoted to several aspects of membrane research have appeared recently. The material presented at the symposium on Drugs and Transport Processes, organized by the Biological Council Coordinating Committee for Symposia on Drug Action, has been published as a book.[12] It contains four chapters on several aspects of thin lipid membranes. Another book[13] also contains material, mostly applications of microelectrodes, presented at a workshop held at Boston University. A book by Ferris,[14] which is probably mistitled, contains a chapter on 'Ion Specific Electrodes' that is very superficial and cursory. It is heavily oriented towards electrical engineering. Three other books on electrodes that have been published during the period under review are by Koryta,[15] Bailey,[16] and Lakshminarayanaiah.[17] One other book, edited by Kessler *et al.*,[18] contains four sections, each composed of several papers written by authorities in the field. Section I contains papers dealing with theoretical and practical aspects of ion-selective electrodes. Section II is composed of papers related to microelectrodes. Section III deals with enzyme electrodes and Section IV deals with applications of electrodes to monitor effects of ion activity in metabolic reactions and in membrane functions.

Several books dealing with membrane principles and applications have also appeared. The book by Tien,[19] although referred to in the earlier review,[3] appeared only in 1974. Volume 3 in the series 'Membranes', edited by Eisenman,[20] contains seven chapters, three of which are related to artificial systems and the rest to biological systems. Three books, one by Lightfoot,[21] the second by Lih,[22] and the

[6] P. Meares, *Pure Appl. Chem.*, 1974, **39**, 99.
[7] 'Biological and Artificial Membranes and Desalination of Water', ed. R. Passino, Elsevier, Amsterdam, The Netherlands, 1976.
[8] K. Sollner, in ref. 7, p. 795.
[9] *Journal of Supramolecular Structure*, Alan R. Liss Inc., New York, 1974, Vol. 2, Nos. 5/6.
[10] *Journal of Dialysis*, ed. K. H. Stenzel and A. L. Rubin, Dekker, New York, 1975—76.
[11] *Journal of Membrane Science*, ed. H. K. Lonsdale, Elsevier, Amsterdam, The Netherlands, 1976.
[12] 'Drugs and Transport Processes', ed. B. A. Callingham, University Park Press, Baltimore, Md., 1974.
[13] 'Ion-selective Microelectrodes', ed. H. J. Berman and N. C. Hebert, Plenum Press, New York, 1974.
[14] C. D. Ferris, 'Introduction to Bioelectrodes', Plenum Press, New York, 1974.
[15] J. Koryta, 'Ion Selective Electrodes', Cambridge University Press, Cambridge, England, 1975.
[16] P. L. Bailey, 'Analysis with Ion Selective Electrodes', Heyden and Son, London, 1976.
[17] N. Lakshminarayanaiah, 'Membrane Electrodes', Academic Press, New York, 1976.
[18] M. Kessler, L. C. Clark, Jr., D. W. Lubbers, I. A. Silver, and W. Simon, 'Ion and Enzyme Electrodes in Biology and Medicine', University Park Press, Baltimore, Md., 1976.
[19] H. T. Tien, 'Bilayer Lipid Membranes (BLM); Theory and Practice', Dekker, New York, 1974.
[20] G. Eisenman, 'Membranes, Vol. 3, Lipid Bilayers and Biological Membranes: Dynamic Properties', Dekker, New York, 1975.
[21] E. N. Lightfoot, 'Transport Phenomena and Living Systems – Biomedical Aspects of Momentum and Mass Transport', Wiley–Interscience, New York, 1974.
[22] M. M. Lih, 'Transport Phenomena in Medicine and Biology', Wiley–Interscience, New York, 1975.

third by Cooney,[23] should prove helpful in understanding the basic principles governing material transport in membranes. The book[24] edited by Sourirajan contains chapters written by authorities in several areas related to reverse osmosis. Similarly, the book[25] edited by Meares has fourteen chapters, very diverse in that each chapter written by an authority deals with a distinct separation or sensing process in which solid and/or liquid membranes are used. Unfortunately, these two books are so heavily priced that they could be found only on the shelves of libraries with money. However, the National Research Council of Canada has recently published a cheaper edition of the book[26] edited by Sourirajan. Two other books, again related to separations using membranes, have appeared. One is by Hwang and Kammermeyer[27] and the other by Madsen.[28] The book by Kotyk and Janacek first published in 1970 has appeared in its second edition[29] and re-appeared (contents almost similar) as Volume 9 in the series 'Biomembranes', edited by Manson.[30] Two serials related to membranes, dealing with several facets of biological membranes, including model systems, are 'Current Topics in Membranes and Transport', edited by Bronner and Kleinzeller,[31] and 'Methods in Membrane Biology', edited by Korn.[32] In another series entitled 'Separation and Purification Methods', edited by Perry et al.,[33] Volumes 3 and 4 contain chapters (three in the former and one in the latter) related to membranes. Similarly, the book 'Biological Membranes'[34] contains some interesting chapters.

A number of extensive reviews of membrane phenomena have appeared. Flynn et al.[35] have reviewed the theoretical principles underlying membrane permeation and transport, placing emphasis on the usefulness of models. Similarly, Bamberg et al.[36] have reviewed ion transport across membranes. Three other reviews of interest are one by Buck,[37] dealing with electroanalytical chemistry of membranes, and the second by Meares,[38] who has considered the mechanism of transport of water in membranes. The third, by Schultz and co-workers,[39,40] contains two parts and deals with facilitated transport that is mediated by carriers in membranes.

[23] D. Cooney, 'Biomedical Engineering Principles: An Introduction to Analysis of Transport Processes', Dekker, New York, 1976.
[24] 'Reverse Osmosis and Synthetic Membranes; Theory, Technology and Engineering', ed., S. Sourirajan, Dekker, New York, 1976.
[25] 'Membrane Separation Processes', ed. P. Meares, Elsevier, Amsterdam, The Netherlands, 1976.
[26] 'Reverse Osmosis and Synthetic Membranes: Theory, Technology and Engineering', ed. S. Sourirajan, National Research Council of Canada, Ottawa, Canada, 1977.
[27] S. T. Hwang and K. Kammermeyer, 'Membranes in Separation', Wiley–Interscience, New York, 1975.
[28] R. F. Madsen, 'Hyperfiltration and Ultrafiltration in Plate and Frame Systems', Elsevier, Amsterdam, The Netherlands, 1977.
[29] A. Kotyk and K. Janacek, 'Cell Membrane Transport, Principles and Techniques', Plenum Press, New York, 1975.
[30] 'Biomembranes', ed. L. A. Manson, Plenum Press, New York, 1977, Vol. 9.
[31] 'Current Topics in Membranes and Transport', ed. F. Bronner and A. Kleinzeller, Academic Press, New York, 1974, Vol. 5; 1975, Vols. 6, 7; 1976, Vol. 8.
[32] 'Methods in Membrane Biology', ed. E. D. Korn, Plenum Press, New York, 1974, Vols. 1, 2; 1975, Vols. 3—5; 1976, Vols. 6, 7; 1977, Vol. 8.
[33] 'Separation and Purification Methods', ed. E. S. Perry, C. J. van Oss, and E. Grushka, Dekker, New York, 1975, Vol. 3; 1976, Vol. 4.
[34] 'Biological Membranes', ed. D. S. Parsons, Oxford University Press, 1975.
[35] G. L. Flynn, S. H. Yalkowsky, and T. J. Roseman, *J. Pharm. Sci.*, 1974, **63**, 479.
[36] E. Bamberg, R. Benz, P. Läuger, and G. Stark, *Chem. Unserer Zeit*, 1974, **8**, 33.
[37] R. P. Buck, *Crit. Rev. Anal. Chem.*, 1976, **5**, 323.
[38] P. Meares, *Phil. Trans. R. Soc. London, Ser. B*, 1977, **B278**, 113.
[39] J. S. Schultz, J. D. Goddard, and S. R. Suchadevo, *AIChE J.*, 1974, **20**, 417.
[40] J. D. Goddard, J. S. Schultz, and S. R. Suchadevo, *AIChE J.*, 1974, **20**, 625.

# Membrane Phenomena

Applications of the principles of irreversible thermodynamics to several membrane phenomena have been outlined by Staverman and Smit.[41] Sollner has published two reviews, in one of which he has outlined the basic electrochemistry of porous membranes;[42] in the other[43] he has traced his links to famous investigators such as Ostwald, Nernst, Haber, Donnan, Michaelis, Freundlich, and Höber, who at various stages in his scientific development inspired him to pursue his own investigations into several membrane phenomena related to the electrochemical behaviour of polymer membranes. Several types of membranes and their properties[44,45] and the use of a membrane as a barrier to control the release of drugs, insecticides, and other chemicals[46] have been summarized. In other reviews, several fields of application of membranes have been treated. Some of these relate to the use of synthetic membranes as models,[47] their use in material separations[48] and in the separation of substances by dialysis and reverse osmosis,[49,50] ion-exchange membranes[51] and their use in electrochemical cells,[52] membranes in the treatment of fermentation waste water,[53] raw sewage, and primary and secondary effluents,[54] desalting in sugar and milk processing,[55] and in the food industry,[56–58] the purification of waste waters from several processing industries,[59–62] and medical uses related to the purification of blood (haemodialysis and haemodiafiltration).[63,64]

## 2 Theoretical Considerations

Transport of ions from one electrolyte solution to another through a membrane involves the following steps: (i) movement through the Nernst stagnant aqueous layer next to the membrane, (ii) transfer through the diffuse double layer adjacent to the membrane, (iii) transfer from the aqueous into the membrane phase, (iv) transport through the membrane, (v) transfer back from membrane into the aqueous phase, and movement again (vi) through the diffuse double layer and (vii) through the stagnant aqueous layer. Steps (i), (ii), and (iii) have been treated

---

[41] A. J. Staverman and J. A. M. Smit, in Phys. Chem: Enriching Top. Colloid Surf. Sci., 1975, 343.
[42] K. Sollner, *J. Dent. Res.*, 1974, **53**, 267.
[43] K. Sollner, in 'Charged Gels and Membranes: I', ed. E. Selegny, Reidel Publishing Co., Dordrecht, Holland, 1976, p. 3.
[44] D. Vofsi and J. Jagur-Grodzinski, *Naturwissenchaften*, 1974, **61**, 21.
[45] B. Richter and R. Voigt, *Pharmazie*, 1974, **29**, 3.
[46] R. W. Baker and H. K. Lonsdale, *Chem. Technol.*, 1975, **5**, 668.
[47] T. Nakagawa, *Kobunshi*, 1974, **23**, 230.
[48] B. Philipp and H. J. Purz, *Z. Chem.*, 1975, **15**, 81.
[49] T. Sakai, *Geppo*, 1975, **28**, 54.
[50] A. Suzuki, *Kagaku (Kyoto)*, 1975, **64**, 33.
[51] W. Pusch, *Chem.-Ing.-Tech.*, 1975, **47**, 914.
[52] G. Richter, *Chem.-Ing.-Tech.*, 1975, **47**, 909.
[53] K. Hashimoto, *Shokuhin Kogyo*, 1974, **17**, 29.
[54] D. C. Sammon and B. Stringer, *Process Biochem.*, 1975, **10**, 4.
[55] R. Ehara, Y. Takatori, N. Okonogi, and M. Tomita, *Shokuhin Kogyo*, 1974, **17**, 58.
[56] S. Kimura, *Shokuhin Kogyo*, 1974, **17**, 20.
[57] K. Hashimoto, *Shokuhin Kogyo*, 1974, **17**, 27.
[58] K. Hashimoto, *Kagaku (Kyoto)*, 1975, **64**, 69.
[59] U. F. Franck, *DECHEMA Monogr.*, 1974, **75**, 9.
[60] A. Watanabe, *Shokuhin Kogyo*, 1974, **17**, 51.
[61] D. C. Sammon, *Pure Appl. Chem.*, 1974, **37**, 423.
[62] E. Staude, *Chem.-Ztg.*, 1975, **99**, 220.
[63] N. Nakabayashi, *Hyomen*, 1974, **12**, 31.
[64] C. F. Gutch, *Ann. Rev. Biophys. Bioeng.*, 1975, **4**, 405.

recently within the framework of Nernst–Planck and Poisson equations and related to step (iv). Current–voltage relations [65] and admittance characteristics[66] have been described, taking into consideration the effects of diffusion of ions in the aqueous phases adjacent to the membrane, of double-layer charges, and of partition equilibrium between water/membrane interfaces. The considerations presented for steps (i), (ii), and (iii) should apply to steps (v), (vi), and (vii), as the latter are the mirror images of the former. The admittance characteristics of membranes containing ion carriers and subject to constancy of electric field have also been described.[67]

Step (iv) has been described by several investigators,[68–70] utilizing an integrated form of Nernst–Planck flux equation [equation (1), where $J_i$ is flux of species $i$ of valence $z_i$ and mobility $u_i$, $C_i$ is concentration/mol l$^{-1}$, and $x$ is the distance in the diffusion layer extending from $x=0$ at one face to $x=d$ at the other face, where $d$ is the thickness of the diffusion layer; $E$ is the electric field and $R$, $T$, and $F$ have their usual meaning].

$$J_i = -u_i RT \frac{dC_i(x)}{dx} - z_i u_i C_i F \frac{dE(x)}{dx} \qquad (1)$$

Equation (1), when applied to a constrained diffusion zone (i.e. a porous membrane), was integrated by Planck (see McInnes[71]). The derivation, however, is complex and applicable to solutions of univalent ions. A less involved derivation has been given by Morf,[72] who showed that the diffusion potential $E_d$ [$E_d = E(d) - E(0)$] is given by

$$E_d = \frac{u'_+ - u'_-}{|z_+|u'_+ + |z_-|u'_-} \left(\frac{RT}{F}\right) \ln\left(\frac{\Sigma C_i(0)}{\Sigma C_i(d)}\right) \qquad (2)$$

where $u'_i$ is the mean mobility for each ion [cation ($+$) or anion ($-$)] and is defined as

$$u'_+ = \frac{\Sigma J_+}{\Sigma(J_+/u_+)} = \text{constant (x)} \qquad (3a)$$

$$u'_- = \frac{\Sigma J_-}{\Sigma(J_-/u_-)} = \text{constant (x)} \qquad (3b)$$

When $J_i$ is constant (steady state), integration of equation (1) gives[72, 73]

$$J_i = -\left(\frac{u_i RT\ C_i(d)\exp[z_i FE(d)/RT] - C_i(0)\exp[z_i FE(0)/RT]}{\int_0^d \exp[z_i FE(x)/RT]dx}\right) \qquad (4)$$

Equation (4) combined with equation (3) gives the value for the mean mobility. Thus

$$u'_i = \frac{\Sigma u_i C_i(d)\exp(z_i E_d F/RT) - \Sigma u_i C_i(0)}{\Sigma C_i(d)\exp(z_i E_d F/RT) - \Sigma C_i(0)} \qquad (5)$$

[65] R. de Levie and N. G. Seidah, *J. Membr. Sci.*, 1974, **16**, 1.
[66] R. de Levie, N. G. Seidah, and H. Moreira, *J. Membr. Biol.*, 1974, **16**, 17.
[67] R. de Levie, *J. Electroanal. Chem. Interfacial Electrochem.*, 1975, **58**, 203.
[68] N. Lakshminarayanaiah, 'Transport Phenomena in Membranes', Academic Press, New York, 1969.
[69] R. de Levie and H. Moreira, *J. Membr. Biol.*, 1972, **9**, 241.
[70] R. de Levie, N. G. Seidah, and H. Moreira, *J. Membr. Biol.*, 1972, **10**, 171.
[71] D. A. McInnes, 'Principles of Electrochemistry', Dover, New York, 1961.
[72] W. E. Morf, *Anal. Chem.*, 1977, **49**, 810.
[73] W. E. Morf, P. Wuhrmann, and W. Simon, *Anal. Chem.*, 1976, **48**, 1031.

With the help of equations (5) (assign a value for $E_d$ to get values for $u'_+$ and $u'_-$) and (2) (use values of $u'_+$ and $u'_-$ to calculate $E_d$), $E_d$ may be evaluated by iterative methods. In most cases, only a few iterative steps are needed to realize the final value.

An interesting aspect of this development is that when $\Sigma C_i(0) = \Sigma C_i(d)$, i.e. the conditions prevailing in biological systems, equation (2) cannot be used. However, the resulting condition, $|z_+|u'_+ + |z_-|u'_- = 0$, when substituted into equation (5) gives, on simplification, the equation for the diffusion potential known in the biological literature as the constant field equation (no assumption of constant field made here). That is

$$E_d = \left(\frac{RT}{F}\right) \ln \left(\frac{\Sigma u_+ C_+(0) + \Sigma u_- C_-(d)}{\Sigma u_+ C_+(d) + \Sigma u_- C_-(0)}\right) \quad (6)$$

for $|z_+| = |z_-| = 1$.

This treatment has been extended by Morf[72] to describe membrane potential, $E_m$, arising across a charged membrane. In accordance with the concepts developed by Teorell and by Meyer and Sievers (see Lakshminarayanaiah[74] for a summary), $E_m$ is considered to arise from two contributions, namely intramembrane diffusion potential and a boundary potential. Thus

$$E_m = E^{II} - E^I = \underbrace{[E^{II} - E(d)] - [E^I - E(0)]}_{E_b = \text{Boundary potential}} + \underbrace{[E(d) - E(0)]}_{E_d = \text{Diffusion potential}} \quad (7)$$

Where I and II indicate the two solutions contacting the membrane boundaries at $x = 0$ and $x = d$.

The distribution of ions at each interface was approximated by the relations

$$\Sigma C_i(0) = \Sigma K_i a_i^I \exp\{-z_i F(E^I - E(0))/RT\} \quad (8a)$$

$$\Sigma C_i(d) = \Sigma K_i a_i^{II} \exp\{-z_i F[E^{II} - E(d)]/RT\} \quad (8b)$$

where $K_i$ is the partition coefficient and the $a_i$'s are activities of outside solutions. The boundary potential $E_b$ (or Donnan potential[75]) is therefore given by

$$E_b = \left(\frac{RT}{z_i F}\right) \ln \left(\frac{\Sigma K_i a_i^{II}}{\Sigma C_i(d)}\right) - \left(\frac{RT}{z_i F}\right) \ln \left(\frac{\Sigma K_i a_i^I}{\Sigma C_i(0)}\right) \quad (9)$$

Combining equation (9) with an appropriate form of equation (2) for $E_d$, viz.

$$E_d = (1 - t_-)\left(\frac{RT}{z_+ F}\right) \ln \left(\frac{\Sigma C_+(0)}{\Sigma C_+(d)}\right) + t_-\left(\frac{RT}{z_- F}\right) \ln \left(\frac{\Sigma C_-(0)}{\Sigma C_-(d)}\right) \quad (10)$$

where $t_-$, the integral anionic transference number, defined as

$$t_- = \frac{|z_-|u'_-}{|z_+|u'_+ + |z_-|u'_-} \quad (11)$$

gives the value for $E_m$. Thus

$$E_m = (1 - t_-)\left(\frac{RT}{z_+ F}\right) \ln \left(\frac{\Sigma K_+ a_+^I}{\Sigma K_+ a_+^{II}}\right) + t_-\left(\frac{RT}{z_- F}\right) \ln \left(\frac{\Sigma K_- a_-^I}{\Sigma K_- a_-^{II}}\right) \quad (12)$$

[74] N. Lakshminarayanaiah, 'Membrane Electrodes', Academic Press, New York, 1976, pp. 65—68.
[75] Y. Shinagawa, *J. Theor. Biol.*, 1977, **64**, 551.

The mean ionic mobility now is defined by

$$u_i' = \frac{\Sigma u_i K_i a_i^{II} \exp(z_i F E_m/RT) - \Sigma u_i K_i a_i^{I}}{\Sigma K_i a_i^{II} \exp(z_i F E_m/RT) - \Sigma K_i a_i^{I}} \qquad (13)$$

Equations (12) and (13) may be utilized to describe membrane selectivity ranging from cationic to anionic.[72]

A classical thermodynamic approach has been applied to calculate the e.m.f. of the galvanic cell

$$\text{Ag, AgCl}|\text{KCl}|C_K^I \ C_{Na}^I \ C_{Cl}^I \ C_R^I||\text{Membrane}||C_K^{II} \ C_{Na}^{II} \ C_{Cl}^{II} \ C_R^{II}|\text{KCl}|\text{AgCl, Ag}$$
$$3\text{ M} \hspace{8cm} 3\text{ M}$$

The membrane is permeable to water, $K^+$, $Na^+$, and $Cl^-$ ions and impermeable to $R^-$ anion, a situation approximating a biological membrane in which separate channels, charged or neutral, exist for transport of cations and anions. The e.m.f. of the cell is given by[76]

$$E_m = \frac{RT}{F}\left[(1-\bar{t}_{Cl})\ln\left(\frac{C_{Na}^I + (\bar{u}_K/\bar{u}_{Na})\,K\,C_K^I}{C_{Na}^{II} + (\bar{u}_K/\bar{u}_{Na})\,K\,C_K^{II}}\right) - \bar{t}_{Cl}\ln\left(\frac{C_{Cl}^I}{C_{Cl}^{II}}\right)\right] \qquad (14)$$

(the overbars refer the parameters to the membrane phase).

When $\bar{t}_{Cl} = 0$, equation (14) roughly approximates equation (6) in which the equilibrium constant, $K_{eq}$, for the exchange reaction

$$K^+ + \bar{N}a^+ \rightleftharpoons Na^+ + \bar{K}^+$$

is not included.

A unified theoretical treatment of ion transport through membranes, combining the Nernst–Planck equations and those of Eyring and Markin, has been presented.[77] Transference of univalent and bivalent cations and of univalent anions through thick carrier membranes has been considered theoretically.[73] A correlation, agreeing with experimental data, between the selectivity of the membrane to the permeating species as measured by ion transport and as observed potentiometrically has been derived. The physico-chemical basis of ion selectivity in biological and model membranes has been discussed at length.[78] It has been shown[79] that a membrane, particularly the uncharged lipid one, will exhibit little selectivity among small ions if the rates of ion permeation are controlled more by interfacial free energies than by activation energies of diffusion. A linear relation between mobility and activity coefficient of small ions in a charged membrane has been derived, using a cell model in which the macro-ion fixed to the membrane matrix is assumed to exist at the centre of the cell.[80] A quantitative study of membrane diffusion to determine molecular size and to monitor changes in molecular size following changes in solvent composition or temperature has been presented.[81] General principles governing irreversible processes in open discontinuous systems have been

---

[76] T. Eørland and T. Østvold, *J. Membr. Biol.*, 1974, **16**, 101.
[77] R. de Levie and K. M. Abbey, *J. Theor. Biol.*, 1976, **56**, 151.
[78] G. Eisenman and S. J. Krasne, in *MTP Int. Rev. Sci. Biochem. Ser.*, 1975, **2**, 27.
[79] R. C. Macdonald, *Biochim. Biophys. Acta*, 1976, **448**, 193.
[80] M. Kamo, *Chem. Pharm. Bull.*, 1975, **23**, 3146.
[81] P. E. O'Connor and M. G. Harrington, *Analyst (London)*, 1976, **101**, 892.

# Membrane Phenomena 47

outlined[82] and the same have been applied to the consideration of electrokinetic effects.[83]

A pore model has been developed to describe the osmotic flow of electrolyte through a charged membrane separating electrolyte solutions of unequal concentration.[84] Transport equations have been developed to describe the non-linear effects observed in membranes that separate two aqueous solutions and are far from equilibrium.[85] Also, simplified flux equations for transport across ion-exchange membranes, particularly to describe the Donnan dialysis process in which $Na^+$ ions are exchanged for $Ca^{2+}$ ions, taking into account the effects due to boundary layers, have been given.[86] The effects of boundary layers on the membrane potential have also been considered.[87] It has been shown that the boundary layer effects were primarily on the Donnan components of the total membrane potential. The effect of electron transfer on membrane potential has been worked out.[88] Time-dependent Nernst–Planck and Poisson equations applied to a membrane separating aqueous solutions containing several ionic species have been solved numerically.[89] Similarly, the transfer of solute through a membrane has been considered from the standpoint of a time-dependent diffusion equation applied to a composite medium.[90] A continuum electrostatic approach, expressing dielectric heterogeneity in membranes and its influence on ion sorption, has been presented.[91] Theoretical calculations show that the membrane/water partition coefficient of ions ($K$) is given by

$$K = \exp\left(-\frac{q^2(\varepsilon_w - \varepsilon_m)}{2bkT\varepsilon_w\varepsilon_m}\right)$$

where $q$ is the charge on the ion of spherical radius $b$, $k$ is the Boltzmann constant, and $\varepsilon$ is the dielectric constant of water (w) and of membrane (m). Again, on the basis of electrostatics, using the method of images, the partition coefficient of ionized solute between membrane and water has been calculated.[92] The partition coefficients were found to be independent of thickness when the membrane thickness exceeded 100 Å. The strong dependence of ionized solute partition on membrane thickness of thin membranes suggests a lower limit to the thickness of the solute-rejecting layer in reverse osmosis membranes. Consideration of ion transport in thin membranes from the precepts of the theory of surface recombination in semiconductors showed that ion flow consisted of two major components: (i) surface-barrier-jumping current and (ii) surface recombination current.[93]

Simulation studies related to electrochemical properties of associated and non-associated liquid ion-exchange membranes[94] and those related to the effects of

[82] R. Haase, *Z. Phys. Chem. (Frankfurt am Main)*, 1976, **103**, 225.
[83] R. Haase, *Z. Phys. Chem. (Frankfurt am Main)*, 1976, **103**, 235, 247.
[84] E. A. Marshall, *J. Theor. Biol.*, 1977, **66**, 107.
[85] R. Schlogl, G. Wiedner, and D. Woermann, *Ber. Bunsenges. Phys. Chem.*, 1975, **79**, 878.
[86] R. P. Wendt, E. Klein, and S. Lynch, *J. Membr. Sci.*, 1976, **1**, 165.
[87] R. J. French, *Biophys. J.*, 1977, **18**, 53.
[88] D. Walz and O. Kedem. *J. Membr. Sci.*, 1977, **2**, 23.
[89] J. P. Meyer and M. D. Kostin, *J. Chem. Phys.*, 1974, **61**, 4067.
[90] J. P. Meyer and M. D. Kostin, *Bull. Math. Biol.*, 1976, **38**, 527.
[91] J. E. Anderson and W. Pusch, *Ber. Bunsenges. Phys. Chem.*, 1976, **80**, 846.
[92] J. E. Anderson and H. W. Jackson, *J. Phys. Chem.*, 1974, **78**, 2259.
[93] B. Y. Woo and L. Y. Wei, *Bull. Math. Biol.*, 1974, **36**, 247.
[94] F. S. Stover and R. P. Buck, *Biophys. J.*, 1976, **16**, 753.

membrane separation parameters using hollow-fibre geometry and liquid/liquid dialysis under ideal conditions[95] have been performed.

Generally, transport across a membrane is considered to be mediated by pores or channels existing in the membrane. Permeability of ions through pores has been considered, taking into account the gradients of both concentration and electric field, *i.e.* the Nernst–Planck flux equation. The current–voltage characteristic of the pore has been shown to attain saturation.[96] When the pore became narrow, the tracer diffusion flow became equal to the osmotic flow. This result has been shown to result from the compensating effects of solvent–solvent interaction on the partitioning of bulk solvent into the pore and on the diffusion rate within the pore.[97] Again, diffusional transport in small pores has been treated from the standpoint of hydrodynamics incorporating the characteristics of Brownian motion.[98] The mechanism of osmotic flow in porous membranes has been considered from the precepts of classical transport and thermodynamic relations.[99] Expressions for the reflection coefficient $\sigma$ as a function of solute dimension and shape have been derived for long cylindrical pores of circular cross-section. The expression for $\sigma$ that is applicable for a spherical macromolecule, *i.e.* $\sigma = (1-K)^2$, where $K$ is the solute distribution coefficient, has been found to be different from those existing in the literature.

The non-electrolyte flux ($J_s$) and volume flux ($J_v$) across a membrane are described in terms of two equations derived from the application of the principles of irreversible thermodynamics.[100] These equations are

$$J_s = \omega RT\Delta C + (1-\sigma_s)J_v\bar{C} \tag{15}$$

$$J_v = L_p(\Delta P - \sigma_v RT\Delta C) \tag{16}$$

where $\Delta C$ is the concentration difference, $\Delta P$ the pressure difference, $\omega$ is the diffusive permeability of solute, $L_p$ is hydraulic permeability, and $\sigma_s$ and $\sigma_v$ are reflection coefficients. The Onsager reciprocity relation leads to equivalence of $\sigma_s$ and $\sigma_v$. This equality between $\sigma_s$ and $\sigma_v$ has been proved, using a continuum analysis of transport through uniform pores.[101] In this development, the expressions (17) and (20) for $J_s$ and $J_v$ have been derived.

$$J_s = (1-\sigma_s)J_v(C_2 e^k - C_1)/(e^k - 1) \tag{17}$$

$$k = -(1-\sigma_s)J_v/\omega RT \tag{18}$$

$\sigma_s$ has been defined by the equation

$$\sigma_s = 1 - 2\lambda^2 \int_0^{(1/\lambda)-1} G_\lambda \beta \, d\beta \tag{19}$$

---

[95] C. H. Lee and E. Perry, *Sep. Sci.*, 1975, **10**, 21.
[96] P. Läuger, *Biochim. Biophys. Acta*, 1976, **455**, 493.
[97] G. S. Manning, *Biophys. Chem.*, 1975, **3**, 147.
[98] J. L. Anderson and J. A. Quinn, *Biophys. J.*, 1974, **14**, 130.
[99] J. L. Anderson and D. M. Malone, *Biophys. J.*, 1974, **14**, 957.
[100] A. Katchalsky and P. F. Curran, 'Nonequilibrium Thermodynamics in Biophysics,' Harvard University Press, Cambridge, Mass., 1965.
[101] D. G. Levitt, *Biophys. J.*, 1975, **15**, 533.

where $\lambda=(a/r)$, $a$ is the solute radius, $r$ is the pore radius, $G_\lambda$ is the drag function, $\beta=(b/a)$, and $b$ is the distance between the pore axis and the particle centre.

$$J_v = L_p[\Delta P - (1 - 2\lambda^2 G_{\lambda(\text{av})})RT\Delta C] \quad (20)$$

and
$$L_p = L_p^0/(1 + P'C_{(\text{av})}) \quad (21)$$

$$P' = -2\eta l N \lambda^4 J_{\lambda(\text{av})} L_p^0/\pi A \quad (22)$$

$\eta$ is viscosity, $l$ is the length of the pore, $N$ is Avogadro's number, $L_p^0$ is the hydraulic permeability in the absence of solute,

$$C_{(\text{av})} = (1/l)\int_0^l C(x)\mathrm{d}x,$$

and average drag functions, $J_{\lambda(\text{av})}$ and $G_{\lambda(\text{av})}$, are given by

$$J_{\lambda(\text{av})} = \int_0^{(1/\lambda)-1} J_\lambda \beta \mathrm{d}\beta; \quad G_{\lambda(\text{av})} = \int_0^{(1/\lambda)-1} G_\lambda \beta \mathrm{d}\beta$$

Comparing equation (20) with equation (16) gives

$$\sigma_v = 1 - 2\lambda^2 G_{\lambda(\text{av})} = 1 - 2\lambda^2 \int_0^{(1/\lambda)-1} G_\lambda \beta \mathrm{d}\beta = \sigma_s$$

This continuum analysis has also been extended to the description of flows in non-uniform pores.[102] The equations are very complex but they approximate equation (16) [in which $\sigma_v = \sigma_s = \sigma$] and equations (17) and (18) when applied to an ideal case of dilute solution.

Equations (15) and (16), based on irreversible thermodynamics, are derived without consideration of membrane structure. These equations have been compared with results derived from considerations of specific membrane models.[103] It is shown that an exact equation for solute flux across an inert porous membrane cannot be given in terms of $\sigma$, $\omega$, and $L_p$ unless the membrane is perfectly homoporous. The effect of heteroporosity on the flux equations for both sieving and non-sieving porous membranes has been considered in detail.[104] The influence of heteroporosity on the kinetics of non-electrolyte tracer flows has also been detailed.[105] It is shown that apparent 'exchange diffusion' by a mobile carrier or 'single file diffusion' through a narrow channel could arise as a consequence of membrane heterogeneity.

When the flow of solvent through a membrane becomes large, equation (15) may not be adequate, and so an alternative equation [equation (23)], applicable to a simple sieving membrane, has been developed,[106] where $P_s$ is the solute permeability and is equal to $(D_s/d)$

$$J_s = \frac{(1-\sigma)J_v\{C_2 - C_1 \exp[(1-\sigma)J_v/P_s]\}}{1 - \exp[(1-\sigma)J_v/P_s]} \quad (23)$$

[102] D. G. Levitt, *Biophys. J.*, 1975, **15**, 553.
[103] E. H. Bresler, E. A. Mason, and R. P. Wendt, *Biophys. Chem.*, 1976, **4**, 229.
[104] R. P. Wendt, E. A. Mason, and E. H. Bresler, *Biophys. Chem.*, 1976, **4**, 237.
[105] J. H. Li and A. Essig, *J. Membr. Biol.*, 1976, **29**, 255.
[106] L. Axel, *Bull. Math. Biol.*, 1976, **38**, 671.

When $\sigma=0$, equation (23) becomes

$$J_s = \frac{J_v[C_2 - C_1 \exp(J_v/P_s)]}{1 - \exp(J_v/P_s)} \quad (24)$$

which is an equation that was derived by Bresler and Wendt.[107] Equation (23) was derived some time ago by Patlak et al.[108] to describe solute flow through a thick homogeneous membrane.

When volume flow $J_v$ is small, equation (23) linearizes approximately to equation (15), which when applied to describe the solute flow across a double membrane made up of two membranes of identical permeabilities but different reflection coefficients ($\sigma_1$ and $\sigma_2$) gave physically unreasonable results. On the other hand, application of equation (23) gave reasonable results. Despite the limited range of applicability of equations (15) and (16), they have been used to describe the flow characteristics of double-membrane systems.[109] As opposed to this approach, simple permeation equations have been used to describe the permeation properties of laminated membranes.[110] In such membranes, rectification of flow would result when the intrinsic permeability constant became concentration-dependent.

The complex admittance characteristics of bipolar membranes (double membranes composed of anion- and cation-exchange membranes) have been described.[111]

## 3 Equilibrium and other Phenomena due to Chemical and Electrochemical Gradients

The state of water in membranes, particularly in reverse osmosis cellulose acetate membranes, is of considerable importance. Differential scanning calorimetry has been employed to study the state of water in the membrane. Based on these studies,[112] it has been suggested that there are four states: completely free water, free water interacting weakly with the membrane matrix, bound water which may contain salt, and bound water which rejects salts. The semipermeability of the membrane has been attributed to the ratio of the four states of water in the membrane. The difficulties in interpreting dye and salt distributions in a cellulose membrane that is in Donnan equilibrium because of the quantity of bound water have been discussed.[113] The relation of water to the stability of pores or channels in ion-exchange membranes has been explored by removal of solvating liquids (including water) in the membrane.[114] Collapse of pores is due to the action of cohesive forces when the solvated polymer chains approach each other by loss of solvent. This effect is stronger in small pores than in larger ones. Porosity is preserved if the rigidity of structures is increased. Consequently, when the membrane matrix material is in its lowest state of hydration, pore stability is high. In ion-exchange resins, collapse of pores is a reversible process.

[107] E. H. Bresler and R. P. Wendt, *Science,* 1969, **163**, 944; **166**, 1438.
[108] C. S. Patlak, D. A. Goldstein, and J. F. Hoffman, *J. Theor. Biol.,* 1963, **5**, 426.
[109] S. W. Rudich, J. B. Wade, and V. A. DiScala, *J. Theor. Biol.,* 1976, **60**, 163.
[110] C. H. Lee, *Sep. Sci.,* 1974, **9**, 479.
[111] R. Simons, *J. Membr. Biol.,* 1974, **16**, 175.
[112] Y. Taniguchi and S. Harigome, *J. Appl. Polym. Sci.,* 1975, **19**, 2743.
[113] R. McGregor and K. H. Ezuddin, *J. Appl. Polym. Sci.,* 1974, **18**, 629.
[114] H. Hilgen, J. DeJong, and W. L. Sederel, *J. Appl. Polym. Sci.,* 1975, **19**, 2647.

Homogeneous and asymmetric membranes of cellulose acetate showed different water sorption properties at higher water activities, with asymmetric membranes showing higher capacity for water sorption.[115] This has been attributed to capillary condensation. A study of solvent contents of perfluorinated sulphonic acid ion-exchange membrane (Nafion) in its sodium and caesium forms (achieved by using different pretreatments) showed that a membrane may be effectively dehydrated by non-aqueous media without prior heating (see Table 1).[116] Rates of exchange of $Na^+$ and $Cs^+$ ions for $H^+$ ions decreased in non-aqueous solvents (see Table 2).

In a number of technical operations, such as reverse osmosis, in which membranes are used as separating devices, it is important to use such polymer material in the preparation of membranes as would bring about the desired separation. Efforts therefore have been made to provide guidelines in that direction.[117] In keeping with this, the influence of added electrolyte upon ion sorption by membranes has been studied.[118] The sorption of permeable ions is considerably increased by the addition of membrane-impermeable salts to the surrounding aqueous solution. The partition coefficient of Na ($0.1M-NaNO_3$) increased from 0.053 to 0.194 upon addition of $1.0M-Mg(NO_3)_2$.

Interpolymer membrane is the name given to those membranes which contain an ionogenic compound (for example, an activating polyelectrolyte) contained in an inert polymeric matrix.[119] Several characteristics of the activating polyelectrolyte that control the properties of the interpolymer membrane formed have been delineated.[119, 120]

Blood purification using membranes (haemodialysis) depends on the prevention of coagulation of blood. For this purpose heparin is used. Attempts have been made

**Table 1** *Drying and solvent composition of a Nafion membrane*

| Cation – Solvent[a] | Treatment[a] | Moles of solvent per $SO_3^-$ site | Moles of water per $SO_3^-$ site |
|---|---|---|---|
| $Na^+$ – PC | Exchanged in PC | 1.9 | 0.3 |
|  | Predried and exchanged in PC | 1.2 | 0.3 |
| $Cs^+$ – PC | Exchanged in PC | 2.0 | 0.3 |
|  | Predried and exchanged in PC | 0.3 | 0.1 |
| $Na^+$ – AN | Exchanged in AN | 1.0 | 0.4 |
|  | Predried and exchanged in AN | 2.0 | 0.3 |
| $Cs^+$ – AN | Exchanged in AN | 1.0 | 0 |
|  | Predried and exchanged in AN | 1.0 | 0.1 |

(a) PC is propylene carbonate and AN is acrylonitrile.

[115] H. G. Burghoff and W. Pusch, *J. Appl. Polym. Sci.*, 1976, **20**, 789.
[116] M. Lopez, B. Kipling, and H. L. Yeager, *Anal. Chem.*, 1976, **48**, 1120.
[117] T. Matsuura, P. Blais, and S. Sourirajan, *J. Appl. Polym. Sci.*, 1976, **20**, 1515.
[118] M. E. Heyde and J. E. Anderson, *J. Phys. Chem.*, 1975, **79**, 1659.
[119] S. R. Caplan and K. Sollner, *J. Colloid Interface Sci.*, 1974, **46**, 46.
[120] S. R. Caplan and K. Sollner, *J. Colloid Interface Sci.*, 1974, **46**, 67, 77.

to incorporate heparin into dialysing membranes instead of adding it to blood. γ-Irradiation of membranes soaked in heparin seems to confer antithrombogenic properties on the membranes.[121]

**Self Diffusion.**—Self diffusion of a number of ionic solutes in several membranes has been determined. In a crosslinked polystyrenesulphonate gel (degree of crosslinking 8%), it was found that $^{14}$C-labelled tetramethylammonium ion (TMA$^+$) diffused less slowly than $^{22}$Na$^+$ ion ($\bar{D}=2.17 \times 10^{-7}$ for TMA$^+$ and $9.61 \times 10^{-7}$ cm$^2$ s$^{-1}$ for Na$^+$) with an energy of activation of 7.0 (TMA) and 5.2 (Na) kcal mol$^{-1}$.[122] Similarly, self diffusion of Na$^+$, Cs$^+$, and I$^-$ in Nafion membrane has been measured using four different solvents.[123] The $\bar{D}$ values shown in Table 3 indicate that the values are high for protic solvents and low for aprotic solvents. Also, the ion-exchange selectivity ratios for Cs$^+$ and Na$^+$ ions showed similar changes between the two kinds of solvents. The effects of tortuosity and electrostatic interactions on ionic mobilities in polystyrenesulphonate resin strips with 2.3% crosslinking when they are in both homo- and hetero-ionic states have been discussed.[124]

Self diffusion of iodide in bentonite clay gel has been measured as a function of increasing clay content.[125] Also, coefficients of transfer of $^{131}$I$^-$ through a cellophane membrane separating aqueous solutions of NaI under different conditions, *viz.* along a gradient (diffusion), no gradient (self diffusion), and against a gradient (interdiffusion) were determined.[126] It was found that the sum of the coefficients of diffusion and interdiffusion was equal to twice the value of the self-diffusion

**Table 2** *Time required for 90% exchange of hydrogen-form Nafion membrane in different solvents*

| Solvent | Water | Acrylonitrile | Propylene carbonate |
|---|---|---|---|
| Time/h | | | |
| Na$^+$ | 0.03 | 5.8 | 19 |
| Cs$^+$ | 0.67 | 47 | 91 |

**Table 3** *Self-diffusion coefficients in Nafion ion-exchange membrane*

| Solvent | $\bar{D}_{ion}$/cm$^2$ s$^{-1}$ | | |
|---|---|---|---|
| | Na$^+$ | Cs$^+$ | I$^-$ |
| Water | $1.2 \times 10^{-6}$ | $3.7 \times 10^{-8}$ | $0.9 \times 10^{-7}$ |
| Methanol | $1.2 \times 10^{-6}$ | $2.0 \times 10^{-8}$ | $1.5 \times 10^{-6}$ |
| Acetonitrile | $4.4 \times 10^{-9}$ | $2.9 \times 10^{-10}$ | $4.8 \times 10^{-10}$ |
| Propylene carbonate | $4.4 \times 10^{-10}$ | $1.0 \times 10^{-10}$ | $1.5 \times 10^{-10}$ |

[121] A. S. Chawla and T. M. S. Chang, *Biomater. Med. Devices Artif. Organs*, 1974, **2**, 157.
[122] G. E. Boyd, *J. Phys. Chem.*, 1974, **78**, 735.
[123] M. Lopez, B. Kipling, and H. L. Yeager, *Anal. Chem.*, 1977, **49**, 629.
[124] R. Fernandez-Prini and M. Philipp, *J. Phys. Chem.*, 1976, **80**, 1976.
[125] J. E. Dufey and H. G. Landelout, *J. Colloid Interface Sci.*, 1975, **51**, 278.
[126] A. Dorabialska, E. Hawlicka, and A. Plonka, *Nukleonika*, 1974, **19**, 65.

coefficient. Diffusion of $Na^+$, $Rb^+$, and $Cl^-$ ions in lecithin–water lamellar phases has been followed at 18 °C, as a function of water content of the phase.[127] At low water content, the diffusion coefficient of $Cl^-$ was greater than that of $Na^+$ or $Rb^+$, whereas at high water content, both cations diffused faster than the anion. This change in relative diffusion occurred at 24% water content (see Table 4), at which value all three ions had the lowest diffusion coefficient. Freeze fracture and polarizing microscopy revealed that a change in long-range organization of the phase occurred at 24% water content. This change, arising from a conformational change of the polar head-groups of the lecithin, was considered responsible for the decrease, in the diffusion coefficients at the level of 24% water content at which the relative diffusion rates inverted.

Factors which govern diffusion in several rubbery polymers have been explored, using radioactively tagged n-hexadecane, heptane-1,7-diol, and ethylene glycol,[128] and n-hexadecane, n-dotriacontane, and a polybutadiene oligomer.[129] Polarity of the penetrant and the penetrant's solubility in the polymer seem to be the dominant factors controlling diffusion. In addition, interaction of the penetrant, as observed in the case of diffusion of $^{35}S$-labelled dye in Nylon-6 film,[130] may complicate the diffusional process by making the diffusion coefficients highly concentration-dependent.

**Diffusion of Electrolyte.**—*Solute and Solvent Fluxes.* Transport of electrolyte across several parchment-supported inorganic precipitate membranes [silver hexacyanoferrate(II), cadmium hexacyanoferrate(II), and barium phosphate;[131] manganese chromate and manganese and cobalt hexacyanoferrate(II)[132]] has been described by the application of the Nernst–Planck flux equation. Similarly, permeation of (1:1)-type electrolyte in a negatively charged membrane (collodion–sulphonated polystyrene interpolymer membrane) has been measured and the results have been compared with theoretical predictions.[133]

A parameter called transference, $T$, to replace permeability, $P$, has been proposed to describe the permeation of solutes through membranes.[134] The essential difference between $T$ and $P$ is that, in the measurement of $T$, a saturated solution (fixed concentration) on one side of the membrane and a low concentration on the other side are used so that the gradient across the membrane is maintained nearly constant.

**Table 4** *Diffusion coefficients in lecithin–water lamellar phases at 18 °C*

| Water content (%) | 13 | 16 | 18 | 23 | 23 | 24 | 26 | 30 | 35 |
|---|---|---|---|---|---|---|---|---|---|
| $\bar{D} \times 10^6/cm^2 \, s^{-1}$ for: $Na^+$ | 0.08 | 0.37 | 1.3 | 1.1 | 0.85 | 0.92 | 0.19 | 0.49 | 0.93 |
| $Rb^+$ | 0.05 | 0.30 | — | — | 0.86 | 0.76 | 0.19 | 0.47 | 0.92 |
| $Cl^-$ | 0.12 | 0.57 | 2.0 | 2.1 | 1.4 | 1.3 | 0.13 | 0.32 | 0.35 |

[127] Y. Lange and C. M. Gary-Bobo, *J. Gen. Physiol.*, 1974, **63**, 690.
[128] C. K. Rhee and J. D. Ferry, *J. Appl. Polym. Sci.*, 1977, **21**, 773.
[129] C. K. Rhee, J. D. Ferry, and L. J. Fetters, *J. Appl. Polym. Sci.*, 1977, **21**, 783.
[130] G. Chantrey and I. D. Rattee, *J. Appl. Polym. Sci.*, 1974, **18**, 105.
[131] F. A. Siddiqi, M. N. Beg, A. Haque, and S. P. Singh, *Bull. Chem. Soc. Jpn*, 1976, **49**, 2858.
[132] F. A. Siddiqi, M. N. Beg, A. Haque, and S. P. Singh, *Electrochim. Acta*, 1977, **22**, 639.
[133] M. Tasaka, N. Aoki, Y. Kondo, and M. Nagasawa, *J. Phys. Chem.*, 1975, **79**, 1307.
[134] F. Theeuwes, R. M. Gale, and R. W. Baker, *J. Membr. Sci.*, 1976, **1**, 3.

A study of the permeability of $^{24}$Na$^+$ and $^{42}$K$^+$ across a protein membrane (serum albumin) containing urease enzyme, measured as a function of pH, showed that, as the pH was increased, the ion permeability decreased, the permeability of K$^+$ being slightly higher than that of Na$^+$.[135] This decrease in permeability with increase in pH was attributed to the Donnan exclusion arising from the ionization of amphoteric sites.

In a cellulose membrane, NaCl, KCl, NaI, and KI were absorbed in an irreversible manner, and the quantity of absorption increased with increase in pH.[136] A study of the diffusion of KCl through the membrane revealed the existence of a barrier on the low-concentration side. Diffusion coefficients of halides in an anion-exchange membrane (Permaplex A20) were around $1 \times 10^{-7}$ cm$^2$ s$^{-1}$ at 0 °C and $1 \times 10^{-6}$ at 60 °C.[137] The overall resistance of the membrane to diffusion in a membrane that was uniform but had low pore density was found to be linearly related to the inverse of the pore area fraction of the membrane, when uniform rates of stirring were employed.[138] Measurements of electrolyte permeation through stressed and unstressed cellophane showed that the relative value of the stressed to unstressed permeability for a particular species depended on its chemical nature and its physical size.[139] Equilibrium dialysis experiments with anionic and cationic dyes revealed the existence of negative charges on poly(vinyl alcohol) membranes.[140] It has been found advantageous to use cellulose acetate membranes for purposes of dialysis in the place of other available membranes.[141] When two or more species of ions of the same charge, coexisting in one solution, exchange at different rates across a permselective membrane, the faster ion (for a short time) may reach a concentration on the other side which will be in excess of the final equilibrium ('overshooting'[142]). The existence of such a phenomenon has been confirmed, and, further, a quantitative analysis of the same has been given for a membrane [collodion–poly(styrenesulphonic acid)] that is permeable to both cations and anions.[143] Interchange of cations with H$^+$ ions across a poly(styrenesulphonic acid) membrane decreased with increase in valence and ion size.[144] A spiral-wound dialyser containing paper-coated poly(vinyl chloride) and dicresyl-butyl phosphate membrane has been found to separate uranium from its fission products and from aluminium.[145]

Water permeability, measured in a variety of membranes as a function of the volume fraction of water absorbed in the membrane, has been interpreted in terms of the effect of different types of absorbed water molecules.[146] Osmosis in permselective membranes has been explained in terms of a coupling effect between ion

---

[135] A. David, M. Metayer, D. Thomas, and G. Broun, *J. Membr. Biol.*, 1974, **18**, 113.
[136] M. Bender, J. K. Moon, J. Stine, A. Fried, R. Klein, and R. Bonjouklian, *J. Chem. Soc., Faraday Trans. 1*, 1975, **71**, 491.
[137] J. D. Lopez-Gonzalez, C. E. Valenzuela, and R. A. Garcia, *An. Quim.*, 1974, **70**, 768.
[138] D. M. Malone and J. L. Anderson, *AIChE J.*, 1977, **23**, 177.
[139] J. D. Rouse and J. Ultman, *Ind. Eng. Chem., Process Des. Dev.*, 1975, **14**, 122.
[140] P. M. Costich and H. W. Osterhoudt, *J. Appl. Polym. Sci.*, 1974, **18**, 831.
[141] J. H. Miller, J. H. Shinaberger, and F. E. Martin, *Med. Instrum.*, 1974, **8**, 214.
[142] R. Neihof and K. Sollner, *J. Phys. Chem.*, 1957, **61**, 159.
[143] N. Takeguchi, I. Horikoshi, and S. Tanaka, *Bull. Chem. Soc. Jpn*, 1975, **48**, 3044.
[144] A. S. Tombalakian, *Can. J. Chem. Eng.*, 1974, **52**, 841.
[145] Z. Ketzinel, Z. Boger, H. Cikurel, D. Vofsi, J. Jagur-Grodzinski, and S. Grassner, *Ind. Eng. Chem., Process Des. Dev.*, 1976, **15**, 524.
[146] Y. J. Chang, C. T. Chen, and A. V. Tobolsky, *J. Polym. Sci., Polym. Phys. Ed.*, 1974, **12**, 1.

streams (difference in Donnan sorption of ions inside the membrane) and water.[147] Similarly, anomalous osmosis in a liquid membrane of potassium polystyrenesulphonate solution confined between two inert cellophane membranes has been attributed to interaction between the flux of solute and that of water.[148] The usual explanation of pressure difference inside the membrane causing anomalous osmosis[149,150] is discarded, as the liquid membrane cannot maintain a pressure difference inside the membrane phase.

A non-aqueous liquid membrane (a layer of solution of copolymer of maleic acid and hexadecyl vinyl ether in n-octanol) separating two aqueous electrolyte solutions has been characterized.[151] The selectivity coefficient $K_{Na,M^{z+}}$, i.e. $K_{Na,M^{z+}} = (\bar{C}_{M^{z+}} C_{Na^+})/(C_{M^{z+}} \bar{C}_{Na^+})$, where $C$'s are concentrations in the membrane (overbars) and aqueous solution containing equinormal $Na^+$ and $M^{z+}$ and $C_{Na^+}$ and $C_{M^{z+}}$ are equal to $10^{-2}$ N, of the membrane for different values of $\alpha$ ($\alpha$ = fractional degree of binding of metal ion by one polymeric acid site) is given in Table 5, from which it is seen that $K_{Na,M^{z+}}$ is independent of $\alpha$, whereas $Na^+$–$Ca^{2+}$ exchange is a function of composition. The affinity of the polymer to $Ca^{2+}$ is high. The cation flux measurements (tracer flux) made using the symmetrical arrangement

| solution (1) | membrane | solution (2) |
|---|---|---|
| pH = 4.9 | | pH = 4.9 |
| $10^{-1}$M-NaCl | | $10^{-1}$M-NaCl |
| $^{22}Na^{36}Cl$ | | |

gave a high flux for $Na^+$ (carrier mediated) compared to $Cl^-$ flux, which was small. Measurement of net flux of $Na^+$ in the arrangement

| solution (1) | membrane | solution (2) |
|---|---|---|
| pH = 5.0 | | pH = 5.0 |
| NaCl | | NaCl ($10^{-2}$M) |
| (i) $10^{-2}$M | | |
| (ii) $10^{-1}$M | | |
| (iii) $10^0$M | | |

showed that $Na^+$ transport was zero in case (i), $\sim 0.5 \times 10^{-7}$ mol h$^{-1}$ in case (ii), and $\sim 2.5 \times 10^{-7}$ mol h$^{-1}$ in case (iii). In the same system, when the pH of solution

**Table 5** *The selectivity constants of a liquid membrane consisting of a solution (1 wt%) of a copolymer of maleic acid and hexadecyl vinyl ether in n-octanol*

| $\alpha$ (%) | $k_{Na, K}$ | $k_{Na, Cs}$ | $k_{Na, Ca}$ |
|---|---|---|---|
| 3 | 3.02 | 4.78 | 21 |
| 6 | 3.02 | 4.78 | 45 |
| 9 | 3.02 | 4.78 | 66 |
| 14 | — | — | 90 |

[147] G. Dickel, *J. Chromatogr.*, 1974, **102**, 31.
[148] M. Tasaka and S. Nagasawa, *Biophys. Chem.*, 1976, **4**, 305.
[149] R. Schlogl, *Z. Phys. Chem. (Frankfurt am Main)*, 1955, **3**, 73.
[150] Y. Toyoshima, Y. Kobatake, and H. Fujita, *Trans. Faraday Soc.*, 1967, **63**, 2828.
[151] E. Pefferkorn and R. Varoqui, *J. Colloid Interface Sci.*, 1975, **52**, 89.

(1) was changed to 3, the flux was from right to left in cases (i) and (ii) and was from left to right in case (iii), being $\sim 1.2 \times 10^{-7}$ mol h$^{-1}$. As H$^+$ ions, in cases (i) and (ii), moved in the direction of decreasing concentration of H$^+$ ion, Na$^+$ ions are driven from the dilute to the concentrated side. In the system

| solution (1) | | solution (2) |
|---|---|---|
| pH = 4.9 | membrane | pH = 4.9 |
| 10$^{-1}$M-NaCl | | 10$^{-1}$M-NaCl |
| $^{22}$Na$^{36}$Cl | | |

tracer flow [$^{22}$Na from solution (1) to solution (2)] was $1.3 \times 10^{-8}$ mol h$^{-1}$ when there was no CaCl$_2$ present in solution (1), but in the presence of CaCl$_2$ the flow was reduced to $0.46 \times 10^{-8}$ mol h$^{-1}$. Ca$^{2+}$ ions occupied the sites in the polymer even at low concentration and impeded the flux of sodium. Both a theoretical and an experimental description of the phenomenon of ion movement against its electrochemical gradient have been given for a number of membrane systems containing carriers[152] such as monensin, cholanic acid,[153,154] and the macrocyclic polyether dibenzo-18-crown-6.[155]

Salinomycin and its derivatives (acetyl and propionyl) show preference to potassium over univalent and bivalent ions in migrating into an organic phase (carbon tetrachloride) in a two-phase aqueous/non-aqueous system[156] (see Table 6). The antibiotic mediated the transport of Na$^+$ and Rb$^+$ as effectively as K$^+$ across the CCl$_4$ bulk phase, but not those of Cs$^+$, Mg$^{2+}$, Ca$^{2+}$, and Sr$^{2+}$. The esters of salinomycin show little activity. Similarly, crown ethers and their polymers[157] and macrocyclic ligands composed of tetrahydrofuran[158] have been used as carriers for the transport of univalent metal picrates or chlorides across a chloroform membrane. Cytochrome $c$ incorporated into cellulose membranes also facilitated the diffusion of ions.[159] The acid metal chloro-complexes of gallium and iron have been demonstrated to be transported across polyurethane membranes by virtue of an acid or chloride gradient.[160] Facilitated transport through liquid membranes may be exploited to effect the separation of the constituents of liquid mixtures.[161] In these operations, facilitation of transport may be caused by (i) allowing an irreversible reaction to occur on the receiving side of the membrane, and thus maximize the concentration gradient, and (ii) adding to membranes additional species that are capable of reversibly reacting with the permeate.[162] As opposed to the aforementioned compounds which facilitated transport, gramicidin A in chloroform (liquid membrane) separating potassium picrate and distilled water allowed little ion transport to occur.[163] This was attributed to the poor complexing abilities of gramicidin A.

[152] D. K. Schiffer, E. M. Choy, D. F. Evans, and E. L. Cussler, *AIChE Symp. Ser.*, 1974, **70**, 150.
[153] E. M. Choy, D. F. Evans, and E. L. Cussler, *J. Am. Chem. Soc.*, 1974, **96**, 7085.
[154] D. K. Schiffer, A. Hochhauser, D. F. Evans, and E. L. Cussler, *Nature (London)*, 1974, **250**, 484.
[155] F. Caracciolo, E. L. Cussler, and D. F. Evans, *AIChE J.*, 1975, **21**, 160.
[156] M. Mitani, T. Yamanishi, and Y. Miyazaki, *Biochem. Biophys. Res. Commun.*, 1975, **66**, 1231.
[157] K. H. Wong, K. Yagi, and J. Smid, *J. Membr. Biol.*, 1974, **18**, 379.
[158] Y. Kobuke, K. Hanji, K. Horiguchi, M. Asada, Y. Nakayama, and J. Furukawa, *J. Am. Chem. Soc.*, 1976, **98**, 7414.
[159] R. Margalit and A. Schejter, *Bioelectrochem. Bioenerg.*, 1976, **3**, 189.
[160] H. D. Gesser, G. A. Horsfall, K. M. Gough, and B. Krawchuk, *Nature (London)*, 1977, **268**, 323.
[161] E. L. Cussler and D. F. Evans, *Sep. Purif. Methods*, 1974, **3**, 399.
[162] E. S. Matulevicius and N. N. Li, *Sep. Purif. Methods*, 1975, **4**, 73.
[163] S. R. Byrn, *Biochemistry*, 1974, **13**, 5186.

**Table 6** Association constants[a] and transport rates[b] for ions and liquid membranes of salinomycin and its derivatives

| Ion | Salinomycin (S) | | Acetyl-S | | Propionyl-S | | Methyl-S | | Bromophenacyl-S | |
|---|---|---|---|---|---|---|---|---|---|---|
| | $K_a$ | $k$ | $K_a$ | $k$ | $K_a$ | $k$ | $K_a$ | $k$ | $K_a$ | $k$ |
| $K^+$ | 3.2 | 67 | 2.4 | 81 | 3.0 | 92 | $<10^{-4}$ | 1 | $<10^{-4}$ | 1 |
| $Na^+$ | 1.7 | 95 | 2.0 | 58 | 2.7 | 114 | $<10^{-4}$ | 1 | $<10^{-4}$ | 1 |
| $Rb^+$ | — | 105 | — | 70 | — | 138 | $<10^{-4}$ | 0.5 | $<10^{-4}$ | 0.5 |
| $Cs^+$ | 0.47 | 11 | 0.05 | 8 | 0.06 | 17 | $<10^{-4}$ | 0.5 | $<10^{-4}$ | 0.5 |
| $Mg^{2+}$ | 0.26 | 6 | 0.007 | 5 | 0.008 | 33 | $<10^{-4}$ | 0.6 | $<10^{-4}$ | 0.6 |
| $Ca^{2+}$ | 0.03 | 11 | 0.004 | 2 | 0.01 | 2 | $<10^{-4}$ | 0.1 | $<10^{-4}$ | 0.1 |
| $Sr^{2+}$ | 0.374 | 12 | 0.05 | 3 | 0.063 | 4 | 0.004 | 0.1 | 0.006 | 0.1 |

(a) $K_a = \dfrac{\text{[antibiotic–metal complex (org)]}}{\text{[antibiotic (org)][M}^+\text{ (aq)]}}$ for univalent ions; $K_a = \dfrac{\text{[antibiotic–metal complex (org)]}}{\text{[antibiotic (org)]}^2\text{[M}^{2+}\text{ (aq)]}}$ for bivalent ions. (b) $k =$ rate of transport in nmol h$^{-1}$.

A number of improvements to existing techniques to follow mass transport through membranes have been introduced.[164] A Raleigh interferometer has been used to follow mass movement in Nucleopore membranes.[165] The existence of Nernst diffusion layers near membrane surfaces has been proved by a laser interferometric method[166] which has been applied to study the characteristics of the diffusion layers. The concentration profiles in these layers at any time followed a quadratic function. The thickness of these layers in the stationary state was estimated to be 575 $\mu$m.[167] Similarly, an experimental apparatus has been developed to estimate the thickness of liquid surfactant membranes;[168] this was found to be of the order of 0.01 cm.

The noise associated with diffusion of ions through pores of known dimension in polycarbonate membranes has been measured[169] and found to correspond approximately to $(1/f)$ noise spectra (where $f$ is the frequency).

Using differential permeabilities, osmotic diffusional salt fluxes and membrane potential have been calculated with the help of a computer.[170]

Permeability of $^{22}$Na$^+$ across simple and layer-type composite membranes of parlodion containing different amounts of poly(styrene sulphonic acid) has been measured as a function of external NaCl concentration and corrected for the effects of aqueous stationary layers present at the membrane/solution interfaces.[171] The permeability of $^{22}$Na$^+$ in opposite directions across the composite membrane was different, whereas it was the same across simple membranes. In both cases, the permeability increased with increase in the concentration of external solution. In the presence of a concentration gradient, the composite membranes generated a potential of $\pm 58$ mV for a ten-fold difference in concentration and gave values for permeability of $^{22}$Na$^+$ which were different in opposite directions. Corrections applied to the two permeability values due to the presence of an electric field across the membrane gave values which agreed qualitatively with those measured in the absence of the gradient. The rectification in the permeability of $^{22}$Na$^+$ was attributed to the presence of a permanent Na$^+$ solubility gradient in the composite membrane which assisted the flow of Na$^+$ in one direction (from the side of high charge density to that of low charge density of the membrane) and opposed it in the opposite direction.

*Membrane Potential.* An experimental cell has been designed to measure diffusion potentials of simple salts and polyelectrolytes. The diffusion potentials generated by concentration gradients of chondroitin sulphate in water and also in the presence of an uncharged polymeric network (dextran) have been measured.[172] The measured potentials agreed with those calculated from the Nernst equation provided that the necessary correction for restricted diffusion or immobilization of the polyions was applied.

[164] R. D. Steele and J. E. Halligan, *Sep. Sci.*, 1974, **9**, 299.
[165] P. H. Bollenbeck and W. F. Ramirez, *Ind. Eng. Chem., Fundam.*, 1974, **13**, 385.
[166] D. Lerche and H. Wolf, *Z. Phys. Chem. (Leipzig)*, 1974, **255**, 126.
[167] D. Lerche, *J. Membr. Biol.*, 1976, **27**, 193.
[168] R. D. Steele and J. E. Halligam, *Sep. Sci.*, 1975, **10**, 461.
[169] M. E. Green, *J. Membr. Biol.*, 1976, **28**, 181.
[170] C. McCallum and P. Meares, *J. Membr. Sci.*, 1976, **1**, 65.
[171] N. Lakshminarayanaiah, *J. Colloid Interface Sci.*, 1975, **50**, 170.
[172] W. D. Comper, W. Lisberg, and A. Veis, *J. Colloid Interface Sci.*, 1976, **57**, 345.

Measurement of electrical potential existing across a Dowex 50W-X2 bead in equilibrium with a KCl solution and penetrated by a microelectrode indicated that the potential was independent of the size of the bead[173] but exhibited Nernstian behaviour in the external KCl concentration range $9.8 \times 10^{-5}$—$7.3 \times 10^{-2}$ mol l$^{-1}$. The activity coefficient of the potassium ion in the bead was found to be approximately constant at 0.27. This was evaluated by measurement of single ion activity, using the Nernst equation, and estimating the amount of KCl in the bead.

Electrochemical cells of the types

SCE|solution (1)||MM||gel membrane||MM||solution (2)|SCE     Type 1
$1 \times 10^{-2}$M

SCE|solution (1)||MM||stearate gel||MM||solution (2)||MM||stearate||MM||solution (3)|SCE     Type 2
octadecanol gel

[SCE is the saturated calomel electrode; potassium stearate gel was made by putting a mixture of potassium stearate and water (1:1 wt. ratio) in a sealed vessel and keeping it in an oven at 100 °C for 2 days and then quickly quenching to room temperature. Potassium stearate octadecanol gel (molar ratio 1:1) was made similarly.]

have been used to follow potentiometrically the polymorphic transitions in the lipoidic structures.[174] In these measurements, solution (1) was kept constant and the concentration of solution (2) was varied. In a cell of Type (2), solutions (1) and (3) were kept constant and solution (2) contained ethanol and KCl of the same concentration as (1) and (3) or varied. MM represents the millipore membrane. Positively charged membranes made of heavy-metal soap (Cu–soap $C_6$—$C_{16}$) by pressing a mixture of the soap with alumina (1:3) at 80 °C and 6000 p.s.i. gave concentration potentials which indicated that, as the number of carbon atoms in the soap molecule was increased, the transport number of the counter-ion increased.[175] This is apparently due to decreased electrolyte solubility and increased crosslinking of membranes. Similarly, regenerated collagen membranes gave membrane potentials which increased with increase in the crosslinking of membrane by formaldehyde.[176] Clay films of montmorillonite suspension (1% or 5% in KCl solution) were formed by filling and draining a small U tube, 4 cm long and of internal diameter 2 mm. These could be prepared as a single zone or as two zones that differed in their concentration of clay particles. In the formation of two homogeneous films, half of the U tube was filled with 5% clay suspension, drained, and dried. The other half was filled with 1% suspension, drained, and dried. The single-zone film, when interposed between two KCl solutions of different concentration, generated potentials which followed the classical theory, with constant ion-transport numbers,[177] whereas the sign and magnitude of potentials observed with bizonal clay films were found to depend on the ratio of KCl concentration inside and outside the film as well as on the difference in the transport numbers of ions in each zone of the film. Membrane potential and apparent transport number of cation have been reported for cured cellulose acetate membranes bounded by electrolyte solutions (0·005—0·05 mol l$^{-1}$). Permselectivity of the membrane to ions increased with

---

[173] M. Goldsmith, D. Hor, and R. Damadian, *J. Phys. Chem.*, 1975, **79**, 342.
[174] C. Botre, C. D. Vecchio, A. Memoli, and M. Mascini, *Anal. Chem.*, 1975, **47**, 1393.
[175] M. A. Beg, F. Ahmed, and A. Razzaq, *J. Am. Oil Chem. Soc.*, 1974, **51**, 439.
[176] M. Sugiura, T. Shimbo, M. Kikkawa, and H. Toyoda, *Nippon Nogei Kagaku Kaishi*, 1974, **48**, 493.
[177] A. K. Helmy, I. M. Natale, and A. M. Grazan, *Colloid Polym. Sci.*, 1976, **254**, 50.

increase in curing temperature. Also, the addition of non-electrolytes to the low-concentration side, which prevented or reversed the direction of flow of water across the membrane, led to higher values for the apparent transport number of the cation.[178] The imposition of a threshold potential across a $BaSO_4$ precipitate membrane ('deconditioning') lowered the membrane potential [$Ba(OH)_2$ and $H_2SO_4$ on either side] existing across it. The rectifying action of the membrane was also lost.[179] On the other hand, when the imposed field was removed, re-adjustment of the adsorbed ions brought back the membrane potential and the rectifying capability of the membrane.

Enzyme (urease)-doped serum albumin membrane gave membrane potentials of the order of 10 mV when it separated electrolyte solutions ($10^{-3}$—$10^{-2}$ M sodium and/or potassium phosphate). The membrane potential increased when a substrate (urea) was introduced.[135] The protein membrane without the enzyme did not show any change in membrane potential on adding urea. The increase in membrane potential of the enzyme-doped protein membrane followed a sigmoid curve with increase in the substrate concentration.

Electrical potentials arising across interpolymer membranes (collodion containing sulphonated polystyrene) have been measured.[133] The effect of water transport on the measured membrane potential has been quantitatively evaluated. The contribution of water flow to membrane potential was estimated to be about 10%.

Bi-ionic potentials (BIP's) measured across anion-exchange membranes [vinyl benzyl chloride–styrene copolymer quaternized with $Me_3N$, acrylonitrile–vinyl benzyl chloride–styrene terpolymer quaternized with $Me_3N$, and butyl acrylate–vinyl benzyl chloride–styrene terpolymer quaternized with $Me_3N$] using 0.1 N solution pairs of $KCl-KNO_3$, $KNO_3-KClO_4$, and $KCl-KClO_4$ gave potentials which followed the additivity rule.[180] Similarly, the BIP's across liquid cation-exchanger (dipicrylaminate) in $o$-nitrotoluene also followed the additivity rule.[181] In addition, the magnitude of BIP was independent of both the concentration of ion-exchanger in the membrane and the concentration of the external solution.

Electrical potentials arising across a membrane whose two faces are differently charged (asymmetric membrane) have been investigated.[182] Identical solutions of (1:1) type electrolyte existing on either side of the asymmetric membrane gave potentials (asymmetry potential) only when the membrane was extremely compact, and its electrical resistance was independent of the concentration of the external solution. These potentials, arising from differences in surface potential, approached zero when the external salt concentration was increased.

A photochromic compound (spiropyran) incorporated with phosphatidyl-choline into an acetylcellulose membrane caused drastic changes in the membrane potential on switching irradiation from the visible to the ultraviolet.[183] This change was reversible and was considered to arise from the change in membrane charge density associated with photoisomerization of spiropyran (spiropyran exists in a coloured form under ultraviolet irradiation and a colourless form under visible

---

[178] S. G. Wong and J. C. T. Kwak, *Desalination*, 1974, **15**, 213.
[179] G. Bahr and P. Hirsch-Ayalon, *J. Membr. Biol.*, 1974, **15**, 405.
[180] C. H. Lee, *J. Appl. Polym. Sci.*, 1977, **21**, 851.
[181] G. M. Shean, *J. Membr. Sci.*, 1977, **2**, 133.
[182] N. Kamo and Y. Kobatake, *J. Colloid Interface Sci.*, 1974, **46**, 85.
[183] S. Kato, M. Aizawa, and S. Suzuki, *J. Membr. Sci.*, 1976, **1**, 289.

irradiation). A similar phenomenon has been noted in the case of a spiropyran-doped triacetylcellulose asymmetric membrane. Photo-induced potential change was dependent on the pH of the electrolyte solution.[184] A complex cardiolipin antigen (cardiolipin, phophatidylcholine, and cholesterol) immobilized in a triacetylcellulose membrane retained its immunochemical reactivity to bind specifically to a Wassermann antibody. This reaction between antigen and free antibody produced a drastic change in membrane potential which was considered to arise from a change in the charge on the membrane.[185]

The theory of membrane potential developed simultaneously by Teorell, Meyer, and Sievers (see Lakshminarayanaiah[186] for a summary) (TMS theory) is used to estimate the quantity of fixed groups in the membrane. Similarly, the theory developed by Kobatake and colleagues and reviewed in the previous Report[3] is also used to estimate the effective charge density $\overline{\phi X}$/mol l$^{-1}$ ($\overline{\phi}$ is a constant, $0 < \overline{\phi} < 1$, and $\bar{X}$ is the stoicheiometric charge density) in the membrane. These methods have been used to estimate the charge on parchment-supported barium sulphate,[187] cobalt and nickel sulphides,[188] and manganese chromate and manganese and cobalt hexacyanoferrate(II)[189] membranes. In all the cases, it was found that $\bar{X} = \overline{\phi X}$, i.e. $\overline{\phi} = 1$. This could be due to the non-existence of fixed groups and to the existence of negatively charged impurities which conferred slight cation selectivity upon the membranes. This comment is based on the fact illustrated in the data given in Table 7 for well-characterized ion-exchange membranes, in which it is seen that the values of $\overline{\phi}$ increase consistently with increase in the external electrolyte concentration. $\overline{\phi X}$ was calculated from the potentiometric data using equations (25) and (26)[190]

$$\xi = (M_1 + M_2)/2\overline{\phi X} = (1 - P_s^2)^{\frac{1}{2}}/2P_s \quad (25)$$

$$P_s = (\bar{t}_{+(\text{app})} - t_+)/[t_+ - (2t_+ - 1)\bar{t}_{+(\text{app})}] \quad (26)$$

$$\bar{t}_{+(\text{app})} = (E_m/2E_{\max}) + 0.5$$
$$E_{\max} = (RT/F) \ln(M_2/M_1)$$

where $M_1$ and $M_2$ are the concentration of (1:1) electrolyte [mol (kg water)$^{-1}$] on either side of the membrane, $E_m$ is the measured membrane potential whose theoretical maximum value is $E_{\max}$, $t$'s are transport number of cation, and overbars refer the parameters to the membrane phase. $\bar{X}$ was determined by analysing the membrane phase for its electrolyte contents.

In the work related to parchment-supported membranes, an average value of $\overline{\phi X}$ at $\xi = 1$ was derived by plotting $P_s$ against log $[(M_1 + M_2)/2]$, in keeping with the original graphical method.[191] The averaging of $\overline{\phi X}$ would be valid when $\bar{X}$ and $\overline{\phi X}$ are independent of concentration, an assumption made by Kamo et al.[191] which is not borne out by experimental facts.

The TMS theory[186] has been extended to liquid ion-exchanger systems in which the concentration of ionized groups may be determined from measurements of

[184] S. Kato, M. Aizawa, and S. Suzuki, J. Membr. Sci., 1977, 2, 39.
[185] M. Aizawa, S. Kato, and S. Suzuki, J. Membr. Sci., 1977, 2, 125.
[186] Ref. 68, pp. 197—203.
[187] F. A. Siddiqi, M. N. Beg, S. P. Singh, and A. Haque, Bull. Chem. Soc. Jpn, 1976, 49, 2864.
[188] M. N. Beg, F. A. Siddiqi, and R. Shyam, Can. J. Chem., 1977, 55, 1680.
[189] F. A. Siddiqi, M. N. Beg, S. P. Singh, and A. Haque, Electrochim. Acta, 1977, 22, 631.
[190] N. Lakshminarayanaiah, J. Membr. Biol., 1975, 21, 175.
[191] N. Kamo, M. Oikawa, and Y. Kobatake, J. Phys. Chem., 1973, 77, 92.

**Table 7** *Values of membrane parameters determined in the evaluation of $\bar{\phi}$*

| Average molality of external soln | $E_m$/mV | $\bar{t}_{+\text{(app)}}$ | $P_s$ | $\xi=(1-P_s^2)^{\frac{1}{2}}/2P_s$ | $\overline{\phi X}$ | $\bar{X}$ | $\bar{\phi}$ |
|---|---|---|---|---|---|---|---|
| *Na–phenol sulphonate membrane in NaCl solutions* ||||||||
| 0.0015 | 17.0 | 0.987 | 0.983 | 0.092 | 0.016 | 1.266 | 0.013 |
| 0.0151 | 16.6 | 0.976 | 0.968 | 0.130 | 0.117 | 1.295 | 0.090 |
| 0.1516 | 11.3 | 0.820 | 0.748 | 0.444 | 0.342 | 1.373 | 0.249 |
| 1.5565 | 1.2 | 0.533 | 0.270 | 1.783 | 0.873 | 1.377 | 0.634 |
| 3.8945 | −2.3 | 0.455 | 0.120 | 4.130 | 0.943 | 1.435 | 0.657 |
| *K–polymethacrylate membrane in KOH solutions* ||||||||
| 0.0062 | 114.5 | 0.997 | 0.998 | 0.035 | 0.175 | 2.846 | 0.062 |
| 0.062 | 108.8 | 0.976 | 0.982 | 0.097 | 0.637 | 2.978 | 0.214 |
| 0.1507 | 31.4 | 0.948 | 0.960 | 0.145 | 1.037 | 3.165 | 0.328 |
| 1.561 | 32.5 | 0.697 | 0.723 | 0.478 | 3.267 | 3.784 | 0.863 |

In the case of the Na–phenol sulphonate membrane system, salt bridges were used in the measurement of membrane potential, whereas in the case of the K–polymethacrylate system, Ag/AgCl electrodes were used.

membrane potential.[192] The theory for estimation of charged sites in the liquid membrane has been illustrated by computer simulation.

*Membrane Electrodes.* This area has attracted a lot of attention, as indicated by the number of papers dealing with the development and evaluation of membrane electrodes. As already mentioned in the Introduction, four books[15–18] deal exclusively with the characteristics and applications of membrane electrodes. There have been a number of reviews dealing with several aspects of membrane electrodes. The review by Covington[193] includes theory, some anion- and cation-responsive electrodes, and discussion of reference electrodes and pX standards. The reviews by Buck[194] cover the literature on ion-selective electrodes and contain tables summarizing electrode preparation, characteristics, and applications. The one by Koryta[195] deals with principles and applications of ion-selective electrodes. Other minor reviews relating to membrane electrodes have appeared.[196,197] Some other summary papers deal with the applications of membrane electrodes in industrial analysis and control,[198] in monitoring of solutes in process plants,[199] in enzymology,[200] and as probes for use in biological systems.[201,202]

---

[192] R. P. Buck, F. S. Stover, and D. E. Mathis, *J. Electroanal. Chem. Interfacial Electrochem.*, 1977, **82**, 345.
[193] A. K. Covington, *Crit. Rev. Anal. Chem.*, 1974, **3**, 355.
[194] R. P. Buck, *Anal. Chem.*, 1974, **46**, 28R; 1976, **48**, 23R.
[195] J. Koryta, *Anal. Chim. Acta*, 1977, **91**, 1.
[196] A. A. Al-Sibaai, *Proc. Anal. Div. Chem. Soc.*, 1975, **12**, 65.
[197] J. D. R. Thomas and G. J. Moody, *Proc. Anal. Div. Chem. Soc.*, 1975, **12**, 48.
[198] E. Hopirtean and I. C. Popescu, *Rev. Chim. (Bucharest)*, 1974, **25**, 679.
[199] D. C. Cornish, *Chimia*, 1975, **29**, 398.
[200] G. J. Moody and J. D. R. Thomas, *Analyst (London)*, 1975, **100**, 609.
[201] G. A. Rechnitz, *Science*, 1975, **190**, 234; *Chem. Eng. News*, 1975, **53**, (4), 29.
[202] K. Cammann, *Z. Anal. Chem.*, 1977, **287**, 1.
[203] W. E. Morf, G. Kahr, and W. Simon, *Anal. Chem.*, 1974, **46**, 1538.

# Membrane Phenomena

*General considerations.* Morf et al.[203] have given a theoretical treatment of the selectivity behaviour and the detection limit of membrane electrodes made of silver compounds. The detection limit is determined either by the solubility of the membrane material or by the activity of the silver defects in the membrane surface, whichever is the larger. It has been shown both theoretically and experimentally that permanent incorporation of lipophilic anions, for example tetraphenylborate, into the membrane prevented the uptake of anions from sample solution and thereby reduced the anion interference.[204, 205] A theoretical investigation of the transient response phenomena of ion-selective electrodes has been carried out by Shatkay[206] and by Morf.[207] It has been shown that transients are a complex phenomenon[206] and that inhomogeneities in the membrane phase may cause the electrode to respond sluggishly.[207] Similarly, Morf et al.[208] have shown that ion-exchange membrane electrodes responded with an exponential time-function (also see Lindner et al.[209]), whereas electrodes constructed of carrier membranes responded at a rate related to the square root of the time-function. In addition, the speed of response of the electrode was dependent not only on the rate of stirring of the solution but also on the direction of change of sample solution, from concentrated to dilute or vice versa.

Values of the response times $t_{95}$ (time taken by an electrode to reach 95% of its equilibrium potential when subjected to a step change in the activity of the ion to which it is responding) of some electrodes[210] are given in Table 8.

Ion-selective electrodes have been used as indicator electrodes in an improved method for determining the equivalence point in potentiometric titrations.[211] Titration rate and the acquisition of data are under the control of a computer. As titration proceeds, the potentiometric curve and Gran plot are displayed on an oscilloscope.

Procedures have been developed for analysing non-Nernstian responses of ion-selective electrodes which may arise from (i) the presence of interfering species, (ii) solubility of the electrode material, and (iii) the presence of the determinand in reagents added to the sample solution.[212] Application of the correct procedure

**Table 8** *Response time ($t_{95}$) of some electrodes*

| Cation-selective | | | Anion-selective | | |
| --- | --- | --- | --- | --- | --- |
| Electrode | Type | $t_{95}$/ms | Electrode | Type | $t_{95}$/ms |
| Calcium | liquid | 2.2 | Chloride | solid | 350 |
| Potassium | liquid | 3.2 | Bromide | solid | 200 |
| Calcium | solid (PVC matrix) | 1.4 | Iodide | solid | 50 |
| Potassium | solid (PVC matrix) | 9.2 | | | |

[204] W. E. Morf, D. Ammann, and W. Simon, *Chimia*, 1974, **28**, 65.
[205] W. E. Morf, G. Kahr, and W. Simon, *Anal. Lett.*, 1974, **7**, 9.
[206] A. Shatkay, *Anal. Chem.*, 1976, **48**, 1039.
[207] W. E. Morf, *Anal. Lett.*, 1977, **10**, 87.
[208] W. E. Morf, E. Lindner, and W. Simon, *Anal. Chem.*, 1975, **47**, 1596.
[209] E. Lindner, K. Toth, and E. Pungor, *Anal. Chem.*, 1976, **48**, 1071.
[210] T. H. Ryan and B. Fleet, *Proc. Anal. Div. Chem. Soc.*, 1975, **12**, 53.
[211] J. W. Frazer, W. Selig, and L. P. Rigdon, *Anal. Chem.*, 1977, **49**, 1250.
[212] D. Midgley, *Anal. Chem.*, 1977, **49**, 1211.

resulted in a linear graph, which gave a slope characteristic of the type of non-ideal behaviour, and an intercept, with the help of which the reagent blank and/or the level of interference can be calculated.

Measurement of exchange current under varying solution conditions has been used to obtain mechanistic information regarding the operation and selectivity of ion-selective membrane electrodes.[213] Procedures have been given to establish the limits of linear[214] and non-linear[215] domain in the responses of ion-sensitive electrodes. Special conditions and procedures to be followed in the evaluation of selectivity coefficients using liquid membrane electrodes have been emphasized.[216] An equation has been derived to describe the e.m.f. of a liquid membrane electrode.[217] This equation

$$E = E_0 - (RT/F) \ln\{C + (C^2 + L_d)^{\frac{1}{2}}]/2\} \tag{27}$$

(where $C$ is the concentration of ion $A^-$ to which the electrode is selective) contains the complex parameter $L_d$, which sets the limit of detection of the electrode to $A^-$ and is given by

$$L_d = \left(\frac{4K}{b_A}\right)\bar{C}_{RA} = \left(\frac{2K^2}{b_A}\right)\{1 + (2X/K) - [1 + (4X/K)]^{\frac{1}{2}}\} \tag{28}$$

where $X = \bar{C}_R + \bar{C}_{RA}$, i.e. the total concentration of ion-exchanger in the membrane, $\bar{C}_R$ is the concentration of exchanger in the membrane, $C_R$ is the concentration of exchanger leaked into the surrounding aqueous solution, and $\bar{C}_{RA}$ is the concentration of ion-exchanger complex, whose dissociation constant, $K$, is given by

$$K = \bar{C}_R C_A / \bar{C}_{RA} \tag{29}$$

The factors $b_A$ and $C_A$ are defined by

$$b_A = \bar{C}_R \bar{C}_A / C_R C_A \tag{30}$$

$$C_A = C + C_R \tag{31}$$

($C_A$ is the total concentration of $A^-$ in the aqueous solution and $C$ is the concentration of $A^-$ initially present in the test solution).

Equations (28)—(31) yield

$$C_A = [C + (C^2 + L_d)^{\frac{1}{2}}]/2$$

$$C_R = L_d / 2[C + (C^2 + L_d)^{\frac{1}{2}}]$$

Equation (27) reduces to the Nernst equation when $L_d = 0$. Theoretical curves of $E$ versus $C$ at a constant value of $L_d$ constructed from equation (27) showed that as $L_d$ became larger, $E$ became independent of $C$ at higher concentrations. When $C$ became infinitely dilute, $E$ reached asymptotically a constant value $E'$ given by

$$E' = E_0 + (RT/F) \ln 2 - (RT/2F) \ln L_d \tag{32}$$

Thus the limit of detection is governed by $L_d$, which from equation (28) simplifies

[213] K. Cammann and G. A. Rechnitz, *Anal. Chem.*, 1976, **48**, 856.
[214] C. Liteanu, I. C. Popescu, and E. Hopirtean, *Anal. Chem.*, 1976, **48**, 2010.
[215] C. Liteanu, E. Hopirtean, and I. C. Popescu, *Anal. Chem.*, 1976, **48**, 2013.
[216] A. Hulanicki and Z. Augustowska, *Anal. Chim. Acta*, 1975, **78**, 261.
[217] N. Kamo, N. Hazemoto, and Y. Kobatake, *Talanta*, 1977, **24**, 111.

to $L_d \simeq (4X^2/b_A)$ when the ion-exchanger exists completely dissociated in the membrane phase, i.e. $(X/K) \ll 1$.

A number of predictions following from equation (27) have been tested for the liquid membrane constructed of salts ($F^-$, $Br^-$, $I^-$, $NO_3^-$, and $ClO_4^-$) of purified Crystal Violet in dichloroethane. Similarly, other salts of Crystal Violet ($Cl^-$, hydrogen maleate, hydrogen phthalate) in nitrobenzene, in dichloroethane, and in chloroform and salts of methyltricaprylammonium ($Cl^-$, $NO_3^-$) in decan-1-ol have been used as liquid membranes and tested for their selectivity to several ions.[218]

Generally, it is found that the selectivity coefficients ($k_{i,j}$'s) are dependent on concentration. Only under ideal conditions are $k_{i,j}$'s of membrane electrodes independent of concentration. This aspect of $k_{i,j}$ dependence has been investigated by computer simulation studies by Buck.[219] Failure of co-ion exclusion, mixed-valence counter-ion transport, and also the existence of equilibrium and non-equilibrium conditions gave electrode responses which generated concentration- or activity-dependent $k_{i,j}$'s, whereas equilibrium conditions for common-valence counter-ions gave $k_{i,j}$ values that were dependent on activity ratios. In general, the $k_{i,j}$'s are influenced by phenomena occurring at and close to the electrode surface. Several uncontrollable factors involved in this led to variability in the values of selectivity coefficients, which is particularly apparent in solid-state electrodes. An attempt to explain this has been made on the basis of a diffusion-layer model.[220]

Chronopotentiometric investigations of diffusion overvoltage at an interface between two immiscible solutions have been reported. The systems investigated were a nitrobenzene solution of tetrabutylammonium tetraphenylborate–aqueous solution of tetrabutylammonium bromide with excess of NaBr[221] and a solution of trimethylammonium picrate in nitrobenzene–aqueous solution of potassium halide.[222] In both cases, diffusion at the interface has been considered to follow the equation due to Sand.[223]

A number of solid and liquid membrane electrodes have been constructed and their characteristics evaluated. Liquid membrane electrodes formed of salts of benzylcetyldimethylammonium cation with anions (toluene-$p$-sulphonate, $NO_3^-$, $I^-$, and $ClO_4^-$) dissolved in nitrobenzene gave Nernstian response to the anions concerned.[224] The strontium salt of Igepol CO-880 (nonylphenoxy-polyethyleneoxyethanol) dissolved in 4-ethylnitrobenzene acted as a liquid membrane electrode that is highly selective to $Sr^{2+}$ ion. The selectivity coefficients of this electrode ($k_{Sr,M}$) were found to be:[225] $Me_4N^+$, $> 10^3$; $Cs^+$, $2 \times 10^2$; $K^+$, $8 \times 10^{-3}$; $Na^+$, $Li^+$, $NH_4^+$, $2 \times 10^{-3}$; $H^+$, $5 \times 10^{-4}$; $Ba^{2+}$, $3 \times 10^2$; $Ca^{2+}$, $Zn^{2+}$, $2 \times 10^{-3}$; $Ni^{2+}$, $Co^{2+}$, $Fe^{2+}$, $8 \times 10^{-4}$; $Mg^{2+}$, $Mn^{2+}$, $7 \times 10^{-4}$; $Fe^{3+}$, $4 \times 10^{-3}$; $Al^{3+}$, $2 \times 10^{-3}$.

Some of the complexes of non-cyclic uncharged ligands (tetra-$N$-substituted dicarboxylic acid diamides of glycol diethers derived from ethane-1,2-diol, cyclohexane-1,2-diol, or benzene-1,2-diol) with alkali-metal or alkaline-earth-metal

---

[218] A. Jyo, M. Torikai, and N. Ishibashi, *Bull. Chem. Soc. Jpn*, 1974, **47**, 2962.
[219] R. P. Buck, *Anal. Chim. Acta*, 1974, **73**, 321.
[220] A. Hulanicki and A. Lewenstam, *Talanta*, 1977, **24**, 171.
[221] C. Gavach and F. Henry, *J. Electroanal. Chem. Interfacial Electrochem.*, 1974, **54**, 361.
[222] C. Gavach and B. D'Epenoux, *J. Electroanal. Chem. Interfacial Electrochem.*, 1974, **55**, 59.
[223] H. J. S. Sand, *Phil. Mag.*, 1900, **1**, 45.
[224] C. Luca, Gh. Semenescu, and C. Nedea, *Rev. Chim. (Bucharest)*, 1974, **25**, 1015.
[225] E. W. Baumann, *Anal. Chem.*, 1975, **47**, 959.

cations dissolved in *o*-nitrophenyl octyl ether or dibutyl sebacate acted as liquid membrane electrodes that are selective to uni- and bi-valent ions.[226] The electrodes have been used in the potentiometric determination of stability constants of several complexes of the non-cyclic ligands with alkali metals and alkaline earth metals at 30 °C in ethanol.[227] Solutions of bis-(*OO'*-di-isobutyl dithiophosphato)nickel(II), bis-(*OO'*-di-isobutyl dithiophosphato)cadmium(II), and bis-(*OO'*-di-isobutyl dithiophosphato)lead(II) in chlorobenzene have been used as liquid membrane electrodes responding selectively to $Ni^{2+}$, $Cd^{2+}$, and $Pb^{2+}$ ions respectively over the concentration range $10^{-1}$—$10^{-4}$ mol $l^{-1}$. The lead electrode has been found to be highly selective in acidic solutions and in solutions containing a number of heavy-metal ions.[228] Quaternary phosphonium salt [tetraoctylphosphonium salt (0.1 mol $l^{-1}$) in decanol or tetraphenylphosphonium salt (0.15 mol $l^{-1}$) in nitrobenzene] was the sensitive element in the formation of liquid membrane electrodes selective to a number of anions ($I^-$, $Br^-$, $Cl^-$, and $NO_3^-$).[229] Tetraoctylphosphonium chloride and nitrate in decanol and tetraphenylphosphonium chloride and nitrate in nitrobenzene are recommended as electrodes for the determination of activity of 1—10M-HCl and 1—9M-$HNO_3$ solutions.[230] Tetraphenylarsonium salts and Crystal Violet act as selective exchangers in nitrobenzene or chloroform for $I^-$ or $SCN^-$ ions.[231] Also, saturated solutions of potassium picrate and potassium tetraphenylborate in several solvents have been tested for their selective response to $K^+$ ions.[232] The behaviour of liquid membrane electrodes constructed of exchange material derived from trioctyl- and trilauryl-amines in $Na_2SO_4$ solutions has been investigated.[233] Chlorobenzene, bromobenzene, nitrobenzene, 1,2-dichlorobenzene, methyl chloride, decyl alcohol, and tributyl phosphate were used as solvents in the preparation of electrodes which were found to be reversible to $SO_4^{2-}$ ion in a certain concentration range. Some organometallic substances have been examined for the preparation of ion-selective electrodes.[234] No useful electrodes for $SO_4^{2-}$, chromate, carbonate, or nitrate were obtained from organometallic salts of lead and thallium. Metal phthalocyanines were found to be selective to anions rather than to cations. Liquid-state ion-selective membranes based on Crystal Violet PAR have been used in the development of electrodes for the determination of acidic and basic dyes.[235] Millipore filters saturated with organic solvents alone (*o*-dichlorobenzene, 1,2-dichloroethane, 2,2-dichlorodiethyl ether and 1,1,2,2-tetrachloroethane) acted selectively to certain ions. Addition of an ion-exchanger to the solvent scarcely affected the potentiometric behaviour of most ions except when there was selective interaction.[236] Niobium oxinate in chloroform, when used as an

[226] D. Ammann, R. Bissig, M. Gueggi, E. Pretsch, W. Simon, I. R. Borowitz, and L. Weiss, *Helv. Chim. Acta*, 1975, **58**, 1535.
[227] N. N. L. Kirsch and W. Simon, *Helv. Chim. Acta*, 1976, **59**, 357.
[228] E. A. Materova, V. V. Muchovikov, and M. G. Grigorjeva, *Anal. Lett.*, 1975, **8**, 167.
[229] Ya. A. Syrchenkov, Yu. I. Urusov, M. A. Geminova, A. F. Zhukov, V. S. Shterman, and A. V. Gordievskii, *Zh. Anal. Khim.*, 1974, **29**, 584.
[230] Ya. A. Syrchenkov, Yu. I. Urusov, M. V. Geminova, A. F. Zhukov, N. I. Savvin, and A. V. Gordievskii, *Zavod. Lab.*, 1974, **40**, 1041.
[231] K. Kina, H. Fukushima, and N. Ishibashi, *Kyushu Daigaku Kogaku Shuho*, 1974, **47**, 787.
[232] R. Geyer and I. Preuss, *Z. Chem.*, 1974, **14**, 29.
[233] E. A. Materova, Z. S. Alagova, and G. R. Mamadieva, *Vestn. Leningr. Univ., Fiz., Khim.*, 1974, 147.
[234] M. Sharp, *Anal. Chim. Acta*, 1975, **76**, 165.
[235] A. G. Fogg, A. A. Al-Sibaai, and K. S. Yoo, *Anal. Lett.*, 1977, **10**, 173.
[236] O. Astrom, *Anal. Chim. Acta*, 1975, **80**, 245.

# Membrane Phenomena 67

electrode, was found to be selective to $ClO_4^-$ ion.[237] The responses of Orion nitrate and perchlorate liquid membrane electrodes to a series of substituted benzoate and phenylacetate ions have been measured.[238] The relative response has been found to depend on the size and position of the substituent. A liquid membrane electrode that is selective to bivalent ions has been evaluated at physiological pH ($\sim$ 7.3) for its responses to a number of protonated amines and polyamines of biochemical interest.[239] All the responses were non-Nernstian in nature. A $10^{-3}$ M solution of ferrous o-phenanthroline dodecyl sulphate in o-dichlorobenzene was used as a liquid membrane to sense dodecylsulphate anion.[240] This electrode could be used in the determination of the critical micellar concentration of anionic surfactants. An ion complex of dodecyl benzenesulphonate with bis(dimethylglyoxime)-o-phenanthrolinecobaltate(III) dissolved in o-dichlorobenzene and n-decanol and held in porous graphite acted as an electrode that is sensitive to soap or surfactant over the pH range 1—13.[241] The role of solvent used in the formation of liquid membranes of tetraheptylammonium nitrate has been explored.[242] The selectivities of the electrode in amyl alcohol and in other benzene and substituted-benzene solvents as membrane solvents have been compared.

A liquid membrane electrode that is sensitive to dibenzyldimethylammonium ion has been constructed.[243] Equal volumes of dibenzyldimethylammonium ion ($10^{-3}$ mol $l^{-1}$) and tetraphenylboron ($10^{-3}$ mol $l^{-1}$) gave a white precipitate which dissolved in dichloroethane to form the liquid ion-exchanger. The response of the electrode was Nernstian, with a slope of 59 mV per decade of concentration, and it has been used to follow the diffusion of dibenzyldimethylammonium ion into biological systems.

The construction of and experimentation with poly(vinyl chloride) (PVC) matrix membranes containing ion-exchangers that are selective to $K^+$ and to $Ca^{2+}$ ions have been described.[244,245] A PVC matrix, plasticized with dibutyl phthalate or dioctyl phthalate and containing tetradecylammonium nitrate, behaved differently in neutral and acidic media.[246] It was highly selective to $HNO_3$ in the presence of HCl, $H_2SO_4$, and $H_3PO_4$ and was reversible to $NO_3^-$ ion in neutral media. Alkylbenzenesulphonate–ferroin complex incorporated into a PVC matrix acted selectively to alkylbenzenesulphonate in the presence of $SO_4^{2-}$, $PO_4^{3-}$, $NO_3^-$, and $Cl^-$ ions.[247] Similarly, Aliquat 336 S in its $Cl^-$, $Br^-$, $I^-$, $NO_3^-$, and $ClO_4^-$ forms, immobilized in a PVC matrix, gave Nernstian response to the respective ions.[248] Metal complexes of the type $M_x[N^{II}L_4]$, incorporated into a PVC matrix, acted as ion-selective electrodes.[249] In the complex, M represents $Ag^I$, $Cu^I$, $Cu^{II}$, $Pb^{II}$, $Hg^{II}$, or Malachite Green, N is Hg, Zn, Co, or Ni, and L is ligand ($SCN^-$ or $I^-$). Some of the complexes

[237] W. Szczepaniak and K. Ren, *Chem. Anal. (Warsaw)*, 1975, **20**, 91.
[238] R. F. Hirsch and G. M. Olderman, *Anal. Chem.*, 1976, **48**, 771.
[239] P. Kent, S. C. Bunce, R. A. Aikens, *Anal. Biochem.*, 1974, **62**, 75.
[240] D. Anghel and N. Ciocan, *Colloid Polym. Sci.*, 1976, **254**, 114.
[241] D. Anghel and N. Ciocan, *Anal. Lett.*, 1977, **10**, 423.
[242] C. Fabiani, *Anal. Chem.*, 1976, **48**, 865.
[243] M. Muratsugu, N. Kamo, K. Kurihara, and Y. Kobatake, *Biochim. Biophys. Acta*, 1977, **464**, 613.
[244] A. Craggs, G. J. Moody, and J. D. R. Thomas, *J. Chem. Educ.*, 1974, **51**, 541.
[245] J. D. R. Thomas, *Proc. Anal. Div. Chem. Soc.*, 1974, **11**, 340.
[246] E. A. Materova, A. L. Grekovich, and N. V. Garbuzova, *Zh. Anal. Khim.*, 1974, **29**, 1900.
[247] T. Tanaka, K. Hiiro, and A. Kawahara, *Anal. Lett.*, 1974, **7**, 173.
[248] A. Hulanicki and R. Lewandowski, *Chem. Anal. (Warsaw)*, 1974, **19**, 53.
[249] D. E. Ryan and M. T. Cheung, *Anal. Chim. Acta*, 1976, **82**, 409.

prepared were Pb[Hg(SCN)$_4$], Cu[Hg(SCN)$_4$], Ag$_2$(HgI$_4$), Cu$_2$(HgI$_4$), and Ag$_{1.14}$ Cu$_{0.86}$(HgI$_4$).

Attempts to replace the PVC matrix by another polymer, poly(vinyl isobutyl ether), revealed that only an electrode that was selective to Ca$^{2+}$ showed good behaviour, whereas others for Cl$^-$ and K$^+$ were unsatisfactory.[250]

A membrane electrode based on Ag$_2$HgI$_4$ crystals showed sensitivity to Ag$^+$ ions similar to that of AgI.[251] Also, the electrode showed linear response for Ag$^+$, I$^-$, and Hg$^{2+}$ ions in the concentration range $10^{-6.5}$—$10^{-1}$ mol l$^{-1}$. The electrode was used in the potentiometric titrations of KCNS and KI with Hg$^{2+}$ ions. AgCl, AgBr, AgI, and Ag$_2$S were precipitated onto finely divided particles of gold and used as electrodes; they showed sensitivity to halides, Ag$^+$, and S$^{2+}$.[252] According to expectations, an AgI/Au electrode was sensitive to CN$^-$ ions. The electrodes had good stability, low impedance, low redox sensitivity, and low response time (0.1—10 s).

A comparative study of the responses of metallic Ag electrodes and Ag$_2$S/AgX solid-state membrane electrodes during rapid changes of concentration in streaming solutions has been presented.[253]

Silver or mercury sulphide, selenide, or telluride matrices to hold active components, the chalcogenides of Cu, Pb, Cd, Zn, Ni, Co, and Bi for cations, and Ag or mercurous halides for anions have been examined[254] and found to be suitable for the preparation of electrodes. Electrodes selective to Ag$^+$, Hg$^{2+}$, Pb$^{2+}$, Cu$^{2+}$, and Cd$^{2+}$ performed well, whereas those that were selective to Zn$^{2+}$, Ni$^{2+}$, Co$^{2+}$, and Bi$^{3+}$ were found to be unsatisfactory.

Solid-state electrodes containing thin ion-selective layers deposited on ionic conductors have been prepared and tested.[255] The supporting ionic conductor materials were Ag$_2$S, Ag$_2$SBr, Ag$_2$SI, and Ag$_{19}$I$_{15}$P$_2$O$_7$. These reacted with gaseous chlorine, bromine, or iodine to form a thin layer of AgCl, AgBr, or AgI on the surface of the supporting material. In the case of Ag$_2$SBr and Ag$_2$SI, the formation of a layer of bromide or iodide was unnecessary since they were already selective to bromide or iodide. The electrode selective to hydrogen orthophosphate ion was made by allowing Ag$_{19}$I$_{15}$P$_2$O$_7$ to react with a concentrated solution of orthophosphate to yield a thin layer of Ag$_3$PO$_4$, P$_2$O$_7^{4-}$ ions being released into the solution. The responses of these electrodes were reversible and fast. Similarly, ion-selective electrodes for bivalent metal ions (Cu$^{2+}$, Pb$^{2+}$, Cd$^{2+}$, and Hg$^{2+}$) have been prepared by covering the above ionic conductors with a thin layer of metal sulphide by heating under a partial pressure of sulphur.[256] The conversion of these and other solid-state ion-selective electrodes into 'combination' type electrodes containing a gel-type reference electrode so as to form a compact unit has been described.[257] Ion-selective electrodes have been prepared from semiconducting salts of 7,7,8,8-tetracyanoquinodimethane (tcnq). The technique used was to apply the

[250] O. F. Schafer, *Anal. Chim. Acta*, 1976, **87**, 495.
[251] A. V. Gordievskii, A. F. Zhukov, V. S. Shterman, N. I. Savvin, and Yu. I. Urusov, *Zh. Anal. Khim.*, 1974, **29**, 1414.
[252] G. W. S. Van Osch and B. Griepink, *Z. Anal. Chem.*, 1975, **273**, 271.
[253] A. Dencks and R. Neeb, *Z. Anal. Chem.*, 1977, **285**, 233.
[254] I. Sekerka and J. F. Lechner, *Anal. Lett.*, 1976, **9**, 1099.
[255] R. E. Van de Leest, *Analyst (London)*, 1976, **101**, 433.
[256] R. E. Van de Leest, *Analyst (London)*, 1977, **102**, 509.
[257] I. Sekerka and J. F. Lechner, *Anal. Lett.*, 1975, **8**, 769.

**Table 9** *Comparison of standard electrode potentials of metal ion|metal tcnq salt and metal ion|metal systems*

| Electroactive salt | $E°[M^{n+}|M_{2/n}(tcnq)_2]/mV$, vs. NHE$^a$ | $E°(M^{n+}|M^0)/mV$, vs. NHE$^a$ |
|---|---|---|
| Ag$_2$(tcnq)$_2$ | 799 | 799 |
| Cu(tcnq)$_2$ | 667 | 337 |
| Pb(tcnq)$_2$ | 404 | −126 |
| Cd(tcnq)$_2$ | 363 | −403 |
| K$_2$(tcnq)$_2$ | 414 | −2952 |
| Na$_2$(tcnq)$_2$ | 474 | −2714 |

(*a*) NHE is the normal hydrogen electrode.

electroactive material manually to a graphite–Teflon conducting surface; the so-called 'Selectrode' technique.[258] The standard electrode potentials measured for metal ion|metal tcnq salts are compared in Table 9 with those of corresponding metal ion|metal systems.[259] There is agreement only in the case of Ag$_2$(tcnq)$_2$, indicating involvement of the cationic component in the redox process. The discrepancies in all other cases point to more complex electrode behaviour. However, the electrodes have been used in determination of the solubility products of the salts of tcnq concerned. Stannic molybdate precipitate incorporated into polystyrene under pressure at 70 °C and used as a membrane electrode gave the order of ion selectivity K$^+$ > Na$^+$ > Li$^+$ > Ba$^+$ > Ca$^{2+}$.[260] Some heavy-metal soaps (palmitates of Zn, Cu, Ni, and Co), when pressed into pellets, served as electrodes and responded in a Nernstian fashion to the cations concerned.[261]

The construction and the performance characteristics of membrane electrodes that sense ammonia, sulphur dioxide, and nitrogen oxide have been described.[262] A graphical method for evaluating the dynamic measuring range of potentiometric gas-sensing devices has been detailed.[263]

Reference standards for the electrometric estimation of potassium and calcium in biological samples have been proposed.[264] The standard solutions should satisfy three conditions: (i) values of $m_K$ and $\gamma$(KCl), (the molality and the activity coefficient) or $m_{Ca}$ and $\gamma$(CaCl$_2$) must be measurable or capable of being calculated, (ii) residual liquid junctions must be small, and (iii) the extrathermodynamic convention to obtain $\gamma$(K) and $\gamma$(Ca) from the mean ionic activity coefficients must be applicable. The following solutions, each having an ionic strength of 0.15 mol (kg of water)$^{-1}$, have been proposed as standards for potassium and calcium in blood serum:

(1) 0.1458m-NaCl + 0.0042m-KCl; pK = 2.504;

(2) 0.1462m-NaCl + 0.00126m-CaCl$_2$; pCa = 3.360;

(3) 0.1422m-NaCl + 0.0041m-KCl + 0.00121m-CaCl$_2$; pK = 2.514, pCa = 3.373, pNa = 0.966

[258] Ref. 17., pp. 114—116.
[259] M. Sharp, *Anal. Chim. Acta*, 1976, **85**, 17.
[260] W. U. Malik, S. K. Srivastava, V. M. Bhandari, and S. Kumar, *J. Colloid Interface Sci.*, 1974, **47**, 1.
[261] M. A. Beg and F. Ahmad, *J. Electronal. Chem. Interfacial Electrochem.*, 1976, **71**, 34.
[262] P. L. Bailey and M. Riley, *Analyst (London)*, 1975, **100**, 145.
[263] E. H. Hansen and N. R. Larsen, *Anal. Chim. Acta*, 1975, **78**, 459.
[264] A. K. Covington and R. A. Robinson, *Anal. Chim. Acta*, 1975, **78**, 219.

In these solutions, the effects of hydrogen carbonate and magnesium (which are generally present in blood plasma) are ignored.

Ion-selective electrodes have found several applications. Alkali-metal-ion-responsive glass and fluoride- or chloride-responsive electrodes have been used to obtain values for the free energies of transfer of alkali-metal fluorides and chlorides from water to methanol and to water mixtures.[265] An $Ag_2S$ crystal membrane electrode has been used to assay sulphur-containing proteins.[266] Fluoride-selective electrode has been used as a reference electrode in cells without liquid junction in mixed solvents for the determination of silver–acetonitrile complexes.[267] A rubber membrane containing 40% by weight of hexadecyltrimethylammonium dodecyl sulphate was found to act as an indicator electrode in potentiometric titrations of RBr (R = hexadecyltrimethylammonium and hexadecylpyridinium ions) with sodium lauryl sulphate.[268] Collodion-coated and uncoated clay or ion-exchange resin membranes have been used in the measurement of mobility ratios of $Na^+$ and $K^+$ ions.[269] $Ag_2S$ deposited electrolytically on a silver rod responded in a Nernstian manner to $Cd^{II}$ in the pCd range 2—10.6. The electrode also gave satisfactory results in (i) titration of $Ag^I$, $Cu^{II}$, $Cd^{II}$, and $Zn^{II}$ with $Na_2S$ solutions, and (ii) titration of $Cu^{II}$ and $Cd^{II}$ with EDTA solutions. It behaved as a third-order self-generating electrode.[270] Dowex 50W X8 in its barium form (200 mesh, and air-dried), commercial graphite powder, and Nujol in the proportion 1:1:1.1, made into a paste and packed into one end of a glass tube and with a platinum wire inserted into the paste, acted as an electrode that is selective to $Ba^{2+}$ ions.[271] This electrode could be used to follow titrations of a variety of metals by addition of EDTA and by back-titrating with a standard $BaCl_2$ solution.[271] Ion-selective electrodes (K electrode, glass electrode, Ca electrode, and bivalent ion electrode) have been used in potentiometric titrations to determine the formation constants of metal ion–ATP and –ADP complexes,[272] and Mg- and Mn-ion-selective electrodes were used to determine the formation constants of $Mg^{II}$–ATP and Mn–ATP.[273] Similarly, a $Cu^{II}$ solid-state electrode has been used as the end-point detector in complexometric back-titrations ($Cu^{II}$ was used as the back titrant) involving alkaline-earth metals with EDTA and EGTA.[274] Also, Cu and Cd electrodes are used as indicator electrodes in the titration of vanadyl ions with EDTA.[275] Again, Cu and Pb electrodes have been used for the measurement of the complexation properties of humic and fulvic acids in natural waters.[276] Stability constants of complexes of silver in sub-micromolar $AgNO_3$ have been determined using the Ag-ion-selective electrode.[277] That the use of silver wire as an indicator electrode in argentometric

---

[265] A. K. Covington and J. M. Thain, *J. Chem. Soc., Faraday Trans. 1*, 1975, **71**, 78.
[266] P. D'Orazio and G. A. Rechnitz, *Anal. Chem.*, 1977, **49**, 41.
[267] K. M. Stelting and S. E. Manahan, *Anal. Chem.*, 1974, **46**, 592.
[268] A. S. Pathan, *Proc. Anal. Div. Chem. Soc.*, 1974, **11**, 143.
[269] M. Adhikari, D. Ghosh, and G. Sen, *J. Indian Chem. Soc.*, 1975, **52**, 408.
[270] J. B. Jensen, *Anal. Chim. Acta*, 1975, **76**, 279.
[271] R. Kuroda and N. Yoshikuni, *Anal. Chim. Acta*, 1976, **87**, 211.
[272] E. J. Fogt and G. A. Rechnitz, *Arch. Biochem. Biophys.*, 1974, **165**, 604.
[273] M. S. Mohan and G. A. Rechnitz, *Arch. Biochem. Biophys.*, 1974, **162**, 194.
[274] J. M. Van der Meer, G. Den Boef, and W. E. Van der Linden, *Anal. Chim. Acta*, 1976, **85**, 309.
[275] A. Napoli and M. Mascini, *Anal. Chim. Acta*, 1977, **89**, 209.
[276] J. Buffle, F. L. Greter, and W. Haerdi, *Anal. Chem.*, 1977, **49**, 216.
[277] A. L. Cummings and K. P. Anderson, *Anal. Chem.*, 1975, **47**, 2310.

titrations has advantages over expensive ion-selective electrodes has been demonstrated.[278] A bivalent-ion-selective electrode ($Mg^{2+}$ or $Zn^{2+}$) has been used in the estimation of tervalent ions.[279] This makes use of the fact that the stability constants of $M^{III}$–EDTA complexes are much larger (for example Bi–EDTA = 1022.9) than those of $M^{II}$–EDTA complexes (Mg–EDTA = 108.7), so that when $M^{III}$– and $M^{II}$–EDTA complexes exist in solution, the amount of $M^{III}$ can be determined by measuring the amount of $M^{II}$ displaced. Stability constants of $Cd^{II}$ with some amino-acids have been determined by three potentiometric methods, in two of which ion-selective electrodes (glass membrane electrode to measure pH and Cd-selective electrode to measure cadmium concentration) have been used.[280] The average values for the stability constants, $\beta(ML)$ and $\beta(ML_2)$, are given in Table 10. Similarly, $Cu^{II}$- and Ca-ion-selective electrodes have been used to investigate the complex formation of these ions with salicylic acid ($H_2$sal).[281] The following species have been found to exist (log $\beta$ values in parentheses): Hsal$^-$ (13.30), $H_2$sal (16.36), CuHsal (15.24), CuH$_2$sal$_2$ (31.23), Cusal (10.34), Cusal$_2$ (19.01); CaHsal (15.24), Casal (5.77), Casal$_2$ (11.88), Hsal (11.22), and $H_2$sal (14.34). The species CuH$_3$sal$_3$ and CuH$_4$sal$_4$ were not detected.

An $Ag_2S$/CuS membrane electrode has been used as an end-point detector in chelometric titrations.[282] A liquid membrane formed of an ion pair, zephiramine (benzyldimethyltetradecylammonium chloride)–Chromazurol S, has been applied to the chelometric titrations of $Cu^{II}$, $Ni^{II}$, $Fe^{III}$, and $Pb^{II}$ with EDTA.[283]

A polycrystalline mixture of HgS and $Hg_2Cl_2$,[284] and another of AgCl, $Ag_2S$, and $Hg_2Cl_2$,[285] have been used as indicator electrodes in the determination of sulphite ion. A method for the microdetermination of sulphate in the presence of phosphate, without separation, has been worked out.[286] A lead-ion-selective electrode and a double-junction reference electrode were used to monitor the potentiometric titrations with 0.01M barium perchlorate in 70% 1,4-dioxan. A buffer system composed of perchloric acid, citric acid, and triethanolamine has been used for the

**Table 10** *Log(stability constant) of cadmium–amino-acid complexes at 25 °C. Ionic strength is 0.1M-KNO$_3$*

| Amino-acid | $\beta(ML)$ | $\beta(ML_2)$ | Amino-acid | $\beta(ML)$ | $\beta(ML_2)$ |
|---|---|---|---|---|---|
| DL-α-Alanine | 4.2 | 7.3 | DL-Serine | 4.0 | 7.3 |
| L-Aspartic acid | 4.8 | 8.2 | L-Threonine | 3.9 | 7.1 |
| L-Glutamic acid | 4.1 | 7.1 | DL-Valine | 3.9 | 6.9 |
| Glycine | 4.5 | 8.0 | | | |

---

[278] M. R. Masson, *Talanta*, 1975, **22**, 933.
[279] F. C. Chang and K. L. Cheng, *Anal. Chim. Acta*, 1975, **76**, 177.
[280] G. J. M. Heijne and W. E. Van der Linden, *Talanta*, 1975, **22**, 923.
[281] P. F. Brun and K. H. Schrøder, *J. Electroanal. Chem. Interfacial Electrochem.*, 1975, **66**, 9.
[282] V. K. Olson, J. D. Carr, R. D. Hargens, and R. K. Force, *Anal. Chem.*, 1976, **48**, 1228.
[283] M. Kataoka, M. Shin, and T. Kambara, *Talanta*, 1977, **24**, 261.
[284] P. K. C. Tseng and W. F. Gutknecht, *Anal. Chem.*, 1976, **48**, 1996.
[285] A. Gordievskii, A. F. Zhukov, V. V. Snakin, Yu. I. Urusov, and V. S. Shterman, *Otkrytiya. Izobret., Prom. Obraztsy, Tovarnye Znaki*, 1975, **52**, 109.
[286] W. Selig, *Mikrochim. Acta*, 1975, **II**, 665.

successive determination of chloride, fluoride, and sodium in a single sample of orthophosphate mineral by means of electrodes that were selective for the appropriate ions.[287] A combination of $Ag_2S$ and $AgI/Ag_2S$ electrodes with different selectivity to the titrated ion has been used in the differential potentiometric titrations of KCN and $AgNO_3$.[288] Also, the effects of ethanol and methanol on solid-state ion-selective electrodes in direct potentiometry have been discussed.[289] Ion-selective electrodes have been used to provide information regarding the mean ion activity coefficient of ions present in sea-water, which is a complex mixture with an ionic strength of $0.7$ mol $l^{-1}$.[290] An electrode that is sensitive to organic cations has been found to sense choline chloride, and so it has been used to determine the mean ionic activity coefficients of choline chloride at molalities up to 4.[291] Values for the solubility product of $BaSO_4$ and $BaCrO_4$ have been derived[292] from measurements of the e.m.f. of cells of the type

| calomel electrode | $M^+$ aqueous solution | cation-exchange membrane | saturated MA ($BaSO_4$ or $BaCrO_4$) | anion-exchange membrane | $A^-$ aqueous solution | calomel electrode |
|---|---|---|---|---|---|---|

Electrodes formed of dodecylsulphate–ferroin or dodecylbenzenesulphonate–ferroin complexes ($10^{-3}$ mol $l^{-1}$) in o-dichlorobenzene–n-decanol solution have been used in the potentiometric analysis of mixtures of sodium octylsulphate and sodium dodecylbenzenesulphonate or sodium di-2-ethylhexyl sulphosuccinate.[293] The titration curves permitted the simultaneous determination of both components of the mixture.

Ion-selective electrodes have been used (i) in a continuous method for calibration and measurement[294] and for the monitoring of effluents,[295] (ii) in a flow-through arrangement for the potentiometric determination of ion ($Na^+$, $K^+$, $Ca^{2+}$, or $Cl^-$) activities,[296] (iii) in the flow injection analysis of soil extracts and blood serum,[297] (iv) in an automated titration system[211] equipped for the oscilloscopic display of some functions,[298] and (v) as detectors in liquid chromatography[299] and in a gas chromatograph for fluorine-, chlorine-, and sulphur-containing organic compounds.[300]

*Air-gap electrodes.* In the preparation of gas-sensing electrodes, gas-permeable hydrophobic membranes separating inner electrolyte from sample solution are used.[301] These membranes are eliminated in the so-called air-gap electrode, and instead an air gap separates the electrolyte layer from the sample solution.[302] The glass and reference (calomel) electrodes are held in a Teflon body. The calomel

[287] D. J. Duff and J. L. Stuart, *Talanta*, 1975, **22**, 823.
[288] I. C. Popescu, C. Liteanu, and A. Mocanu, *Rev. Roum. Chim.*, 1975, **20**, 397.
[289] A. M. Elbakai, G. J. Kakabadse, M. N. Khayat, and D. Tyas, *Proc. Anal. Div. Chem. Soc.*, 1975, **12**, 83.
[290] M. Whitfield, *Proc. Anal. Div. Chem. Soc.*, 1975, **12**, 56.
[291] J. B. Macaskill, M. S. Mohan, and R. G. Bates, *Anal. Chem.*, 1977, **49**, 209.
[292] M. Adhikari and D. Gangopadhyay, *J. Indian Chem. Soc.*, 1975, **52**, 654.
[293] N. Ciocan and D. F. Anghel, *Anal. Lett.*, 1976, **9**, 705.
[294] G. Horval, K. Toth, and E. Pungor, *Anal. Chim. Acta*, 1976, **82**, 45.
[295] M. E. Hofton, *Proc. Anal. Div. Chem. Soc.*, 1975, **12**, 308.
[296] H. F. Osswald, R. E. Dohner, T. Meier, P. C. Meier, and W. Simon, *Chimia*, 1977, **31**, 50.
[297] J. Ruzicka, E. H. Hansen, and E. A. Zagatto, *Anal. Chim. Acta*, 1977, **88**, 1.
[298] J. W. Frazer, A. M. Kray, W. Selig, and R. Lim, *Anal. Chem.*, 1975, **47**, 869.
[299] F. A. Schults and D. E. Mathis, *Anal. Chem.*, 1974, **46**, 2253.
[300] E. Pungor, K. Toth, Zs. Feher, G. Nagy, and M. Varadi, *Anal. Lett.*, 1975, **8**, ix.
[301] Ref. 17, pp. 315–331.
[302] J. Ruzicka and E. H. Hansen, *Anal. Chim. Acta*, 1974, **69**, 129.

electrode makes contact with the electrolyte layer through a ceramic pin. For sensing $CO_2$, the electrolyte layer used was a solution ($2 \times 10^{-2}$ mol l$^{-1}$) of $NaHCO_3$ saturated with a non-ionic wetting agent. The sample solution (50 μl) was kept in a polyethylene cup to which 100 μl of 0.1M-lactic acid solution was added. The cup enclosed the membrane of the glass electrode, leaving an air gap between the sample solution and the bicarbonate layer on the glass membrane. The $CO_2$ evolved from the well-stirred sample solution brought about a change in pH of the electrolyte layer which was sensed by the glass membrane. An improved version[303] in which the calomel electrode is replaced by a Ag/AgCl reference electrode is shown in Figure 1.

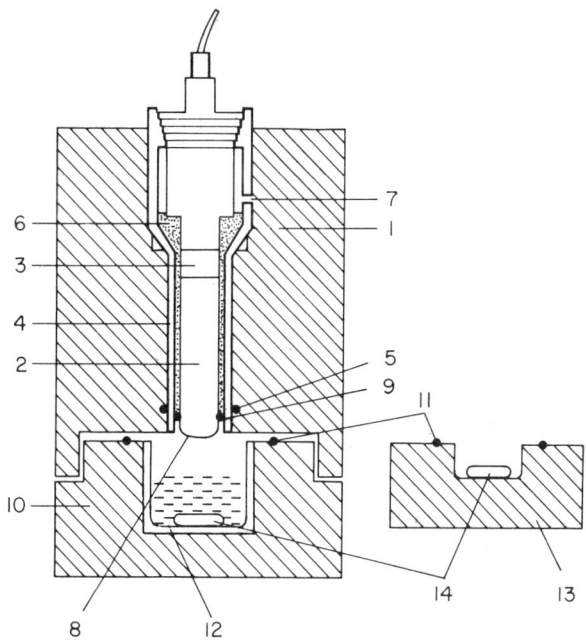

**Figure 1** *An air-gap electrode (after Hansen and Ruzicka[303]): 1 is the cylindrical Teflon body holding the glass 2 and reference 3 electrodes in a common Perspex tube 4; 5 is the O-ring; and 6 is the electrolyte solution surrounding the Ag/AgCl reference electrode and contained in the Perspex tube, the solution being introduced through a small hole 7. The same solution covers the pH-sensitive surface 8 of the glass electrode. 9 is an O-ring that has been degreased and made hydrophilic in a strong solution of detergent, and it prevents leakage of electrolyte from the reference compartment. 10 is the sample chamber (macrochamber, and 13 is the microchamber) with a smooth coaxial hole and fitted to the main body of the equipment through the O-ring 11, which serves as a gasket. 12 is a polythene vial, fitting closely into the cavity of the macrochamber. The microchamber can be used by itself as a sample holder. Both chambers are equipped with Teflon-coated stirring bars 14.*

(Reproduced by permission from 'Membrane Electrodes', Academic Press, New York, 1976)

---

[303] E. H. Hansen and J. Ruzicka, *Anal. Chim. Acta*, 1974, **72**, 353.

The air-gap electrode will sense ammonia gas if a solution of $NH_4Cl$ is used as the electrolyte on the glass membrane. Similarly, other gases such as $SO_2$ or oxides of nitrogen may be sensed when a layer of the appropriate electrolyte on the glass membrane is used. The advantages and disadvantages of both the air-gap and polymeric-membrane-based sensors[262] have been pointed out.[302] An even simpler arrangement of the air-gap electrode has been described.[304] In this construction, a combined glass–calomel electrode is used. To hold the electrolyte on the glass membrane, a 1% methylcellulose was added to the electrolyte. This made contact through a porous pin with the calomel reference electrode. The accuracy of measurements with this electrode compared favourably with similar measurements made with its earlier prototypes.

The air-gap electrode has been used in the determination of the ammonia content of waste waters,[305] the hydrogen sulphite content of wine,[306] the total inorganic and total organic carbon contents in water,[307] and total inorganic nitrogen.[308] The air-gap electrode has been used to assay methylamine in the concentration range $10^{-5}$—$10^{-2}$ mol $l^{-1}$.[309] In this procedure, a conical polyurethane sponge, soaked in the proper amount of electrolyte solution (in most assays, $5 \times 10^{-3}$M-methylamine and $5 \times 10^{-2}$M-KCl plus wetting agent) has been used for depositing the electrolyte layer on the membrane of the glass electrode. The pH of the sample was maintained at 12.

*Coated wire electrodes.* A coated platinum wire electrode that is selective to anionic detergents has been prepared.[310] The tip of the Pt wire was fused to make a ball, 1.5 mm in diameter, and dipped repeatedly in the coating mixture [3:1 mixture of PVC in cyclohexanone and a solution of ion-association complex in decanol which is prepared by shaking a mixture (50% v/v) of 0.1M aqueous solution of methyltricaprylammonium chloride]. Similarly, Orion perchlorate liquid ion-exchanger mixed with PVC and coated on to Pt wire acted as an electrode selective to perchlorate ion.[311] This electrode had a Nernstian response to perchlorate in the concentration range $10^{-1}$—$10^{-4}$ mol $l^{-1}$.

A platinum wire coated with a membrane of potassium tetraphenylborate, PVC, and a plasticizer was evaluated for its response to potassium ion over a wide concentration range, using three plasticizers; 3-nitro-*o*-xylene (A), dibutyl phosphate (B), and di-iso-octyl phthalate (C).[312] The electrode with (A) gave a linear response over a large concentration range and exhibited high selectivity to $K^+$ ions over $Ca^{2+}$, $Mg^{2+}$, and $Na^+$ ions, but its lifetime was short, compared to electrodes containing (B) and (C), which had lifetimes of 3 weeks. Evaluation of the electrode containing (C) showed[312a] that it gave Nernstian responses to quinine, brucine, and cinchonine in the concentration ranges $10^{-1}$—$10^{-7}$ mol $l^{-1}$ (slope per decade of

---

[304] J. Fligier and Z. Gregorowicz, *Anal. Chim. Acta*, 1977, **90**, 263.
[305] J. Ruzicka, E. H. Hansen, P. Bisgaard, and E. Reymann, *Anal. Chim. Acta*, 1974, **72**, 215.
[306] E. H. Hansen, H. Bergamin Filho, and J. Ruzicka, *Anal. Chim. Acta*, 1974, **71**, 225.
[307] U. Fielder, E. H. Hansen, and J. Ruzicka, *Anal. Chim. Acta*, 1975, **74**, 423.
[308] E. H. Hansen, J. Ruzicka, and N. R. Larsen, *Anal. Chim. Acta*, 1975, **79**, 1.
[309] K. P. Hsiung, S. S. Kuan, and G. G. Guilbault, *Anal. Chim. Acta*, 1976, **84**, 15.
[310] T. Fuginaga, S. Okazaki, and H. Freiser, *Anal. Chem.*, 1974, **46**, 1842.
[311] T. J. Rohm and G. G. Guilbault, *Anal. Chem.*, 1974, **46**, 590.
[312] E. Hopirtean, C. Liteanu, and E. Stefaniga, *Rev. Roum. Chim.*, 1974, **19**, 1651.
[312a] E. Hopirtean and E. Stefaniga, *Rev. Roum. Chim.*, 1976, **21**, 305.

concentration was 55 mV, $10^{-1}$—$3 \times 10^{-6}$ mol $l^{-1}$ (54 mV), and $10^{-1}$—$10^{-5}$ mol $l^{-1}$ (61 mV), respectively.

In the aqueous phase, zephiramine cation ($Z^+$) formed ion pairs with dodecylbenzenesulphonate anion ($DBS^-$). These ion pairs were extracted into a nitrobenzene phase. A melt of naphthalene with nitrobenzene extract was solidified on a clean platinum wire. This, when used as an electrode, showed good linearity and Nernstian slope for pDBS from 3 to 7 and for pZ from 4 to 7.[313] It has been found to be useful in potentiometric titrations. Similarly, the zephiramine–Eriochrome Black T ion pair, dissolved in hot nitrobenzene–naphthalene and coated on to a platinum wire, has been used in chelometric titrations of metal ions with EDTA.[314]

Aliquat 336 S (tricaprylammonium) salts and PVC in either tetrahydrofuran or cyclohexanone have been used in the preparation of several coated wire electrodes that are selective to the ions concerned. For example, Aliquat tetrachloroferrate(III),[315] chlorocuprate(II),[316] chlorozincate(II),[317] chloro- and iodomercurates(II),[318] and chlorocadmate(II)[319] have been used to prepare electrodes selective to iron(III), chlorocuprate(II), tetrachlorozincate(II), $Hg^{II}$, and chlorocadmate(II) respectively. Some of these have been used in the estimation of $Fe^{III}$ in mineral samples[320] and in complexometric titrations.[317,318] Similarly, Aliquat–phenobarbitol-coated platinum wire showed (at pH 9.6) a fast response time with interference (moderate) from $NO_3^-$, $Cl^-$, and salicylate ions.[321] However, assay of tablets compared very well with USP methods.

Coated wire electrodes that are selective to $Ca^{2+}$ ions have been prepared.[322] In this investigation, several alkyl phosphoric acid esters and PVC in tetrahydrofuran have been used to form the electrodes. Of the several electrodes tested, some were superior to others. These superior electrodes, whose characteristics are given in Table 11, had membranes with the following composition:

| mixture type | composition |
|---|---|
| A | 30% PVC + (10:1) DEHEHP–$CaH_2X_4$ |
| B | 40% PVC + (10:1) DEHEHP–$CaH_2X_4$ |

DEHEHP [= di-(2-ethylhexyl) 2-ethylhexylphosphonate] was the plasticizer. The calcium salts of diesters ($CaH_2X_4$) were prepared by mixing calcium salt ($CaX_2$) with the phosphoric acid ester (HX) in mole ratio 1:2.

The selectivities ($k_{i,j}$) of these superior electrodes have also been evaluated.[323] This evaluation was made by measuring the cell potential $E_1$ for a pure $CaCl_2$ solution ($10^{-3}$ mol $l^{-1}$) followed by measurement of the cell potential $E_2$ for a

---

313 M. Kataoka and T. Kambara, *Denki Kagaku Oyobi Kogyo Butsuri Kagaku*, 1975, **43**, 209.
314 A. R. Rajaput, M. Kataoka, and T. Kambara, *J. Electroanal. Chem. Interfacial Electrochem.*, 1975, **66**, 67.
315 R. W. Cattrall and C. P. Pui, *Anal. Chem.*, 1975, **47**, 93.
316 R. W. Cattrall and C. P. Pui, *Anal. Chim. Acta*, 1976, **83**, 355.
317 R. W. Cattrall and C. P. Pui, *Anal. Chim. Acta*, 1976, **87**, 419.
318 R. W. Cattrall and C. P. Pui, *Anal. Chem.*, 1976, **48**, 552.
319 R. W. Cattrall and C. P. Pui, *Anal. Chim. Acta*, 1977, **88**, 185.
320 R. Cattrall and C. P. Pui, *Anal. Chim. Acta*, 1975, **78**, 463.
321 G. D. Carmack and H. Freiser, *Anal. Chem.*, 1977, **49**, 1577.
322 R. W. Cattrall, D. M. Drew, and I. C. Hamilton, *Anal. Chim. Acta*, 1975, **76**, 269.
323 R. W. Cattrall and D. M. Drew, *Anal. Chim. Acta*, 1975, **77**, 9.

**Table 11** Response characteristics of some coated-wire electrodes that are selective to $Ca^{2+}$ ions,[322] and their selectivity coefficients[323]

| Reagent[a] | Composition | Concentration range/mol l$^{-1}$ | Slope (mV per decade of activity) | Response time/min | Selectivity coefficient, $k_{i,j} \times 10^3$ | | | | |
|---|---|---|---|---|---|---|---|---|---|
| | | | | | Mg | Sr | Ba | Na | K |
| HDEHP | A | $3 \times 10^{-2}$—$4 \times 10^{-4}$ | 28 | 5—10 | 50 | 24 | 9 | 3 | 23 |
| HDOP | A | $3 \times 10^{-2}$—$5 \times 10^{-4}$ | 29 | 5—10 | 6 | 12 | 11 | 3 | 1 |
| OPPA | A | $3 \times 10^{-2}$—$4 \times 10^{-4}$ | 29 | 10 | 26 | 21 | 9 | 1 | 1 |
| HDOPP | A | $2.5 \times 10^{-2}$—$6 \times 10^{-5}$ | 29 | 5 | 28 | 36 | 8 | 1 | 61 |
| HDEHP | B | $1 \times 10^{-2}$—$2 \times 10^{-4}$ | 29.5 | 5—10 | 68 | 28 | 15 | 4 | 1 |
| IOPA | B | $3 \times 10^{-2}$—$2.5 \times 10^{-4}$ | 34 | 10 | 54 | 7 | 5 | 1 | 1 |
| OPPA | B | $3 \times 10^{-2}$—$1.4 \times 10^{-5}$ | 30 | 10 | 52 | 25 | 38 | 1 | 6 |

(a) HDEHP is di-(2-ethylhexyl) hydrogen phosphate; HDOP is di-n-octyl hydrogen phosphate; OPPA is octylphenyl dihydrogen phosphate; HDOPP is di-(n-octylphenyl) hydrogen phosphate; IOPA is iso-octyl dihydrogen phosphate

Membrane Phenomena 77

solution containing a mixture of $CaCl_2$ and the interfering salt $(10^{-3}M\text{-}CaCl_2 + 10^{-2}M\text{-}MCl_n)$. Thus, using the equation

$$E_2 - E_1 = \left(\frac{2.303RT}{nF}\right) \log \left(1 + k_{i,j} \frac{a_j}{a_i}\right) \qquad (33)$$

where $a_i$ is the activity of the primary ion and $a_j$ is that of the interfering ion, the value of $k_{i,j}$ was derived. Some of the values so derived are given in Table 11. Platinum wire coated with a solution of PVC, a plasticizer (either DEHEHP or di-n-decyl phthalate, DDP), and valinomycin dissolved in tetrahydrofuran acted as an electrode that is selective to $K^+$ ions.[324] A DDP-containing electrode gave better response to $K^+$ ions than one containing DEHEHP. $Cu^{II}$ and $Zn^{II}$ ions poisoned the electrode, which, however, has been used to determine the potassium content of whole blood and of sea-water.

*Glass membrane electrodes.* The theory and applications of glass membrane electrodes have been summarized recently by Schwabe[325] and also by Lakshminarayanaiah.[326]

Electrodes made from chalcogenide glass, $CuAs_2S_3$, have been found to be selective to $Cu^{II}$ ions in concentrated solutions of chloride, bromide, nitrate, and acetate over the range $10^{-1}$—$10^{-6}$ mol $l^{-1}$ of $Cu^{II}$ ions.[327] Slopes of 30 mV per decade of concentration for $NO_3^-$ and acetate solutions, 50 mV for 1.0M chloride or bromide electrolyte, and a slope of 40 mV for 0.1M chloride solution have been reported. No interferences from Ca, Pb, Zn, Ni, or Mn and interferences from Ag and $Fe^{III}$ (when present in 10-fold excess), and no response to $Cu^I$, have been found.

It is generally believed that a swollen layer of hydrous silica which is necessary for the proper functioning of the pH electrode exists on the surface of the conditioned glass membrane. In the gel layer of a pH glass, protons are bound to the negative charges in a silicon network crosslinked with bi- or ter-valent metal ions. The outer part of the gel layer contains water. Between the gel layer and dry glass, there is a transition layer that acts as a barrier to transport of ions and prevents rapid corrosion of glass.[328,329] By etching away the gel layers and the transition layers on both sides of the glass membrane and then starting hydration simultaneously on both sides, two symmetric layers can be produced. This will eliminate any asymmetry potential.[329]

Glass electrodes previously used in alkaline solutions undergo structural transformations within the gel layer and generally give rise to large potential drifts in neutral or acidic test solutions.[330] Consequently alternating transfers between acidic and basic solutions should be avoided to get stable electrode-potential response. The changes in potential either during the hydration of freshly etched electrode or the drying of hydrated electrodes have been observed to be slight.[330] Replacement of silicon in the network by aluminium causes the degree of $H^+$ and $M^+$ ion exchange to decrease[328] and favours selectivity of the glass to alkali-metal ion.

[324] R. W. Cattrall, S. Tribuzio, and H. Freiser, *Anal. Chem.*, 1974, **46**, 2223.
[325] K. Schwabe, *Adv. Anal. Chem. Instrum.*, 1974, **10**, 495.
[326] Ref. 17, pp. 277—311.
[327] R. Jasinski, I. Trachtenberg, and G. Rice, *J. Electrochem. Soc.*, 1974, **121**, 363.
[328] A. Wikby, *Talanta*, 1975, **22**, 663.
[329] G. Johansson, B. Karlberg, and A. Wikby, *Talanta*, 1975, **22**, 953.
[330] B. Karlberg, *Talanta*, 1975, **22**, 1023.

In order to study further the properties of surface layers of pH glass electrodes, the ion-sputtering method has been used.[331] The profiles of ion concentration formed under several conditions (change of temperature, pH, and hydrodynamic conditions) were found to be reversible. The rate process governing complete ion exchange in the gel layer in the steady state was found to be diffusion of the ion that is present in deficit in the gel layer. Estimates of ratios of self-diffusion coefficient for interdiffusing ions have been made by using the results obtained from e.m.f. measurements.[332] Typical e.m.f. changes observed after transferring the glass electrode from acidic to isopropyl alcohol solution and *vice versa* have been measured as a function of time. From these curves, $t_{10}$ and $t_{20}$, *i.e.* the times at which the steady-state e.m.f. value was within 10 and 20 mV, were determined. The difference, $\Delta t_{10,20}$, was considered to be inversely proportional to the diffusion coefficient of the ion. Thus $D_H(\Delta t_{10,20})_H = \text{constant} = D_{ion}(\Delta t_{10,20})_{ion}$. The diffusion coefficient ratio is given by

$$\frac{D_H}{D_{ion}} = \frac{\Delta t_{10,20(ion)}}{\Delta t_{10,20(H)}}$$

Some of the values determined for two glass electrodes are given in Table 12.

The factors that affect the limit of detection of sodium-responsive glass electrodes have been examined, and it has been found that the ultimate limit is set by the dissolution of alkali-metal ions from the glass membrane itself.[333] The limit of detection of sodium of an EIL GEA 33 electrode in an EIL flow cell with a flow rate of 4 ml min$^{-1}$ was about 0.07 $\mu$g l$^{-1}$ at 20 °C, when the pH was maintained with diethylamine at 11.0. Several sodium-selective glass membrane electrodes have been evaluated for their selectivity in mixed solutions, their response times,[334] and their use in automatic monitoring systems employing concentrated solutions.[335] By using a flow channel of special design which enhanced the flow of solution past the

**Table 12** *Characteristics of some commercially available glass electrodes*

| | | $t_{10,20}$/s for transfer from: | | | | |
|---|---|---|---|---|---|---|
| Electrode | State of the electrode | acid to isopropyl alcohol | isopropyl alcohol to acid | Ion | $\dfrac{D_{H+}}{D_{ion}}$ | Apparent exchange capacity/nmol cm$^{-2}$ |
| Beckmann E-2 | hydrated | 120 | 765 | Na | 6.4 | — |
| Ingold 201 | hydrated | 25 | 240 | Li | 9.6 | 1.5 |
| | etched | 15 | 105 | Li | 7.0 | 0.8 |
| | dried | 190 | 550 | Li | 2.9 | — |
| | hydrated | 35 | 220 | Na | 6.3 | 8.0 |
| | etched | 4 | 8 | Na | 2 | 2.4 |
| | etched and then hydrated (1 h) | 4 | 7 | Na | 2 | 1.8 |
| | dried | 120 | 405 | Na | 3.4 | — |

---

[331] F. G. K. Baucke, *J. Non-cryst. Solids*, 1975, **19**, 75.
[332] B. Karlberg, *J. Electroanal. Chem. Interfacial Electrochem.*, 1974, **49**, 1.
[333] G. I. Goodfellow, D. Midgley, and H. M. Webber, *Analyst (London)*, 1976, **101**, 848.
[334] M. F. Wilson, E. Haikala, and P. Kivalo, *Anal. Chim. Acta*, 1975, **74**, 395, 411.
[335] K. Bergner, *Anal. Chim. Acta*, 1976, **87**, 1.

sensor bulb, low levels of sodium have been estimated.[336] When determining low levels of sodium (about 1 p.p.b.), an unusual temperature effect has been noted, probably due to the response of the electrode to $H^+$ ions produced by the splitting of water at elevated temperatures.[337]

A new method for evaluating the performance of glass electrodes in the pH range 6.5—12.6 has been given.[338] Cells without liquid junction with Ag/AgCl reference electrodes and amine buffers and their hydrochlorides with test solutions were used. Also, the method has been extended at both ends of the pH scale to cover the range 0—14.[339]

A differential pH electrode system incorporating a new type of reference electrode, a conventional pH-measuring glass electrode, and a dual high-impedance input amplifier has been described.[340] The reference electrode was a conventional pH electrode (matched to the pH-measuring electrode) whose filling solution was similar in composition to the solution under test. Thus a symmetrical cell (potentiometric output nearly zero) which reduced or eliminated the usual sources of error of conventional pH cells, such as variation in liquid junction potential, temperature coefficients, ionic strength, and activity coefficients, may be used conveniently in situations where acidity had to be controlled.

A pH-dependent electrode to monitor hydrogen ion concentration has been prepared by thermomoulding a mixture of powdered antimony and a thermoplastic polymer.[341] Its behaviour has been claimed to be highly reproducible and reliable, and considered suitable for use in gas-sensing probes for $NH_3$, $CO_2$, and $SO_2$.

The behaviour of a glass electrode in heavy water ($D_2O$) has been evaluated. Cation-selective glass electrodes (CSGE) in $D_2O$ were used in cells of the type[342]

CSGE|NaCl($m$ in $H_2O$)|AgCl, Ag           Type (1)

CSGE|NaCl($m'$ in $D_2O$)|AgCl, Ag           Type (2)

The CSGE's showed little drift when they were transferred from one solution in $H_2O$ to another in $D_2O$, and steady potentials were quickly observed. The e.m.f.'s measured in the selectivity ($k_{H,M}$) determination experiments were fitted to the equation

$$E = E° - (RT/F) \ln\{(a_H + K_{H,M}a_M)(a_{Cl})\} \quad (34)$$

where M is Na or K and $a_H$, $a_M$, and $a_{Cl}$ are activities of $H^+$, $M^+$, and $Cl^-$. The values of $k_{H,M}$ derived on the assumption that the activity coefficients of $H^+$ and $Na^+$ were equal are given in Table 13. The electrodes, in view of the large selectivity coefficients, were not affected by the contributions of $H^+$ and $D^+$ ions. The value of $k_{K,Na}$ ($=k_{H,Na}/k_{H,K}$) is low for the GKN electrode, indicating a slight response to $K^+$ ions in NaCl solutions in the above cells.

Also, the changes in e.m.f. following the transfer of the glass electrode in cell (1) from a solution of molality $m_1$ to another of molality $m_2$ and similar changes in cell

---

[336] E. L. Eckfeldt and W. E. Proctor, Jr., *Anal. Chem.*, 1975, **47**, 2307.
[337] E. L. Eckfeldt, *Anal. Chem.*, 1975, **47**, 2309.
[338] M. Filomena, G. F. C. Camoes, and A. K. Covington, *Anal. Chem.*, 1974, **46**, 1547.
[339] A. K. Covington and M. I. A. Ferra, *Anal. Chem.*, 1977, **49**, 1363.
[340] T. S. Light and J. L. Swartz, *Anal. Chem.*, 1977, **49**, 1009.
[341] M. Mascini and C. Cremisini, *Anal. Chim. Acta*, 1977, **92**, 277.
[342] B. M. Lowe and D. G. Smith, *J. Electroanal. Chem. Interfacial Electrochem.*, 1974, **51**, 295.

(2) have been measured. Similarly, the changes in e.m.f. following transfer from cell (1) to cell (2), [$\Delta E$(H→D)] and then return to cell (1) [$\Delta E$(D→H)] have been measured. From these measurements, and using appropriate equations, values for the aqua-molal activity coefficient $\gamma$ for NaCl solutions in $D_2O$ have been derived. Some of these values are given in Table 14.

The temperature dependence of the response of the glass electrode in $D_2O$ has also been studied.[343] Solutions of NaOH of equal molality in $H_2O$ and $D_2O$ at constant ionic strength were prepared. At a given temperature, concentrations of $H^+$ and $D^+$ were calculated and also measured using a glass membrane electrode. The difference between $\Delta_{calc}$ and $\Delta_{exp}$ ($\Delta = pD - pH$), designated $\Delta pD$, determined as a function of temperature, conformed to the relation

$$\Delta pD = (2.02 \times 10^2/T) - 0.204$$

provided the electrode showed low Na error.

The impedance characteristics of two ion-selective glass electrodes (Beckmann E-2 and General Purpose) have been investigated in the low-frequency range.[344] The low-frequency responses deviated from responses expected of a simple finite Warburg model but conformed to an empirical relation based on Cole–Cole distribution of time constants.[345]

A new solid-state theory of glass electrode responses applied to a homogeneous-site glass membrane gave the equation[346]

$$E = \text{Const} + (RT/F) \ln\{[a'_H + K_{H/M}(K_1 a'_H a'_M)^{1/2}](1 + K_1 a'_M/a'_H)^{1/2}\} \quad (35)$$

where the quantities $a$ are activities of $H^+$ and $M^+$ ions in the sample at (′) side, $K_1$ is the ion-exchange equilibrium constant for metal ions replacing protons on silicate sites, while $K_{H/M}$ is a mobility and defect generation ratio term given by

$$K_{H/M} = \frac{u_i^M K_{Ii}^{M^-} \bar{\gamma}_{1M}}{u_i^H K_{Ii}^{H^-} \bar{\gamma}_{1H}} \quad (36)$$

where $u_i^M$ and $u_i^H$ are interstitial mobilities of ions $M^+$ and $H^+$, $K_{Ii}^M$ and $K_{Ii}^H$ are equilibrium constants for the generation of vacancies and interstitials for M and H

**Table 13** *Selectivity coefficients of cation-selective glass electrodes*

| Electrode | $k_{H,Na}$ ($H_2O$) | $k_{D,Na}$ ($D_2O$) | $k_{H,K}$ | $k_{K,Na}$ |
|---|---|---|---|---|
| GEA 33 (Na-selective) | 330 | 90 | 1.4 | 230 |
| GKN 33 (K-selective) | 780 | 210 | 15 500 | 0.05 |

**Table 14** *Values of $\gamma$ for solutions of sodium chloride in $D_2O$*

| Aqua-molality, m | 0.1 | 0.3 | 0.5 | 0.7 | 1.0 |
|---|---|---|---|---|---|
| $-\log \gamma$ | 0.110 | 0.151 | 0.170 | 0.179 | 0.188 |

[343] R. K. Force and J. D. Carr, *Anal. Chem.*, 1974, **46**, 2049.
[344] J. R. Sandifer and R. P. Buck, *J. Electroanal. Chem. Interfacial Electrochem.*, 1974, **56**, 385.
[345] K. S. Cole, 'Membranes, Ions and Impulses', University of California Press, Berkeley, California, 1968.
[346] R. P. Buck, *Anal. Chem.*, 1973, **45**, 654.

# Membrane Phenomena 81

in silica sites (type 1), and $\gamma_{1M}$ and $\gamma_{1H}$ are the activity coefficients of lattice-site $M^+$ and $H^+$ ions in type 1 sites.

The non-ideal parameter that appears in the Eisenman theory[347] has been shown to be given by[346]

$$n \approx \frac{59.14}{17.7} \log(1 + K_{H/M}^{\frac{1}{2}})\left(1 + \frac{1}{K_{H/M}}\right)^{\frac{1}{2}} \tag{37}$$

For cation-selective glass membranes containing heterogeneous sites, the potential has been shown to be given by (e.g. for the $Na^+$-selective electrode)[346,348]

$$E = \text{Const} + (RT/F)\ln[a'_{Na} + K_{Na/M}^{(1)}(a'_{Na}a'_M)^{\frac{1}{2}}] \times [1 + K_{Na/M}^{(2)}(a'_M/a'_{Na})^{\frac{1}{2}}] \tag{38a}$$

$$= \text{Const} + (RT/F)\ln[K_{Na/M}^{(1)}K_{Na/M}^{(2)}] + (RT/F)\ln[a'_M + K_{M/Na}^{(1)}(a'_{Na}a'_M)^{\frac{1}{2}}]$$

$$\times [1 + K_{M/Na}^{(2)}(a'_{Na}/a'_M)^{\frac{1}{2}}] \tag{38b}$$

$K_{Na/M}^{(2)}$ refers $K_{Na/M}$ to sites of type 2 in the membrane, so that

$$K_{Na/M}^{(1)} = \frac{1}{K_{M/Na}^{(1)}} = \left(\frac{K_{2i}^M \bar{\gamma}_{2M}}{K_{2i}^H \bar{\gamma}_{2H}}\right) K_2 \tag{39}$$

$$K_{Na/M}^{(2)} = \frac{1}{K_{M/Na}^{(2)}} = \left(\frac{u_i^M}{u_i^{Na}}\right) K_{Na/M}^{(1)} \tag{40}$$

where $K_2$ is the ion-exchange constant applied to type 2 sites and $K_{2i}$ is the equilibrium constant for the generation of vacancies and interstitials in sites of type 2: $a_M$ is the activity of interfering ion. In this case $n$ is given by

$$n \approx \frac{59.14}{17.7}\log\left[1 + \left(\frac{u_i^M}{u_i^{Na}}\right)^{\frac{1}{2}}\right] \times \left[1 + \left(\frac{u_i^{Na}}{u_i^M}\right)^{\frac{1}{2}}\right] \tag{41}$$

Using several electrodes, the potential responses were measured as a function of pH.[346,348] In the case of pH electrodes, the experimental curves followed equation (35). The responses of a sodium-selective electrode were computed with the help of equations (38)—(40) and compared with experimental e.m.f. vs. pH curves at constant sodium concentration. The results realized showed that the selectivity coefficients were concentration-dependent. This aspect of the dependence of selectivity coefficient on concentration, or on the ratio of concentrations in mixtures, for glass and other ion-selective electrodes has also been described.[219]

Glass membrane electrodes have been used in the determination of the equilibrium constant of the reaction[349]

$$HClO_4 + LiOAc \underset{}{\overset{K_f}{\rightleftharpoons}} LiClO_4 + HOAc$$

and in the potentiometric titration of zinc and cadmium acetates with $HClO_4 +$ LiOAc, enabling the determination of formation constants for the acetate complexes of zinc and cadmium. Also, the electrodes are used to determine the dissociation constants of weak acids in various solvents.[350]

---

[347] G. Eisenman, *Biophys. J.*, 1962, **2**, Part 2, 259.
[348] R. P. Buck, J. H. Boles, R. D. Porter and J. A. Margolis, *Anal. Chem.*, 1974, **46**, 255.
[349] K. Sawada, M. Nakano, H. Mori, and M. Tanaka, *Bull. Chem. Soc. Jpn*, 1975, **48**, 2282.
[350] M. Bos and W. Lengton, *Anal. Chim. Acta*, 1975, **76**, 149.

Cation-selective glass microelectrodes made from $NAS_{27\text{-}04}$ have been evaluated for their electrochemical properties.[351] A marked fall in resistance was noted when microelectrodes were stored in 3M-KCl. This fall in resistance was accompanied by an increase in selectivity of the microelectrode to $K^+$ ions. These changes were attributed to the hydration of the glass membrane. Similarly, microelectrodes with sensing bulbs (100—500 μm range) have been prepared[352] and used in routine pH measurements.[353] pH-Sensitive microelectrodes have been used to estimate intracellular pH.[354] A serious problem in these measurements is the elimination of tip potentials, which has been studied in depth.[355] An increase in tip potential following a decrease in electrode resistance was noted when microelectrodes were studied in 3M-KCl at 37 °C. Measurement of e.m.f. across the thin glass wall in the tip region indicated that the thin glass wall contributed to the generation of tip potential. After a study of the effects of acid treatment of glass it was suggested that the negative charges on the glass wall played a significant role in the generation of tip potential, the magnitude being governed by both diffusion and interfacial potentials.

CATION-SELECTIVE MEMBRANE ELECTRODES. *Ammonium- and ammonia-selective membrane electrodes.* Membranes containing 3% sodium tetraphenylborate, ammonium tetraphenylborate, or potassium tetraphenylborate, 35% PVC, and 62% dibutyl phthalate gave responses which were linear in the $pNH_4$ ranges 0—3.8, 0—3.5, or 0—3.2 respectively, with a slope of 54 mV per unit of $pNH_4$.[356] Sodium interfered with the response of the electrode, whereas calcium or magnesium did not.

Nonactin held in a silicone rubber matrix acted selectively towards $NH_4^+$ ion.[357] Earlier, a liquid membrane [saturated solution of nonactin (72%) and monactin (28%) in tris-(2-ethylhexyl) phosphate][358] was shown to act as an electrode that was selective to $NH_4^+$ ions, exhibiting the selectivity sequence $NH_4^+ > K^+ > Rb^+ > H^+ > Cs^+ > Li^+ > Na^+ > Ca^{2+}$. The selectivity of this electrode towards organic onium ions has been determined.[359]

The ammonia probe and $NH_4^+$-selective membrane electrodes have been used in the estimation of $NH_4^+$ ions present in airborne particulates after filtering the air and then collecting and extracting the particles. Both electrodes, as expected, gave identical results.[360] The electrodes have been used to detect $NH_4^+$ ion produced in the urea–urease reaction, which has been used in an automated process to determine serum urea.[361] Similarly, it is used to assay nitrate and nitrite reductases in stationary and flow stream conditions.[362,363] The reactions involved are:

[351] C. O. Lee and H. A. Fozzard, *Biophys. J.*, 1974, **14**, 46.
[352] J. D. Czaban and G. A. Rechnitz, *Anal. Chem.*, 1975, **47**, 1787.
[353] J. D. Czaban and G. A. Rechnitz, *Anal. Chem.*, 1976, **48**, 277.
[354] Y. Okada and A. Inouye, *Biophys. Struct. Mechanism*, 1976, **2**, 21.
[355] Y. Okada and A. Inouye, *Biophys. Struct. Mechanism*, 1976, **2**, 31.
[356] E. Hopirtean and E. Stefaniga, *Rev. Roum. Chim.*, 1975, **20**, 863.
[357] G. G. Guilbault and G. Nagy, *Anal. Chem.*, 1973, **45**, 417.
[358] R. P. Scholer and W. Simon, *Chimia*, 1970, **24**, 372.
[359] E. F. Vansant, H. Deelstra, and R. Dewolfs, *Talanta*, 1974, **21**, 608.
[360] M. L. Eagan and L. Dubois, *Anal. Chim. Acta*, 1974, **70**, 157.
[361] R. A. Llenado and G. A. Rechnitz, *Anal. Chem.*, 1974, **46**, 1109.
[362] W. R. Hussein and G. G. Guilbault, *Anal. Chim. Acta*, 1974, **72**, 381.
[363] W. R. Hussein and G. G. Guilbault, *Anal. Chim. Acta*, 1975, **76**, 183.

Membrane Phenomena 83

$$NO_3^- \xrightarrow{\text{nitrate reductase}} NO_2^-$$

$$NO_2^- + 6e^- + 8H^+ \xrightarrow{\text{nitrite reductase}} NH_4^+ + 2H_2O$$

The increase in the $NH_4^+$ concentration is monitored with the help of the electrode. The gas-permeable ammonia electrode has been used to determine ammonia in waste water,[364] ammonia, nitrate, and organic nitrogen in a number of waters and effluents,[365] ammonia in natural and waste waters, on an automated continuous-flow system with on-line minicomputer and printer,[366] nitrogen in tobacco and tobacco smoke,[367] mobile nitrogen in steel,[368] and bound nitrogen in uranium hexafluoride.[369]

The use of an improved cell and pretreatment of the ammonia electrode by storing it in a solution containing 1 p.p.m. of $NH_4Cl$ has been claimed to improve the performance of the electrode.[370] The multiple-addition method, in combination with Gran plots, has been used for the determination of 2.8 p.p.m.—14 p.p.b. of ammonia nitrogen.[371]

With the ammonia probe, the change in pH of the internal electrolyte is measured, to estimate the quality of ammonia in a sample. Instead of measuring the change in pH, the change in pM (where M is $Ag^+$, $Cu^{2+}$, or $Hg^{2+}$, which form complexes with $NH_3$) may be measured, by means of ion-selective electrodes, to estimate ammonia.[372] The ion-selective electrodes used are the precipitate-based Ag, Cu, or Hg electrodes. The internal electrolyte [$NH_4NO_3$, $M(NO_3)_n$, and NaF] is separated from the sample by a hydrophobic gas-permeable membrane. The sensor is immersed in a sample solution containing $NH_3$, which diffuses across the membrane until the concentration of $NH_3$ reaches equilibrium. The response of the electrode is governed by the following reactions.

Ammonia diffusing into the internal solution forms complexes. Thus

$$M^{n+} + iNH_3 \rightleftharpoons M(NH_3)_i^{n+} \qquad (42)$$

$$\beta_i = \frac{[M(NH_3)_i^{n+}]}{[M^{n+}][NH_3]^i} \quad (i=1, 2, \ldots N) \qquad (43)$$

Total metal ion concentration is given by $[M_t] = [M_{free}] + [M_{combined}]$. Thus

$$[M_t] = [M^{n+}]\left(1 + \sum_{i=1}^{N} \beta_i [NH_3]^i\right) \qquad (44)$$

The concentration of free metal ion is given by

$$[M^{n+}] = [M_t] / \left(1 + \sum_{i=1}^{N} \beta_i [NH_3]^i\right) \qquad (45)$$

[364] W. H. Evans and B. F. Patridge, *Analyst (London)*, 1974, **99**, 367.
[365] L. R. McKenzie and P. N. W. Young, *Analyst (London)*, 1975, **100**, 620.
[366] I. Sekerka and J. F. Lechner, *Anal. Lett.*, 1974, **7**, 463.
[367] C. H. Sloan and G. P. Morie, *Anal. Chim. Acta*, 1974, **69**, 243.
[368] J. B. Headridge and G. D. Long, *Analyst (London)*, 1976, **101**, 103.
[369] O. A. Vita, *Anal. Chim. Acta*, 1976, **81**, 45.
[370] N. Shibata, *Anal. Chim. Acta*, 1976, **83**, 371.
[371] W. Selig, J. W. Frazer, and A. M. Kray, *Mikrochim. Acta*, 1975, **II**, 675.
[372] T. Anfalt, A. Graneli, and D. Jagner, *Anal. Chim. Acta*, 1975, **76**, 253.

The indicator electrode measures $[M^{n+}]$ and so the potential is given by

$$E = E° + (RT/nF) \ln[M^{n+}] \qquad (46)$$

Substituting for $[M^{n+}]$ from equation (45) gives, when $\Sigma\beta_i[NH_3]^i \gg 1$, the equation

$$E = E°' - (RT/nF) \ln\sum\beta_i[NH_3]^i \qquad (47)$$

where $E°' = E° + (RT/nF) \ln 10 \log [M_t]$.
Thus, according to equation (47), the electrode responds to changes in the concentration of $NH_3$.

*Barium-selective membrane electrode.* Both solid and liquid membranes that are selective to $Ba^{2+}$ ions have been prepared. The sensor was the tetraphenylborate salt of the barium complex of a nonylphenoxy-poly(ethyleneoxy)ethanol. Electrodes with the sensor and solvent mediator in liquid membranes have been made by using a number of nitro-aromatic solvents.[373] Similarly, these solvents have been used in the formation of PVC matrix membrane electrodes. It was found that a more viscous solvent mediator than 4-nitroethylbenzene was required for long life of the electrode. Both 2-nitrophenyl octyl ether and di-2-nitrophenyl ether in PVC matrix gave functional electrodes, but the latter mediator was found to be superior. These electrodes could be used in the determination of $Ba^{2+}$ and $SO_4^{2-}$ ions in potentiometric titrations.

Another barium-selective electrode based on a neutral carrier, $NNN'N'$-tetraphenyl-3,6,9-trioxaundecane diamide, in a PVC matrix has been described.[374] This electrode discriminated against other cations by factors in the range $10^2$—$10^3$ for alkali-metal cations and 30—$10^5$ for alkaline-earth cations.

*Cadmium- and caesium-selective membrane electrodes.* Three types of commercial cadmium-selective electrodes (Orion 94-48A, Selectrode, and TOA Cd-125, which is Japanese) have been evaluated.[375] Orion and TOA are solid-state electrodes whose membranes are based on homogeneous precipitates. The selectivities of these three electrodes are given in Table 15.

Caesium tetraphenylborate dissolved in nitrobenzene (saturated solution) acted as a liquid membrane electrode whose selectivity coefficients ($k_{Cs,M}$) were: $Cs^+$, 1; $Rb^+$, 0.19; $K^+$, 0.037; $NH_4^+$, 0.025; $Na^+$, 0.0023; and $Mg^{2+}$, 0.0009.[376] A liquid membrane electrode with 4-ethylnitrobenzene gave a nearly Nernstian response (slope = 52 mV per decade concentration) for caesium concentrations of from $10^{-1}$ to $< 10^{-4}$ mol $l^{-1}$.[377] The electrode has been used in the determination of parts per million of caesium in simulated nuclear waste. Caesium tetrakis($m$-trifluoromethylphenyl)borate dissolved in ethylnitrobenzene has been tested as a liquid membrane electrode that is selective to caesium ions.[378]

*Calcium-selective membrane electrode.* The calcium salts of di-(n-octylphenyl) hydrogen phosphate, di-($p$-n-octyl-$o$-nitrophenyl)hydrogen phosphate, and di-($p$-n-octyl-$o$-bromophenyl) hydrogen phosphate have been studied as sensors for

---

[373] A. M. Y. Jaber, G. J. Moody, and J. D. R. Thomas, *Analyst (London)*, 1976, **101**, 179.
[374] M. Guggi, E. Pretsch, and W. Simon, *Anal. Chim. Acta*, 1977, **91**, 107.
[375] P. Kivalo, R. Virtanen, K. Wickstrom, M. Wilson, E. Pungor, K. Toth, and G. Sundholm, *Anal. Chim. Acta*, 1976, **87**, 387.
[376] C. J. Coetzee and A. J. Basson, *Anal. Chim. Acta*, 1976, **83**, 361.
[377] E. W. Bauman, *Anal. Chem.*, 1976, **48**, 548.
[378] C. J. Coetzee and A. J. Basson, *Anal. Chim. Acta*, 1977, **92**, 399.

# Membrane Phenomena

**Table 15** *Selectivity coefficients of cadmium-selective membrane electrodes*

| Interfering ion | $[Cd]/mol\ l^{-1}$ | Orion | Electrode Selectrode | TOA |
|---|---|---|---|---|
| $Zn^{2+}$ | $10^{-2}$ | no interference | no interference | no interference |
|  | $10^{-4}$ | — | — | $7.8 \times 10^{-4}$ |
|  | $10^{-5}$ | $9.5 \times 10^{-5}$ | $2.3 \times 10^{-4}$ | — |
|  | $10^{-6}$ | $6.8 \times 10^{-5}$ | $1.1 \times 10^{-4}$ | $1.5 \times 10^{-4}$ |
| $Fe^{2+}$ | $10^{-4}$ | $2 \times 10^{-3}$ | $3 \times 10^{-3}$ | $3 \times 10^{-3}$ |
| $Pb^{2+}$ | $10^{-2}$ | — | — | 0.5 |
|  | $10^{-3}$ | 7.1 | 5.4, 21.8 | — |
|  | $10^{-4}$ | 0.6 | 0.4 | 4.8 |

calcium.[379] Simple electrophilic substitution in the n-octylphenyl group of the calcium salt of di-(n-octylphenyl) hydrogen phosphate neither improved the detection limit of the calcium electrode nor its useful pH range. However, they were used in potentiometric titrations of the calcium present in sea-water.

The dynamic response of the calcium-selective liquid membrane electrode has been evaluated by measuring the time ($t_{95}$) required for the electrode to reach 95% of its steady-state potential after a rapid change in ion activity.[380] In pure calcium solution, the $t_{95}$ was 2.25 s, whereas in the presence of interfering ion it was 2.2 ($Me_4N^+$), 2.3 ($H^+$), 4 ($Mg^{2+}$), 8.5 ($Sr^{2+}$), and 22.4 s ($Ba^{2+}$).

Use of a complexation buffer (40.4 g of $KNO_3$, 3.6 g of disodium iminodiacetate, 160 ml of aqueous 0.5M acetylacetone solution, 2 ml of 10M ammonia, and 1.07 g of $NH_4Cl$; made up to a litre) has been recommended for the estimation of calcium in water samples using the calcium electrode.[381] The electrode has been used to determine the solubility of calcium molybdate in water.[382]

The role of solvent in determining the selectivity of the calcium liquid membrane electrode [sensor: bis(di-n-decyl phosphate)] has been explored by using three groups of solvents. The selectivity sequences realized with the three groups of solvents, *viz.* $Cu > Ni \approx Mg > Ca$ for n-alcohols, octan-2-ol, and 2-methylheptan-2-ol; $Cu > Ca > Ni \approx Mg$ for octan-3-ol, octan-4-ol, and methyl-heptanols; and $Ca > Cu > Ni \approx Mg$ for all tri-n-alkyl phosphates, indicate that tri-n-alkyl phosphates are the best.[383] Similarly, the influence of the dielectric constant of the solvent on the selectivity of the calcium- and potassium-selective liquid membrane electrodes has been investigated, using neutral carrier ligands; $NN'$-di-(11-ethoxycarbonyl)undecyl-$N,N',4,5$-tetramethyl-3,6-dioxaoctanediamide (1) for calcium and valinomycin for potassium.[384] Selectivity for univalent cations was favoured by a solvent of low dielectric constant and that for bivalent ions by a solvent of high dielectric constant. The ligand (1) showed high selectivity for $Ca^{2+}$ over $H^+$, $Mg^{2+}$, $Na^+$, and $Zn^{2+}$ ions when it was used either as a microelectrode

---
[379] D. Jagner and J. P. Østergaard-Jensen, *Anal. Chim. Acta*, 1975, **80**, 9.
[380] B. Fleet, T. H. Ryan, and M. J. D. Brand, *Anal. Chem.*, 1974, **46**, 12.
[381] A. Hulanicki and M. Trojanowicz, *Anal. Chim. Acta*, 1974, **68**, 155.
[382] N. A. A. Mumallah and W. J. Popicol, *Anal. Chem.*, 1974, **46**, 2055.
[383] K. Garbett, *Proc. Anal. Div. Chem. Soc.*, 1975, **12**, 60.
[384] U. Fiedler, *Anal. Chim. Acta*, 1977, **89**, 111.

$$\text{EtO}_2\text{C}(\text{H}_2\text{C})_{11} \underset{N-C}{\overset{Me}{\diagdown}} \overset{O}{\diagup} \text{H}_2\text{C}-\text{O} \overset{H\ H}{\underset{Me\,C-C\,Me}{}} \text{O}-\text{CH}_2 \overset{O}{\diagup} \underset{C-N}{\overset{Me}{\diagdown}} (\text{CH}_2)_{11}\text{CO}_2\text{Et}$$

(1)

(10% solution, 1% sodium tetraphenylborate in o-nitrophenyl n-octyl ether)[385] or incorporated into a PVC matrix (0.9% ligand, 34.4% PVC, 0.4% sodium tetraphenylborate, and 64.3% o-nitrophenyl n-octyl ether; cast into a membrane).[386] Poly(vinyl chloride) (PVC) has proved to be the best matrix to hold ion-selective sensors and to form membranes. Other matrices, such as mixtures of PVC and partially hydrolysed vinyl chloride–vinyl acetate copolymer (copolymer VAGH), and phosphorylated copolymer VAGH, have been used to form membranes.[387] The PVC–copolymer VAGH membranes contained Orion 92-20-02 Ca liquid ion-exchanger, whereas the phosphorylated copolymer VAGH contained di(octylphenyl) phosphonate and the calcium salt of phosphorylated VAGH. The latter showed no advantages over the former but gave fast and steady response. PVC matrix membranes containing Orion 92-20-02 liquid ion-exchanger are permselective to cations.[388] A study of diffusional transport of $^{45}$Ca showed that $Mg^{2+}$, $Sr^{2+}$, $Ba^{2+}$, and $Be^{2+}$ ions inhibited its migration through the membrane. The inhibition by the first three ions was attributed to the low affinity of those ions to the ion-exchanger sites, whereas that by $Be^{2+}$ was attributed to site blockage by adsorption of $Be^{2+}$ to the dialkyl phosphate groups.[389] The role of solvent mediators in these calcium-selective membranes has also been studied.[390] Di-(n-octylphenyl) phosphonate, whose preparation is also described,[391] served as the most useful solvent mediator for ion-selective membrane electrodes based on calcium bis(dialkyl phosphate) sensors trapped in PVC. Decan-1-ol was found to be acceptable as a solvent mediator for electrodes that are selective for bivalent ions and based on bis(di-2-ethylhexyl phosphate).

A PVC-based membrane containing di-(n-octylphenyl) hydrogen phosphate has been placed in direct contact with the internal electrode made of Ag, Ag/AgCl, or Teflonized graphite.[392] Silver or Teflonized graphite contacts conferred greater sensitivity and gave stable potentials.

The responses of the calcium electrode to $Ca^{2+}$ ions have been affected by the presence of surfactants [linear alkylbenzenesulphonate (LAS) or di-isobutylphenoxyethoxyethyl(dimethyl)benzylammonium chloride (Hyamine)] in the sample solution.[393] Interference by LAS has been attributed to competing reactions between organophilic membrane and LAS and $Ca^{2+}$ ions, whereas interference by

---
[385] M. Oehme, M. Kessler, and W. Simon, *Chimia*, 1976, **30**, 204.
[386] D. Ammann, M. Guggi, E. Pretsch, and W. Simon, *Anal. Lett.*, 1975, **8**, 709.
[387] L. Keil, G. J. Moody, and J. D. R. Thomas, *Analyst (London)*, 1977, **102**, 274.
[388] A. Craggs, G. J. Moody, and J. D. R. Thomas, *Proc. Anal. Div. Chem. Soc.*, 1975, **12**, 64.
[389] A. Craggs, G. J. Moody, and J. D. R. Thomas, *Talanta*, 1976, **23**, 799.
[390] A. Craggs, L. Keil, G. J. Moody, and J. D. R. Thomas, *Talanta*, 1975, **22**, 907.
[391] A. Craggs, L. Keil, G. J. Moody, and J. D. R. Thomas, *J. Inorg. Nucl. Chem.*, 1975, **37**, 577.
[392] A. Hulanicki and M. Trojanowicz, *Anal. Chem. Acta*, 1976, **87**, 411.
[393] R. A. Llenado, *Anal. Chem.*, 1975, **47**, 2243.

## Membrane Phenomena

Hyamine has been attributed to displacement of $Ca^{2+}$ from the ion-exchange site. The electrode is $10^3$ times more selective to Hyamine than to $Ca^{2+}$, and so the surfactant interference cannot be overcome. On the other hand, the LAS interference can be overcome by loading the membrane phase with an equilibrium amount of LAS; this desensitizes the sensing element for its response to LAS.

The carboxylic antibiotic A23187, dissolved in nitrobenzene and supported on cellulose ester membranes, has been evaluated as a liquid membrane electrode for $Ca^{2+}$.[394] The selectivity order was found to be $Ba^{2+} > Sr^{2+} > Ca^{2+} > Mg^{2+}$. This is in reverse order to that determined by phase-extraction studies.[395]

Microelectrodes using calcium salts of di(n-octylphenyl) hydrogen phosphate[396] and of di[p-(1,1,3,3-tetramethylbutyl)phenyl] hydrogen phosphate[397] as sensors have been prepared. The mediator used was di(octylphenyl) phosphonate in both cases. A mixture of sensor and mediator (in 10:1 wt ratio) was dissolved with known weights of PVC in tetrahydrofuran to form the solution, which was allowed to form a membrane at the tip of the microelectrode. The selectivity coefficients of the microelectrode were[397] $k_{Ca,K} = 5 \times 10^{-7}$, $k_{Ca,Na} = 1\text{---}4 \times 10^{-7}$, and $k_{Ca,Mg} = 10^{-3}$. This electrode has been used to estimate the intracellular calcium activity $[a(Ca^{2+}) = 5.4 \times 10^{-7}$ mol $l^{-1}]$ in *Aplysia* neurone and the stability constant of Ca-EGTA ($10^{11}$ l mol$^{-1}$).[398]

*Copper-selective membrane electrode.* Powdered crystals of $Cu_2HI_4$ mixed with incompletely polymerized epoxy-resin ED-6 (7% by weight of powder) and polyethylene polyamine (10% by weight of resin), compacted at 90 °C under a pressure of 150 kg cm$^{-2}$ for 4 h, gave membranes which were used in the fabrication of electrodes that acted selectively to $Cu^{II}$ ions with little interference from other ions such as those of K, Na, Mg, Ca, Al, Zn, Co, Ni, Pb, and Cd. However, those ions, *e.g.* $I^-$ and $CNS^-$, which formed weakly dissociated complexes with $Cu^{2+}$ affected the electrode potential.[399]

A single crystal of synthetic chalcocite ($Cu_2S$), when used as an electrode, responded primarily to $Cu^I$ ion in solution.[400] The potential of the electrode is given by

$$E = E° + (RT/F) \ln\{a(Cu^I) + k(Cu^I, Cu^{II})[a(Cu^{II})]^{\frac{1}{2}}\} \quad (48)$$

When $a(Cu^{II}) \ll k(Cu^I, Cu^{II})[a(Cu^{II})]^{\frac{1}{2}}$, equation (48) becomes

$$E = E° + (RT/F) \ln k(Cu^I, Cu^{II}) + (RT/2F) \ln a(Cu^{II}) \quad (49)$$

The selectivity coefficient may be evaluated from the solubility products ($k_s$'s)

$$k(Cu^I, Cu^{II}) = \left(\frac{k_s(Cu_2S)}{k_s(CuS)}\right)^{\frac{1}{2}} = 10^{-6.6} = \text{constant}$$

[394] A. K. Covington and N. Kumar, *Anal. Chim. Acta*, 1976, **85**, 175.
[395] P. R. Reed and H. A. Lardy, *J. Biol. Chem.*, 1972, **247**, 6970.
[396] G. R. J. Christoffersen and E. S. Johansen, *Anal. Chim. Acta*, 1976, **81**, 191.
[397] H. M. Brown, J. P. Pemberton, and J. D. Owen, *Anal. Chim. Acta*, 1976, **85**, 261.
[398] J. D. Owen, H. M. Brown, and J. P. Pemberton, *Anal. Chim. Acta*, 1977, **90**, 241.
[399] A. F. Zhukov, A. V. Gordievskii, Yu. I. Urusov, and M. N. Voronovskaya, *Tr. Mosk. Khim. Tekhnol. Inst.*, 1974, **81**, 69.
[400] A. Hulanicki, M. Trojanowicz, and M. Cichy, *Talanta*, 1976, **23**, 47.

On this basis, equation (49) becomes

$$E = E^{\circ\prime} + (RT/2F) \ln a(\text{Cu}^{II}) \quad (50)$$

Accordingly, the electrode should respond to $\text{Cu}^{II}$ ions with a slope of 29 mV and to $\text{Cu}^{I}$ ions with a slope of 59 mV per decade of concentration. Experimental data confirmed these predictions. The ion-exchange reactions at the surface of this electrode have been discussed in terms of a diffusion-layer model.[401]

Several methods for the preparation of $\text{CuS-Ag}_2\text{S}$ precipitates have been investigated for their suitability in the construction of $\text{Cu}^{II}$-selective membrane electrodes. Homogeneous membranes with satisfactory performance were obtained from precipitates formed by the addition of metal salt solutions to sodium sulphide.[402] The electrode from this gave Nernstian response to $\text{Cu}^{II}$ solutions, and the response did not depend on the ionic strength, provided that the background electrolyte had no chloride. If the background electrolyte contained chloride, the electrode response, although linear, was not Nernstian.[403] Both chloride and fluoride ions produced shifts in standard potentials of the electrode on prolonged exposure.[404] Chloride has been found to tarnish the shiny surface of a $\text{Cu}^{II}$-selective electrode.[405] Interferences by mineral acids depended on the nature of the anion present. The magnitudes of these effects are determined by the condition of the membrane surface. The electrode behaviour in solutions containing submicro-levels of $\text{Cu}^{II}$ ions ($10^{-6}$—$10^{-9}$ mol l$^{-1}$) has been described.[406]

A microelectrode that is selective to $\text{Cu}^{II}$ ions has been developed, and, in combination with a microcell and nanolitre burette, it has been used in the microdetermination of $\text{Cu}^{II}$ ions.[407]

Four types of copper-selective electrodes, *i.e.* Orion 94-29 A (polycrystalline membrane of $\text{Ag}_2\text{S-CuS}$), Ruzicka Selectrode ($\text{Ag}_2\text{S-CuS}$-impregnated graphite–Teflon rod), Tacussel PCU 2 (polycrystalline membrane of $\text{Ag}^{I}$ and $\text{Cu}^{II}$ selenides), and Radiometer F 1112 Cu (single crystal of $\text{Cu}_{1.8}\text{Se}$), showed different behaviour, especially at copper concentrations below $10^{-6}$ mol l$^{-1}$.[408] None of the electrodes gave Nernstian response to $\text{Cu}^{II}$ at concentrations below $10^{-6}$ mol l$^{-1}$. The causes for this deviation are (i) oxidation of membrane in the case of $\text{Ag}_2\text{S-CuS}$ membrane electrodes, (ii) a redox effect in the case of the single-crystal electrode, and (iii) a combination of (i) and (ii) in the case of the Selectrode. Consequently, great care must be exercised in using these electrodes to estimate microquantities of $\text{Cu}^{II}$. However, in a flowing system[409] the electrode ($\text{Ag}_2\text{S-CuS}$) response was Nernstian down to $10^{-8}$ mol l$^{-1}$; but below $10^{-6}$ mol l$^{-1}$, steady potentials were attained only after long periods of contact of the flowing solution with the electrode.

The copper-selective electrodes have been used in the determination of copper in sea-water[410] and of n-butyl-*l*-biguanide,[411] citrates and 8-hydroxyquinoline,[412] com-

---

[401] A. Hulanicki and A. Lewenstam, *Talanta*, 1976, **23**, 661.
[402] G. J. M. Heijne, W. E. Van der Linden, and G. Den Boef, *Anal. Chim. Acta*, 1977, **89**, 287.
[403] G. B. Oglesby, W. C. Duer, and F. J. Millero, *Anal. Chem.*, 1977, **49**, 877.
[404] D. Midgley, *Anal. Chim. Acta*, 1976, **87**, 19.
[405] D. J. Crombie, G. J. Moody, and J. D. R. Thomas, *Talanta*, 1974, **21**, 1094.
[406] W. J. Blaedel and D. E. Dinwiddie, *Anal. Chem.*, 1974, **46**, 873.
[407] J. A. Van der Meer, G. Den Boef, and W. E. Van der Linden, *Anal. Chim. Acta*, 1976, **85**, 317.
[408] D. Midgley, *Anal. Chim. Acta*, 1976, **87**, 7.
[409] W. J. Blaedel and D. E. Dinwiddie, *Anal. Chem.*, 1975, **47**, 1070.
[410] R. Jasinski, I. Trachtenberg, and D. Andrychuck, *Anal. Chem.*, 1974, **46**, 364.

plexometric titrations,[400,413] determination of stability constants of $Cu^{II}$ complexes with acetate, ammonia, ethylenediamine, glycine, iminodiacetic acid, and 1,10-phenanthroline,[414] and determination of ferric iron at pH 2—3.[415]

*Lead-selective membrane electrode.* A Selectrode that is selective to lead ions has been prepared by using Pbs–$Ag_2S$ as the electro-active material deposited on a graphite rod.[416] It exhibited a Nernstian response in the concentration range $10^{-2}$—$10^{-11}$ mol $l^{-1}$. It could be used for measuring the activity of $Pb^{II}$ ion and for indicating the end point in potentiometric titrations with EDTA. This electrode has been compared to the Orion 94-82 $Pb^{II}$-selective electrode, whose response was found to be 5—10 times faster than that of the Selectrode.[417] $Zn^{2+}$ had insignificant interference with both the electrodes, whereas $Cd^{2+}$ showed the following selectivity: $k_{Pb,Cd}$ ($10^{-6}$M-$Pb^{2+}$) 0.12 for the Orion electrode and 0.012 for the Selectrode, and at a $Pb^{2+}$ concentration of $10^{-4}$ mol $l^{-1}$ the selectivities were 0.18 and 0.064 respectively.

A lead-selective electrode has been used indirectly to estimate sulphate ion in natural waters.[418] This was accomplished by titration with lead nitrate solution containing 80% isopropyl alcohol. In titrations with lead(II) perchlorate solutions, addition of 75% methanol eliminated interference from 200-fold excess of nitrate.[419] However, calcium interfered, and this interference was overcome by increasing the ionic strength with sodium perchlorate.

A lead-selective electrode has been used in the study of chemical equilibria in lead–halide systems.[420] The solubility products ($k_s$'s) of lead salts [$k_s = 3.07 \times 10^{-13}$ for $PbCO_3$, $4.44 \times 10^{-13}$ for $PbCrO_4$, $3.7 \times 10^{-10}$ for $PbC_2O_4$, $2.09 \times 10^{-13}$ for $Pb(IO_3)_2$, and $2.48 \times 10^{-6}$ for $Pb_2Fe(CN)_6$] and the stability constants of lead halides have been determined. Similarly, the solubility products of lead molybdate ($1.2 \times 10^{-13}$), lead tungstate ($8.4 \times 10^{-11}$), and lead perrhenate ($6.9 \times 10^{-9}$) have been determined.[421] The electrode has been used to determine, (i) micro- and semimicro-amounts of fluoride and sulphate or of fluoride and phosphate, without prior separation,[422] (ii) inorganic sulphate, organic sulphur, and oxalate,[423,424] and (iii) micro-estimation of orthophosphate in the presence of large amounts of fluoride.[425]

*Lithium-selective membrane electrode.* A liquid membrane electrode based on n-decanol and selective to lithium over other alkali-metal ions has been described, and its behaviour was comparable to that of LAS 15-25 glass electrode (selective to

---

[411] G. E. Baiulescu, V. V. Cosofret, and F. G. Cocu, *Talanta,* 1976, **23**, 329.
[412] M. F. El-Taras and E. Pungor, *Anal. Chim. Acta,* 1976, **82**, 285.
[413] J. M. Van der Meer, G. Den Boef, and W. E. Van der Linden, *Anal. Chim. Acta,* 1975, **76**, 261; **79**, 27; 1976, **85**, 309.
[414] G. Nakagawa, H. Wada, and T. Hayakawa, *Bull. Chem. Soc. Jpn.,* 1975, **48**, 424.
[415] Y. S. Fung and K. W. Fung, *Anal. Chem.,* 1977, **49**, 497.
[416] E. H. Hansen and J. Ruzicka, *Anal. Chim. Acta,* 1974, **72**, 365.
[417] P. Kivalo, R. Virtanen, K. Wickstrom, M. Wilson, E. Pungor, G. Horvai, and K. Toth, *Anal. Chim. Acta,* 1976, **87**, 401.
[418] E. P. Scheide and R. A. Durst, *Anal. Lett.,* 1977, **10**, 55.
[419] A. Hulanicki, R. Lewandowski, and A. Lowenstam, *Analyst (London),* 1976, **101**, 939.
[420] C. Birraux, J.-Cl. Landry, and W. Haerdi, *Anal. Chim. Acta,* 1977, **90**, 51.
[421] E. E. Chao and K. L. Cheng, *Talanta,* 1977, **24**, 247.
[422] W. Selig, *Mikrochim. Acta,* 1974, 515.
[423] W. Selig and A. Salomon, *Mikrochim. Acta,* 1974, 663.
[424] W. Selig, *Mikrochim. Acta,* 1976, **II**, 9.
[425] W. Selig, J. W. Frazer, and A. M. Kray, *Mikrochim. Acta,* 1975, **II**, 581.

univalent cations).[426] Another membrane electrode using a synthetic neutral carrier (2) (5.8%) with tris-(2-ethylhexyl) phosphate (62.8%) in a PVC matrix has been described.[427] Lithium activity in the range $10^{-5}$—$1.0$ mol $l^{-1}$ may be determined with the help of the electrode. Its selectivity to $Li^+$ over $Na^+$, $K^+$, $Mg^{2+}$, and $Ca^{2+}$ was 2, 140, 600, and 1900 respectively.

A microelectrode using the ligand (2) has been constructed and used to measure the accumulation of lithium by snail neurones.[428]

$$\underset{Me(H_2C)_6}{\overset{Me}{\diagdown}} N-\underset{\underset{O}{\parallel}}{C}-OCH_2-\underset{CH_2}{\overset{Me_2}{\overset{|}{C}}}-CH_2-O-CH_2-\underset{\underset{O}{\parallel}}{C}-N\underset{(CH_2)_6Me}{\overset{Me}{\diagup}}$$

(2)

*Mercury-selective membrane electrode.* A liquid membrane electrode that is selective to $Hg^{II}$ ions has been constructed. The membrane is formed by impregnating a hydrolysed graphite rod with the mercuric chelate of diketohydrindylidene-diketohydrindamine in chloroform.[429] The electrode showed linear response in the concentration range $10^{-1}$—$10^{-5}$ mol $l^{-1}$ of $Hg^{II}$, with a slope of 31 mV per decade of concentration and a response time of a few seconds. It could be used in potentiometric titrations.

It is difficult to construct electrodes that are selective to $Hg^I$ ions owing to the instability of insoluble mercurous compounds. However, an electrode that is sensitive to mercurous ions has been constructed and found suitable for the determination of the equivalence point in mercurometric titration.[430] This electrode contained graphite coated with a solution of palladium dithizonate [$Pd(Hdz)_2$]. When it came into contact with mercurous ions, $Hg_2(Hdz)_2$ appeared in the membrane, and made it selective to mercurous ions.

Generally, iodide-selective membranes respond to mercuric ions, and so an $AgI/Ag_2S$ membrane electrode has been used as an indicator electrode in the complexometric titrations of tervalent cations ($Bi^{3+}$, $Fe^{3+}$, $Cr^{3+}$) for which there are no selective electrodes.[431]

*Molybdenum-selective membrane electrode.* A solution of bis-tetraethy-lammonium pentathiocyanato-oxomolybdate in a mixture of nitrobenzene and o-dichlorobenzene and absorbed by lightly crosslinked natural rubber membrane has been used to sense the molybdenum present in the complex pentathiocyanato-molybdate(v) in the concentration range $10^{-2}$—$5 \times 10^{-8}$ mol $l^{-1}$.[432] Iron,

---

[426] W. A. Hildebrandt and K. H. Pool, *Talanta*, 1976, **23**, 469.
[427] M. Guggi, U. Fiedler, E. Pretsch, and W. Simon, *Anal. Lett.*, 1975, **8**, 457.
[428] R. C. Thomas, W. Simon, and M. Oehme, *Nature*, 1975, **258**, 754.
[429] G. E. Baiulescu and V. V. Cosofret, *Talanta*, 1976, **23**, 677.
[430] G. E. Baiulescu and N. Ciocan, *Talanta*, 1977, **24**, 37.
[431] E. Hopirtean, C. Liteanu, and R. Vlad, *Talanta*, 1975, **22**, 912.
[432] A. G. Fogg, J. L. Kumar, and D. T. Burns, *Anal. Lett.*, 1974, **7**, 629.

## Membrane Phenomena

vanadium, tungsten, rhenium, and niobium (which form thiocyanate complexes) interfered with the response of the electrode.

*Potassium-selective membrane electrode.* A silicone rubber membrane containing potassium zinc ferrocyanide has been constructed and found to be selective to $K^+$ ions over $Na^+$ ions.[433,434] However, other ions ($NH_4^+$, $Rb^+$, and $Cs^+$) interfered with the response of the electrode to $K^+$ ions.[433]

Of the several neutral ion carriers, valinomycin and some crown compounds have proved their usefulness as sensors for $K^+$ ions.[435] Several studies have been reported with valinomycin, employing a variety of methods to determine the role of several components used in the construction of potassium-selective electrodes. A plasticized polymer matrix containing valinomycin [PVC+dibutyl phthalate (1:3) + valinomycin dissolved in cyclohexanone and cast into a membrane] detected $K^+$ ions in the presence of 1000-fold excess of $Na^+$, $H^+$, $Li^+$, $Ca^{2+}$, and $Mg^{2+}$ ions and 10- to 20-fold excess of $NH_4^+$ ions.[436] The electrode could be used in the pH range 0—9, with a lifetime of 6 months. Other charged polymeric membranes used to hold valinomycin and plasticizers also functioned selectively to $K^+$ ions.[437] A long-lived (3—5 years) potassium-selective electrode has been fabricated from membranes incorporating potassium valinomycin tetraphenylborate salt into a block copolymer of poly(bisphenol-A carbonate) and poly(dimethylsiloxane) with sufficient cyanoethyl substitution in the latter to provide a dielectric constant of 5.2. Nearly ideal Nernstian response was observed over more than 3 years.[438] A liquid membrane electrode that is selective to $K^+$ ions in the presence of $Na^+$ ions was made from valinomycin and tetraphenylborate dissolved in bis-2-ethylhexyl adipate and nitrobenzene.[439] Another liquid membrane electrode (valinomycin in diphenyl ether) was used to follow its responses to various inorganic and organic ions. It was found that membrane-soluble tetraphenylborate caused interference.[440] Also, cetyltrimethylammonium bromide, a positively charged surfactant, caused interference, whereas a negatively charged surfactant (sodium dodecyl sulphate), a non-ionic surfactant (Tween 80), and a positively charged non-surfactant (tetramethylammonium bromide) caused no interference, thereby indicating that the response of the potassium-selective electrode was affected by only those compounds that had positive charge and were surface-active. The effects of several oil-soluble anions on the behaviour of a valinomycin-based electrode have been investigated both theoretically and experimentally.[441,442] Equations have been derived to explain the reduction of interference by lipid-soluble anions in the sample solution.[441] The interference of tetraphenylborate with the response of the potassium-selective electrode already referred to may be reduced or eliminated by incorporation of the anion into a less polar membrane phase, so as to shift the extraction equilibrium of

---

[433] A. G. Fogg, A. S. Pathan, and D. T. Burns, *Anal. Lett.*, 1974, **7**, 539.
[434] P. A. Rock, T. L. Eyrich, and S. Styer, *J. Electrochem. Soc.*, 1977, **124**, 530.
[435] Ref. 17, pp. 237—244.
[436] B. P. Nikol'skii, E. A. Materova, A. L. Grekovich, and V. E. Yurinskaya, *Zh. Anal. Khim.*, 1974, **29**, 205.
[437] M. Perry, E. Lobel, and R. Bloch, *J. Membr. Sci.*, 1976, **1**, 223.
[438] O. H. LeBlanc, Jr., and W. T. Grubb, *Anal. Chem.*, 1976, **48**, 1658.
[439] D. M. Band and J. Kratochvil, *J. Physiol.*, 1974, **239**, 10P.
[440] S. M. Hammond and P. A. Lambert, *J. Electroanal. Chem. Interfacial Electrochem.*, 1974, **53**, 155.
[441] W. E. Morf, G. Kahr, and W. Simon, *Anal. Lett.*, 1974, **7**, 9.
[442] W. E. Morf, D. Ammann, and W. Simon, *Chimia*, 1974, **28**, 65.

the system to a state in which the response of the electrode to cations was improved. This was accomplished, according to the theoretical principles enunciated,[441] by incorporating valinomycin and tetraphenylborate into a membrane formed of PVC dissolved in 2-nitro-$p$-cymene.[441,442] Efforts have been made to discover a new formula regarding solvent, support, and surface coating which would improve the linearity and response time of a valinomycin-based potassium-selective electrode.[443] The use of a solvent of low dielectric constant (1-bromoheptane), potassium salts with oil-soluble anions, and a very hydrophobic support such as polycellulose triacetate that is coated on one side with Teflon gave electrodes which had short response times and long lifetimes, and were free from 'memory effects'.

In order to investigate the effect of the structure of crown compounds on the selectivity they exhibit toward $K^+$ and $Na^+$ ions, a large number of crown compounds were prepared and used to form membranes,[444] from 10 mg of crown compound, 1 ml of dipentyl phthalate, and 10 ml of 5% (w/V) solution of PVC in cyclohexanone. These membranes, when used as electrodes, showed the selectivities given in Table 16, in which are also shown the Nernstian slopes obtained for some membrane electrodes used in an earlier study.[445] Crown-containing membranes, as can be seen from Table 16, do not display that high selectivity to $K^+$ over $Na^+$ shown by valinomycin-based membranes. In another study[446] both valinomycin and dimethyldibenzo-30-crown-10 have been introduced into a PVC matrix and the membranes so formed have been used as electrodes. Both graphical and numerical

**Table 16** *Selectivity coefficients* ($k_{K,Na}$) *of PVC membranes containing crown compounds and valinomycin*

| Crown compound | $k_{K,Na}$[444] | Slope[445] (mV per p$K^+$ unit) |
|---|---|---|
| Dicyclohexyl-18-crown-6 | $1.1 \times 10^{-2}$ | 58 |
| Dibenzo-18-crown-6 | $7.7 \times 10^{-2}$ | 51 |
| Dimethyldibenzo-18-crown-6 | $6.7 \times 10^{-2}$ | 60 |
| Di-n-propyldibenzo-18-crown-6 | $6.3 \times 10^{-2}$ | 59 |
| Dimethyldicyclohexyl-18-crown-6 | $1.1 \times 10^{-2}$ | 58 |
| Di-n-propyldicyclohexyl-18-crown-6 | $1.6 \times 10^{-2}$ | 60 |
| Dibenzo-16-crown-5 | 1.0 | — |
| Dibenzo-19-crown-6 | $2.2 \times 10^{-2}$ | — |
| Dicyclohexyl-19-crown-6 | $1.5 \times 10^{-2}$ | — |
| Dibenzo-22-crown-7 | $3.3 \times 10^{-2}$ | — |
| 4-Methylbenzo-15-crown-5 | $6.7 \times 10^{-2}$ | — |
| Benzo-18-crown-6 | $0.53 \times 10^{-2}$ | — |
| 4-Nitrobenzo-18-crown-6 | $0.28 \times 10^{-2}$ | — |
| Dimethyldibenzo-24-crown-8 | $1.0 \times 10^{-1}$ | 59 |
| Dibenzo-30-crown-10 | $0.85 \times 10^{-2}$ | — |
| Dimethyldibenzo-30-crown-10 | $0.22 \times 10^{-2}$ | — |
| Di-t-butyldibenzo-30-crown-10 | $0.24 \times 10^{-2}$ | — |
| Dinitrodibenzo-30-crown-10 | $0.95 \times 10^{-2}$ | — |
| Di-t-butyldibenzo-36-crown-12 | $1.2 \times 10^{-2}$ | — |
| Dibenzo-36-crown-12 | $1.6 \times 10^{-2}$ | — |
| Valinomycin[445] | $0.03 \times 10^{-2}$ | 60 |

[443] S. B. Lewis and R. P. Buck, *Anal. Lett.*, 1976, **9**, 439.
[444] J. Petranek and O. Ryba, *Anal. Chim. Acta*, 1974, **72**, 375.
[445] O. Ryba, E. Knizakova, and J. Petranek, *Collect. Czech. Chem. Commun.*, 1973, **38**, 497.
[446] M. Semler and H. Adametzova, *J. Electroanal. Chem. Interfacial Electrochem.*, 1974, **56**, 155.

methods were used to calculate values for the selectivity coefficient. Although both electrodes were selective to $K^+$ ions over $NH_4^+$ and $Na^+$ ions, the valinomycin-based electrode was superior to the crown-compound-based electrode. The role of plasticizers used with valinomycin and crown compounds has been investigated.[447] Membranes contained 0.15—0.2 g of PVC, 5—10 mg of neutral carrier (valinomycin, dicyclohexyl-18-crown-6, or dibenzo-18-crown-6 cyclic polyethers), and 0.1—0.4 g of plasticizer (tributyl phosphate, dibutyl phthalate, or diphenyl ether). Polymers other than PVC and dicyclohexyl-18-crown-6 were found unsatisfactory. Similarly, the responses of potassium-selective electrode with valinomycin and dimethyldibenzo-30-crown-10 in PVC that was plasticized with dipentyl phthalate in solutions containing lipophilic anions such as perchlorate and picrate have been evaluated.[448] Other $C_{12}$—$C_{14}$ phthalates used as plasticizers increased the concentration range of cation response with a Nernstian slope. Using dipentyl phthalate with 5% of high-molecular-weight PVC in cyclohexanone and the above neutral carriers, miniature electrodes (membrane area $< 0.2$ mm$^2$) with a Nernstian response to $K^+$ ions and long-term stability have been constructed.[449] Similarly, microelectrodes with valinomycin as the sensor, both commercial (Crytur 19-15)[450] and experimental,[451] have been used in the determination of $K^+$ ions. Also, a valinomycin-based liquid membrane electrode has been used in the determination of $K^+$ ions in natural and waste waters on an automated continuous-flow system with on-line minicomputer and printer.[452]

*Selenium-selective membrane electrode.* This is a liquid membrane electrode containing a saturated solution of 3,3'-diaminobenzidine in hexane.[453] The electrode response was linear to selenium(IV) at pH 2.5 up to a concentration of $10^{-4}$ mol l$^{-1}$, with a slope of 60—65 mV per decade change in selenium concentration. The interferences of other ions with the response of the electrode to selenium(IV) at a concentration of $10^{-4}$ mol l$^{-1}$ have been determined. The interference decreased in the order $V^V > Te^{IV} > Sb^{III} > Mo^{VI} > ClO_4^- > Ag^+ > As^{III} > Na^+ > Br^- > Hg^{II} > Cl^- > SO_4^{2-}$.

*Silver-selective membrane electrode.* A liquid membrane electrode that is selective to silver ions was obtained by impregnating a graphite bar with the silver chelate of 2-(2′,3′,5′-tri-*o*-benzoyl-$\beta$-D-ribofuranosyl)-5-thio-6-benzylthio-*as*-triazine-3,5-dione in chloroform. The response in $KNO_3$ was Nernstian in the $Ag^+$ concentration range $10^{-6}$—$10^{-1}$ mol l$^{-1}$, with $Hg^{2+}$ and $Cu^{2+}$ causing interference. When silver ions are involved, the electrode could be used in potentiometric titrations.[454] The $Ag_2S$-based electrode has been used in the potentiometric estimation of silver in mixtures containing copper and cadmium[455] and in the microtitration of chloride,[456] using $10^{-3}$ N silver perchlorate in acetone with a titration system controlled by a conventional PDP-8/I minicomputer processor.[298]

[447] M. Mascini and F. Pallozzi, *Anal. Chim. Acta*, 1974, **73**, 375.
[448] O. Ryba and J. Petranek, *J. Electroanal. Chem. Interfacial Electrochem.*, 1976, **67**, 321.
[449] O. Ryba and J. Petranek, *Talanta*, 1976, **23**, 158.
[450] M. Semler and H. Adametzova, *Chem. Prum.*, 1975, **25**, 377.
[451] M. Oehme and W. Simon, *Anal. Chim. Acta*, 1976, **86**, 21.
[452] I. Serkerka and J. F. Lechner, *Anal. Lett.*, 1974, **7**, 463.
[453] T. L. Malone and G. D. Christian, *Anal. Lett.*, 1974, **7**, 33.
[454] G. E. Baiulescu, V. V. Cosofret, and C. Cristescu, *Rev. Chim. (Bucharest)*, 1975, **26**, 429.
[455] M. Geissler, *Anal. Chim. Acta*, 1977, **90**, 249.
[456] W. Selig, *Microchem. J.*, 1976, **21**, 291.

*Sodium-selective membrane electrode.* The most popular Na$^+$-selective electrode is the glass membrane electrode. The ligand (3) prepared by Ammann *et al.*,[457] when used as a membrane electrode (solid or liquid), responded selectively to Na$^+$ ions.[458] The PVC membrane formed by using a solvent of low dielectric constant (dibenzyl ether, $\varepsilon=4.0$) showed better selectivity response than the one formed by using a solvent of higher dielectric constant (*o*-nitrophenyl n-octyl ether, $\varepsilon=24$). This behaviour is dramatically illustrated by using several solvents of increasing dielectric constant.[459] The solvents used were dibutyl sebacate ($\varepsilon=4$), di-(2-ethylhexyl) sebacate, di-(2-ethylhexyl) adipate, di-(2-ethylhexyl) phthalate, and *o*-nitrophenyl n-octyl ether ($\varepsilon=24$). The first solvent gave the best results.

(3)

A new liquid membrane microelectrode containing the sodium-selective exchanger monensin in nitrobenzene (10%) has been constructed and evaluated.[460] The electrode exhibited a selectivity ratio against K$^+$ ion of 15:1 at 0.1 M concentration and of 13:1 at 1.0 M concentration. This selectivity ratio at 1.0 M concentration for other ions was 143:1 for Ca$^{2+}$, 154:1 for NH$_4^+$, 14:1 for H$^+$, and 7:1 for Mg$^{2+}$.

The electrode has been used to follow changes in extracellular concentration of sodium during spreading depression in the catfish cerebellum. The Orion sodium-selective electrode has been used to estimate sodium in alumina.[461]

*Thallium- and antimony-selective membrane electrodes.* An electrode that is selective to thallium(I) and is precipitate-based has been described. Thallium iodide precipitate was added to silver iodide precipitate in the construction of the electrode.[462] The electrode showed good selectivity to Tl$^I$, with little interference from ordinary bivalent ions. It has been used in the potentiometric titration of thallium with K$_2$CrO$_4$.

Thallium(I) *O,O'*-didecyl dithiophosphate in chlorocyclohexane acted as the sensor for thallium in a liquid membrane electrode with a Nernstian behaviour in the pTl range 1—5.5 and slope equal to 57.6 mV per decade of activity at an ionic strength of 0.1 mol l$^{-1}$.[463] In the pH range 5—12, the electrode was selective to Tl$^I$,

[457] D. Ammann, E. Pretsch, and W. Simon, *Helv. Chim. Acta*, 1973, **56**, 1780.
[458] D. Ammann, E. Pretsch, and W. Simon, *Anal. Lett.*, 1974, **7**, 23.
[459] U. Fiedler, *Anal. Chim. Acta*, 1977, **89**, 101.
[460] R. P. Kraig and C. Nicholson, *Science*, 1976, **194**, 725.
[461] D. E. Thompson and R. S. Danchik, *Anal. Lett.*, 1975, **8**, 699.
[462] A. V. Gordievskii, A. F. Zhukov, Yu. I. Urusov, and V. S. Shterman, *Zh. Anal. Khim.*, 1974, **29**, 1298.
[463] W. Szczepaniak and K. Ren, *Anal. Chim. Acta*, 1976, **82**, 37.

## Membrane Phenomena

with little interference from alkali-metal and alkaline-earth cations. It can be used in direct potentiometry and in potentiometric titrations with iodide or tetraphenylborate solutions. Use of EDTA masked the interferences from several ions.

Thallium tetrakis($m$-trifluoromethylphenyl)borate dissolved in ethylnitrobenzene has been tested as a liquid membrane electrode that is selective to thallium ions.[378]

Tetrachlorothallium(III) and hexachloroantimonate(V) salts of Sevron Red L, Sevron Red GL, Flavinduline O, and phenazinduline O, dissolved in $o$-dichlorobenzene and saturating a natural rubber membrane, functioned as liquid membrane electrodes.[464] Saturated solutions of tetrachlorothallate and hexachloroantimonate in 2M-HCl were used for the calibration of the electrodes. Full Nernstian responses in the range $10^{-6}$—$10^{-2}$ mol $l^{-1}$ (Tl) and $10^{-8}$—$10^{-2}$ mol $l^{-1}$ (Sb) were noted. They were used to determine thallium and antimony. In the case of the latter, the electrode responded to both $Sb^{III}$ and $Sb^{V}$.

*Uranyl-ion-selective membrane electrode.* Uranyl organophosphorus complexes held in PVC matrices acted as sensors for uranium. Of the several complexes used in a suitable diluent, PVC, and tetrahydrofuran to form membranes,[465] the best electrodes were obtained from di-(2-ethylhexyl) hydrogen phosphate in diamyl phosphonate, di-(2-ethylhexyl) ethylphosphonate, or tri-(2-ethylhexyl) phosphate as the diluent. Electrodes made from each of these diluents gave Nernstian responses to uranyl ions with little interference from common bi- and ter-valent ions.

*Zinc-selective membrane electrode.* The zinc salt of di-(n-octylphenyl) hydrogen phosphate (8%) and PVC (30%) in di(octylphenyl) phosphonate (62%), when cast into a membrane, was found to be selective to zinc ions, with little interference from alkali-metal ions, $NH_4^+$, $Ba^{2+}$, $Cu^{2+}$, and $Cd^{2+}$.[466] The ions $H^+$, $Ca^{2+}$, $Sr^{2+}$, and $Pb^{2+}$ showed some interference. The interference by calcium may be prevented by using a calcium-precipitating buffer composed of 0.02M-HOAc and 0.02M-NaOAc and 0.04M-NaF with ionic strength 0.06 and pH 4.8.

ANION-SELECTIVE MEMBRANE ELECTRODES. *Carbonate- and carbon-dioxide-selective membrane electrodes.* Aliquat 336 (tricaprylmethylammonium chloride) in chloride form, or after conversion into the bicarbonate form, dissolved in trifluoroacetyl-$p$-butylbenzene, served as a liquid membrane electrode that is selective to carbonate ions in the presence of bicarbonate.[467] The internal reference solution was a mixture of NaCl and $NaHCO_3$, both of concentration 0.1 mol $l^{-1}$. The electrode responded rapidly to carbonate in the concentration range $10^{-2}$—$10^{-6}$ mol $l^{-1}$ and at pH 5.5—8.5. The selectivity coefficients $(k_{i,j})$[468] of the electrode are given in Table 17. The electrode has been used in a continuous flow system to determine the $CO_2$ levels in human serum.[469] Usually a $CO_2$-sensing membrane electrode[470] or the air-gap electrode already described is used to estimate

---
[464] A. G. Fogg, A. A. Al-Sibaai, and C. Burgers, *Anal. Lett.*, 1975, **8**, 129.
[465] D. L. Manning, J. R. Stokely, and D. W. Magouyrk, *Anal. Chem.*, 1974, **46**, 1116.
[466] L. Gorton and U. Fiedler, *Anal. Chim. Acta*, 1977, **90**, 233.
[467] H. B. Hermans and G. A. Rechnitz, *Science*, 1974, **184**, 1074.
[468] H. B. Hermans and G. A. Rechnitz, *Anal. Chim. Acta*, 1975, **76**, 155.
[469] H. B. Hermans and G. A. Rechnitz, *Anal. Lett.*, 1975, **8**, 147.
[470] Ref. 17, pp. 316—322.

**Table 17** Values of selectivity coefficient $(k_{i,j})$ for the carbonate-ion-selective liquid membrane

| Ion | Interferent concentration /mol l$^{-1}$ | $k_{i,j}$ | Ion | Interferent concentration /mol l$^{-1}$ | $k_{i,j}$ |
|---|---|---|---|---|---|
| $CO_3^-$ | $10^{-3}$ | 1 | Acetate | $10^{-1}$ | $2.6 \times 10^{-2}$ |
| $SO_4^{2-}$ | $10^{-1}$ | $1.5 \times 10^{-4}$ | Borate | $10^{-2}$ | $4.9 \times 10^{-2}$ |
| $HPO_4^{2-}$ | $2.5 \times 10^{-2}$ | $2.6 \times 10^{-4}$ | Hydrogen phthalate | $5 \times 10^{-2}$ | 83 |
| $Cl^-$ | $10^{-1}$ | $1.9 \times 10^{-4}$ | | | |
| $NO_3^-$ | $10^{-1}$ | $3.0 \times 10^{-1}$ | | | |
| $ClO_4^-$ | $10^{-1}$ | 25 | | | |

$CO_2$ levels in blood and other biological fluids. The Radiometer type E5036 membrane electrode has been used to determine $CO_2$ in power-station waters.[471]

*Chlorate-selective membrane electrode.* The Corning 477316 nitrate liquid ion-exchanger [tri-dodecyl(hexadecyl)ammonium nitrate/n-octyl o-nitrophenyl ether; sensor/mediator] was converted into the chlorate form and incorporated into a PVC matrix membrane.[244] The internal reference solution was a mixture of KCl ($5 \times 10^{-2}$ mol l$^{-1}$) and $KClO_3$ ($5 \times 10^{-2}$ mol l$^{-1}$). The electrode showed nearly Nernstian response which was fast under dynamic conditions. The selectivity coefficients, $k_{i,j}$, were: $4.7 \times 10^{-4}$ ($F^-$), $2.5 \times 10^{-3}$ ($Cl^-$), $8.9 \times 10^{-2}$ ($Br^-$), 4.2 ($I^-$), $3.6 \times 10^{-3}$ ($NO_2^-$), 0.66 ($NO_3^-$), $3.3 \times 10^{-4}$ ($SO_4^{2-}$), 220 ($ClO_4^-$), $7.2 \times 10^{-2}$ ($IO_3^-$), 0.1 ($BrO_3^-$), and 370 ($IO_4^-$). The serious interferent ion is periodate.[472]

*Chromate-selective membrane electrode.* The liquid membrane was formed by using chromate–ferroin ion-association complex {1,10-phenanthroline (=phen) and $FeSO_4$ with chromate solution at pH 3.5 gave $[Fe(phen)_3][HCrO_4]_2$}, extracted into nitrobenzene.[473] The electrode gave a linear response to log[chromate] in the concentration range $10^{-4}$—$10^{-2}$ mol l$^{-1}$ at pH 2—4.

*Cyanide-selective membrane electrode.* Any halide membrane electrode can be used as an electrode that is sensitive to cyanide ions, but in practice an AgI-based electrode is preferred because of its high selectivity to cyanide ions. Polyethylene-moulded silver halide electrodes have been used to study their responses to cyanide ions in solutions containing metal ions.[474] Titrations of solutions of cyanide with solutions of $Cd^{II}$, $Zn^{II}$, or $Ni^{II}$ at a pH that is regulated between 6 and 8 were followed potentiometrically. A plot of $log[X^-]$ vs. $log[M]_t$, where $[X^-]$ is halide concentration and [M] that of metal ion, was made and compared with the theoretical (calculated) curve. Only in the case of the AgI electrode and the $Cd^{II}$—$CN^-$ system was good agreement found between experimental and theoretical curves.

The electrode has been used to determine thiocyanate present in water and other samples.[475] The cationic impurities in water, e.g. $Fe^{2+}$, $Fe^{3+}$, and $Co^{2+}$, were removed by passing the sample through a cation-exchange column. Then CNS was

---

[471] D. Midgley, *Analyst (London)*, 1975, **100**, 386.
[472] K. Hiiro, G. J. Moody, and J. D. R. Thomas, *Talanta*, 1975, **22**, 918.
[473] T. Tanaka, K. Hiiro, and A. Kawahara, *Osaka Kogyo Gijyutsu Shikensho Kiho*, 1974, **25**, 241.
[474] M. Mascini and A. Napoli, *Anal. Chem.*, 1974, **46**, 447.
[475] G. Nota, *Anal. Chem.*, 1975, **47**, 763.

*Membrane Phenomena* 97

converted into CN by treating the water sample with 20% $H_3PO_4$. Bromine water was added and the excess bromine was removed by phenol. The CNBr generated in the reaction

$$SCN^- + 4Br_2 + 4H_2O \rightleftharpoons CNBr + SO_4^{2-} + 7Br^- + 8H^+$$

was reduced by a saturated aqueous solution of $SO_2$ to HCN:

$$CNBr + SO_2 + 2H_2O \rightleftharpoons HCN + Br^- + SO_4^{2-} + 3H^+$$

Addition of 4M-NaOH hydrolysed HCN to $CN^-$, which was sensed by the cyanide-selective electrode. The electrode has been used in the determination of cyanide in silver-plating baths.[476]

The $Ag_2S$ membrane electrode also responds to cyanide ions and so may be used to determine cyanide in pure solutions[477] and in effluents from steel works,[478] and also in the microdetermination of *m*-dinitro-compounds. The last determination is based on a reaction of the nitro-compound with excess KCN solution. Using the cyanide electrode, the excess cyanide was potentiometrically estimated by titration with $AgNO_3$.[479]

*Fluoroborate-selective membrane electrodes.* An *o*-phenanthroline-based liquid membrane electrode has been found to be selective to $BF_4^-$.[480] Recently, a solution of Brilliant Green tetrafluoroborate in chlorobenzene adsorbed on natural rubber sheeting has been used as a membrane electrode that shows Nernstian response to $BF_4^-$ in the concentration range $10^{-3}$—$10^{-1}$ mol $l^{-1}$, with a slope of 58.5 mV per decade of concentration.[481] It was selective to $BF_4^-$ over halides, nitrate, acetate, and borate ions, with some interference from perchlorate ions. *o*-Phenanthroline-based electrodes have been used to determine boron in silicon[482] after conversion of boron into fluoroborate ion by treatment with 4M-HF, 2M-$NH_4F$, and $H_2O_2$ in the presence of $Cu^{II}$ ion as the catalyst.

*Halide-selective membrane electrodes.* In the preparation of halide-selective membrane electrodes, the precipitate of silver halide is held in a silver sulphide matrix. Instead, mercuric sulphide matrices have been used to hold appropriate mercurous or silver compounds; pressure, and elevated temperature, were used to form pellets.[483] In this way, precipitates of $Hg_2Br_2$, $Hg_2I_2$, $Hg_2(CN)_2$, AgI, AgBr, and AgCNS have been incorporated into HgS matrices to form bromide-, iodide-, cyanide- and thiocyanate-selective solid-state membrane electrodes. These electrodes showed improved selectivity over the corresponding ones based on $Ag_2S$ matrices. However, HgS–$Hg_2I_2$ did not give satisfactory performance.

Halide-selective electrodes of the second kind and other solid-state electrodes formed of the same materials have been used in comparative studies, and their standard potentials have been computed.[484]

[476] L. N. Lapatnik, *Anal. Chim. Acta*, 1974, **72**, 430.
[477] H. Clysters, F. Adams, and F. Verbeek, *Anal. Chim. Acta*, 1976, **83**, 27.
[478] P. J. Cusbert, *Anal. Chim. Acta*, 1976, **87**, 429.
[479] S. S. M. Hassan, *Anal. Chem.*, 1977, **49**, 45.
[480] J. W. Ross, Jr., in 'Ion Selective Electrodes', ed R. A. Durst, National Bureau of Standards, Special Publication 314, Washington, D.C., 1969, Ch. 2.
[481] A. G. Fogg, A. S. Pathan, and D. T. Burns, *Anal. Lett.*, 1974, **7**, 545.
[482] P. Lanza and P. L. Buldini, *Anal. Chim. Acta*, 1975, **75**, 149.
[483] I. Sekerka and J. R. Lechner, *J. Electroanal. Chem. Interfacial Electrochem.*, 1976, **69**, 339.
[484] R. P. Buck and V. R. Shepard, Jr., *Anal. Chem.*, 1974, **46**, 2097.

Polycrystalline halide-selective electrodes have been used in flowing systems to evaluate their dynamic response time, $t_{95}$.[485]

A method to determine halides ($F^-$, $Cl^-$, $Br^-$, $I^-$) in a single sample of 100 mg of orthophosphate mineral has been described.[486] In successive determinations, mutual halide interference did not occur if the level of iodides was less than $\frac{1}{20}$ of that of bromide and if $Cl^-$ concentration was between 50- and 200-fold in excess of that of bromide. Similarly, halogens in marine algae have been determined by using an iodide-selective electrode.[487] This involved either alkali fusion and ashing of dry algal tissue or ether extraction of algal tissue. Halides in aqueous solution or those that were soluble in organic solvents were determined by argentimetric titration, using an iodide-selective membrane electrode.

A new type of fluoride-selective electrode based on a so-called ceramic membrane has been described.[488] This is formed by sintering lanthanum fluoride, europium fluoride, and calcium fluoride at a temperature greater than 1200 °C for 3—15 h in a stream of HF gas. The electrode gave a Nernstian response and the equilibrium potential was established in less than a minute; $H^+$, $Al^{III}$, and $Fe^{III}$ interfered with the response of the electrode.

A mechanistic study of a fluoride solid-state electrode has been made[489] to elucidate the mechanism of its response. A fast flow technique and impedance measurements have been presented. Analysis of the data indicated that the response of the electrode arose from heterogeneous ion-exchange reactions of fluoride at two types of sites, one involving electronic conduction and the other ionic conduction.

The estimation of the lower limit of detection of $F^-$ ion by a fluoride-selective electrode, discussed by several investigators, is subject to some uncertainties related to the variable liquid junction and the activity coefficient of fluoride ion. These factors and others (such as membrane solubility, impurities in supporting electrolyte, and adsorption of test ions on the walls of the container) governing the behaviour of the electrode have been considered in detail.[490] It has been shown that the solubility of the membrane did not interfere with the electrode function, which, however, was affected by impurities in the supporting electrolyte. Adsorption of fluoride ions, rather than solubility of the crystal membrane, determined the lower limit of detection of the electrode.[491] The study of the effect of solution acidity on the $LaF_3$ single-crystal electrode showed that the electrode response was determined by the competitive adsorption of $OH^-$ and $F^-$ ions and of various fluorine-containing species in the film formed on the electrode surface. The electrode was most sensitive to $F^-$ ions at pH 5.5.[492] Nitrate and chloride showed no interference with the response of the electrode, whereas monofluorophosphate ion did.[493] A kinetic study of the electrode in a fast flow and an automatic system showed that the behaviour of the electrode did not conform to any of the equations proposed for ion-selective

---

[485] R. Rangarajan and G. A. Rechnitz, *Anal. Chem.*, 1975, **47**, 324.
[486] E. J. Duff and J. L. Stuart, *Analyst (London)*, 1975, **100**, 739.
[487] J. N. C. Whyte and J. R. Englar, *Analyst (London)*, 1976, **101**, 815.
[488] H. Hirata and M. Ayuzawa, *Chem. Lett.*, 1974, 1451.
[489] P. Van den Winkel, J. Mertens, T. Boel, and J. Vereecken, *J. Electrochem. Soc.*, 1977, **124**, 1338.
[490] N. Parthasarathy, J. Buffle, and D. Monnier, *Anal. Chim. Acta*, 1974, **68**, 185.
[491] J. Buffle, N. Parthasarathy, and W. Haerdi, *Anal. Chim. Acta*, 1974, **68**, 253.
[492] J. Vesely and K. Stulik, *Anal. Chim. Acta*, 1974, **73**, 157.
[493] A. F. Berndt and R. I. Stearns, *Anal. Chim. Acta*, 1975, **74**, 446.

# Membrane Phenomena

electrodes.[494] A discrepancy between dynamic responses measured in the injection method and in autoanalysers has been noted. This behaviour of the electrode in the latter method is controlled by film diffusion.

Methods for separation and determination of fluoride in samples that cannot be directly analysed for fluoride have been described. In one method,[495] applied to determine fluoride in toothpaste, hexamethyldisiloxane with $HClO_4$ is used to form hydrophobic fluorosilanes, which are hydrolysed to fluoride and silanols in an alkaline trapping solution. In another method, complexing agents for interfering ions have been used.[496] Thus, using Tiron as the masking agent for aluminium, fluoride in samples containing aluminium as an impurity has been estimated with the help of the fluoride-selective electrode. The electrode has been used in the determination of fluorine in alloys of Th—U and U—Zr,[497] in coal,[498] in silicates,[499] in plant ashes (by using sodium citrate to complex Al, Fe, Mg, or silicate),[500] in boron-containing materials,[501] in organic and inorganic compounds,[502] in soil and vegetation (using alkali fusion),[503] and in biological samples, by using a reverse extraction technique.[504]

A sensitive method to determine sub-nanogram amounts of fluoride in sample solutions with the Orion flouride electrode has been described.[505] Similarly, the electrode has been used to monitor fluoride in air[506, 507] and to determine nanomole amounts of aluminium in aqueous solution[508] and in paper-making machine white water.[509] The electrode has been used, under computer control, in flowing systems to monitor the response of $F^-$ ions.[510] It has been used (i) as a reference electrode in mixed solvents in electrochemical cells without transport,[267] and in the study of formation constants of complexes of acetonitrile and allyl alcohol with AgI,[511] and (ii) to investigate the complex formed between fluoride and lanthanum alizarin complexone ($H_4A$), the formula of the complex being $La(LaA)_4F_2$.[512]

A chloride-selective membrane electrode has been constructed from a pellet formed by pressing 30—70 mole % HgS and 70—30 mole % $Hg_2Cl_2$.[513] The electrode conducted well, had a good response time, and was used in precipitation titrations.

Membranes made of pure AgCl single crystal, annealed in partially evacuated

---

[494] J. Mertens, P. Van den Winkel, and D. L. Massart, *Anal. Chem.*, 1976, **48**, 272.
[495] R. Sara and E. Wanninen, *Talanta*, 1975, **22**, 1033.
[496] S. Tanikawa, H. Kirihara, N. Shiraishi, G. Nakagawa, and K. Kodama, *Anal. Lett.*, 1975, **8**, 879.
[497] F. C. Chang, H. T. Tsai, and S. C. Wu, *Anal. Chim. Acta*, 1974, **71**, 477.
[498] J. Thomas, Jr., and H. J. Gluskoter, *Anal. Chem.*, 1974, **46**, 1321.
[499] J. B. Bodkin, *Analyst (London)*, 1977, **102**, 409.
[500] B. Vickery and M. L. Vickery, *Analyst (London)*, 1976, **101**, 445.
[501] B. Schreiber and R. W. Frei, *Mikrochim. Acta*, 1975, **I**, 219.
[502] E. E. M. Hussain, *Mikrochim. Acta*, 1974, 889.
[503] N. R. McQuaker and M. Gurney, *Anal. Chem.*, 1977, **49**, 53.
[504] P. Venkateswarlu, *Anal. Chem.*, 1974, **46**, 878.
[505] A. S. Hallsworth, J. A. Weatherell, and D. Deutsch, *Anal. Chem.*, 1976, **48**, 1660.
[506] A. Hrabeczy-Pall, K. Toth, E. Pungor, and F. Vallo, *Anal. Chim. Acta*, 1975, **77**, 278.
[507] M. Mascini, *Anal. Chim. Acta*, 1976, **85**, 287.
[508] Nj. Radic, *Analyst (London)*, 1976, **101**, 657.
[509] A. Homola and R. O. James, *Anal. Chem.*, 1976, **48**, 776.
[510] J. J. Zipper, B. Fleet, and S. P. Perone, *Anal. Chem.*, 1974, **46**, 2111.
[511] K. M. Stelting and S. E. Manahan, *Anal. Chem.*, 1974, **46**, 2118.
[512] T. Anfalt and D. Jagner, *Anal. Chim. Acta*, 1974, **70**, 365.
[513] J. F. Lechner and I. Sekerka, *J. Electroanal. Chem. Interfacial Electrochem.*, 1974, **57**, 317.

glass ampoules at 320 °C, have been used in the construction of 90 ion-selective chloride electrodes.[514] Annealing decreased the resistance of the electrode and improved its potential response.

Entirely solid-state Ag/AgCl ring membrane electrodes with pressed-in silver foil have been used to measure their standard potentials at 25, 50, and 75 °C with respect to the hydrogen electrode by means of cells without[515] and with[516] transference. The thermodynamic behaviour of these membrane electrodes was identical to that of the corresponding Ag/AgCl electrodes of the second kind.

The chloride-selective electrode has been used to determine chloride in inorganic orthophosphates[517] and in yttrium compounds containing silicon.[518]

A new solid-state microelectrode has been prepared by sealing the tips of tapering glass capillaries, coating them under vacuum with a thin layer of pure silver, and sealing them inside tapered glass shields.[519] Some 106 microelectrodes gave an average slope of 55 mV per decade change in chloride activity. Tip resistance was $77 \times 10^9$ Ω and electrode response was rapid and unaffected by $HCO_3^-$, $H_2PO_4^-$, $HPO_4^{2-}$, or protein. These electrodes have been used in the measurement of intracellular chloride activities.

A combination chloride-selective electrode for potentiometric automated measurements has been proposed.[520] This type of electrode eliminated oscillations even at high flow rates by earthing the flow-through cell. This has been used to estimate chloride in mineral water samples.

A membrane electrode that is selective to bromide ions was formed from pellets made out of a mixture of $Hg_2Br_2$ and HgS (2:1 wt ratio) by subjecting the mixture to a pressure of 25 270 p.s.i.[521] Internal contact was made through mercury. The limit of detection of bromide was $10^{-6}$ mol l$^{-1}$. The ions Cl$^-$, I$^-$, and $SO_4^{2-}$ interfered with the response of the electrode.

A PVC gel containing a solution of tridodecylammonium iodide in nitrobenzene as the sensor for the iodide and a lead wire inserted in it gave a calibration curve that was linear up to $10^{-5}$ mol l$^{-1}$ of metal iodide solution.[522] Similarly, another liquid ion-exchange membrane electrode, based on the extraction of halides from the $LiNO_3$–$KNO_3$ eutectic at 160 °C by tetraoctylphosphonium nitrate, has been constructed.[523] The iodide electrode gave a linear response to iodide ions in the presence of Br$^-$ and Cl$^-$ ions.

An interesting type of iodide-selective electrode, based on the capacity of an anion-exchange resin to adsorb tri-iodide ions selectively and irreversibly, has been constructed.[524] Dowex 2-X8 was prepared, using agar-agar as the binder, and fixed to the end of a glass tube. A platinum wire dipping in the iodide solution and

---

[514] H. Adametzova and R. Vadura, *J. Electroanal. Chem. Interfacial Electrochem.*, 1974, **55**, 53.
[515] F. G. K. Bauke, *J. Electroanal. Chem. Interfacial Electrochem.*, 1976, **67**, 277.
[516] F. G. K. Bauke, *J. Electroanal. Chem. Interfacial Electrochem.*, 1976, **67**, 291.
[517] E. J. Duff and J. L. Stuart, *Talanta*, 1975, **22**, 901.
[518] N. Shibata, K. Oshima, and H. Kojima, *Anal. Chim. Acta*, 1975, **76**, 452.
[519] W. McD. Armstrong, W. Wojtkowski, and W. R. Bixenman, *Biochim. Biophys. Acta*, 1977, **465**, 165.
[520] M. Vandeputte, L. Dryon, and D. L. Massart, *Anal. Chim. Acta*, 1977, **91**, 113.
[521] P. K. C. Tseng and W. F. Gutknecht, *Anal. Lett.*, 1976, **9**, 795.
[522] Y. Shijo, *Bull. Chem. Soc. Jpn.*, 1975, **48**, 1647.
[523] A. Rouchouse, J. Mesplede, and M. Porthault, *Anal. Chim. Acta*, 1975, **74**, 155.
[524] M. Novkirishka and R. Christova, *Anal. Chim. Acta*, 1975, **78**, 63.

# Membrane Phenomena 101

contacting the membrane formed the half cell [*i.e.* Pt ($I_3^-$, $I^-$)] which was connected to a standard calomel electrode. The electrode reaction was

$$(I_3^-)_m + 2e^- \rightleftharpoons 3(I^-)_m$$

Thus the electrode potential is given by

$$E = E° + (RT/2F) \ln\left(\frac{[(I_3^-)_m]}{[(I^-)_m]^3}\right)$$

Changes in $E$ were proportional to the change in the iodide concentration. A linear response over the $pI^-$ range 1—5, with excellent selectivity over other halide ions, was observed.[524]

An extensive study of the properties of both AgI and $AgI/Ag_2S$ membrane electrodes under a variety of conditions has been made.[525] The properties of AgI electrode depend on the ratio of $\gamma$- and $\beta$-AgI modifications in the membrane; $\gamma$-AgI is recommended for the preparation of highly sensitive electrodes. During prolonged contact with $I^-$ solutions, $\gamma$-AgI changed into $\beta$-AgI, and the supersensitivity effect gradually disappeared. Contrarily, electrodes made from $Ag_2S$–AgI mixture do not change their properties with time. These are highly conducting, owing to the presence of $Ag_3SI$. Both types of electrodes were affected by light, and the photosensitivity was governed by the changes in the concentration of interstitial $Ag^+$ ions.

A variation of the pressed $Ag_2S$–AgI precipitate membrane where the membrane is split into two pieces, to serve as two identical iodide-selective electrodes, has been described.[526] Both these electrodes, on calibration, gave identical slopes of 57.6 mV per decade of concentration. These electrodes could be used in a differential mode to estimate iodide concentrations.

The Orion 94-53A solid-state iodide-/silver-ion-selective electrode has been used in studies related to the detection limits of the two ions.[527] The detection limit could vary, depending on the way it is defined, and could depend ultimately on the solubility product of the membrane sensor material. The other factors contributing to the detection limits are absorption of the primary ion on container walls, contamination of the sensor surface, interference from supporting electrolyte, and solid-state defects.

Electron and scanning microscopy have been applied to study the surface morphology of AgI-based silicone membranes[528] and $AgI$–$Ag_2S$ membranes.[529] The electrode surface is attacked upon prolonged use, and the active material is removed from the membrane.

The iodide-selective electrode is used to determine formaldehyde by potentiometric titration.[530] It is based on the reactions

$$HCHO + I_2 + 2KOH \rightleftharpoons 2KI + HCO_2H + H_2O$$
$$KI + AgNO_3 \rightleftharpoons AgI + KNO_3$$

*i.e.* $[HCHO] = 2[KI] = 2[AgNO_3]$

[525] J. Vesely, *Collect. Czech. Chem. Commun.*, 1974, **39**, 710.
[526] R. Wawro and G. A. Rechnitz, *Anal. Chem.*, 1974, **46**, 806.
[527] J. Kontoyannakos, G. J. Moody, and J. D. R. Thomas, *Anal. Chim. Acta*, 1976, **85**, 47.
[528] M. Malissa, M. Grasserbauer, E. Pungor, K. Toth, M. K. Papay, and L. Polos, *Anal. Chim. Acta*, 1975, **80**, 223.
[529] M. H. Sorrentino and G. A. Rechnitz, *Anal. Chem.*, 1974, **46**, 943.
[530] S. Ikeda, *Anal. Lett.*, 1974, **7**, 343.

The electrode is used as an indicator electrode in the titration of poly(vinyl sulphate) with tetradecyl(dimethyl)benzylammonium chloride,[531] and in the potentiometric determinations of hydrazine and hydroxylamine,[532,533] arsenite, sulphite, ascorbic acid,[533] nitrate, and nitramine in micro-quantities.[534]

*Nitrate-selective membrane electrode.* A universal ion-selective electrode, based on Aliquat 336 in the anion form of interest (for example, the nitrate form) and mixed with commercial graphite powder, has been described.[535] The paste, packed into a tubing, served as the electrode.

Silver diethyldithiocarbamate powder compressed at a pressure of 7 ton cm$^{-2}$ has been used as a nitrate-selective electrode.[536] Before use, the electrode was soaked in 0.1M-NaNO$_3$ for a day. Adsorbed nitrate ion probably made the membrane selective to nitrate ions. Cations except Ag$^+$, Hg$^{II}$, and H$^+$ showed no interference with the response of the electrode. Other common anions (except cyanide), at pH > 6, showed slight or no interference.

Quaternary ammonium compounds in the nitrate form [*e.g.* tetraheptylammonium, trioctyl(methyl)ammonium, tetraoctylammonium, tetra(dodecyl)ammonium, and tetra(tetradecyl)ammonium], incorporated into PVC with a plasticizer (di-isodecyl phthalate, dioctyl phthalate, dibutyl phthalate, or diethyl phthalate), served as nitrate-selective membranes which when used as electrodes were found to be superior to other electrodes in stability and sensitivity.[537] This superiority was attributed to the increased distribution ratio of electroactive material between the membrane and the aqueous solution.

A liquid nitrate-selective membrane electrode based on Nitron (1,4-diphenyl-3,5-endanilo-4,5-dihydro-1,2,4-triazole) in nitrobenzene has been evaluated.[538] The selectivity coefficients ($k_{i,j}$) were: $1.9 \times 10^2$ (ClO$_4^-$), 14 (I$^-$), 1.7 (ClO$_3^-$), 0.13 (Br$^-$), $6.4 \times 10^{-2}$ (NO$_2^-$), $4.8 \times 10^{-2}$ (Ac$^-$), $6 \times 10^{-3}$ (Cl$^-$), $2.2 \times 10^{-3}$ (CrO$_4^{2-}$), and $1.4 \times 10^{-3}$ (SO$_4^-$). The electrode has been used as a detector in liquid chromatography [539] and to determine nitrate nitrogen in soils,[537,540] pickling baths,[541] plants,[542] and in grass and clover.[543] It has been used in an automatic apparatus developed to determine nitrate in soils.[544] Also, the electrode has been modified by replacing the membrane with a wick of natural or synthetic porous polymer. The wick is soaked with the electro-active material.[545] This electrode has been used to determine nitrate in different waters.

The bacillus *Escherichia coli,* during growth under anaerobic conditions, uses

---

[531] N. Ishibashi, K. Kina, and K. Tamura, *Anal. Lett.,* 1975, **8**, 867.
[532] S. Ikeda and J. Motonaka, *Anal. Chim. Acta,* 1977, **90**, 257.
[533] R. Christova, M. Ivanova, and M. Novkirishka, *Anal. Chim. Acta,* 1976, **85**, 301.
[534] S. S. M. Hassan, *Talanta,* 1976, **23**, 738.
[535] J. P. Sapio, J. F. Colaruotolo, and J. M. Bobbitt, *Anal. Chim. Acta,* 1974, **71**, 222.
[536] T. Nomura and G. Nakagawa, *Anal. Lett.,* 1975, **8**, 873.
[537] H. J. Nielsen and E. H. Hansen, *Anal. Chim. Acta,* 1976, **85**, 1.
[538] A. Tateda and H. Murakami, *Bull. Chem. Soc. Jpn,* 1974, **47**, 2885.
[539] F. A. Schultz and D. E. Mathis, *Anal. Chem.,* 1974, **46**, 2253.
[540] V. Simeonov, I. Asenov, and V. Diadov, *Talanta,* 1977, **24**, 199.
[541] J. O. Burman and G. Johansson, *Anal. Chim. Acta,* 1975, **80**, 215.
[542] G. R. Smith, *Anal. Lett.,* 1975, **8**, 503.
[543] A. W. M. Sweetsur and A. G. Wilson, *Analyst (London),* 1975, **100**, 485.
[544] D. Goodman, *Analyst (London),* 1976, **101**, 943.
[545] A. Hulanicki, R. Lewandowski, and M. Maj, *Anal. Chim. Acta,* 1974, **69**, 409.

Membrane Phenomena                                                                 103

only nitrate, which is reduced to nitrite by dissimilatory nitrate reductase. This growth of *E. coli* has been followed using the nitrate-selective electrode.[362]

*Perchlorate-selective membrane electrode.* Commercially available Orion perchlorate liquid ion-exchanger, mixed with PVC and formed into a disk or coated on a platinum wire, acted as an electrode that is selective to perchlorate ions.[311] It showed a Nernstian response in the concentration range $10^{-1}$—$10^{-4}$ mol $l^{-1}$ of perchlorate solution. Little interference was observed from $OH^-$, $I^-$, $Br^-$, or $NO_3^-$ ions.

A liquid membrane electrode based on the perchlorate salt of tetrakis(triphenylphosphine)silver(I) in nitrobenzene showed improved characteristics, exhibiting little interference from $OH^-$, $NO_3^-$, $Cl^-$, $SO_4^{2-}$, and acetate ions.[546] The electrode has been found to be superior to Orion electrode, which showed poor selectivity in the presence of $OH^-$ ions. Another liquid membrane electrode based on Brilliant Green perchlorate in chlorobenzene has been developed.[547] It showed a Nernstian response which was constant in the pH range 4.5—8.0. A quaternary phosphonium salt has also been shown to respond selectively to $ClO_4^-$ ions over a wide pH range.[548]

The perchlorate electrode has been used as a periodate sensor in a number of reactions. Vicinal glycol may be oxidized by periodate, and the kinetics of the reaction may thus be followed with the help of a perchlorate-selective electrode. Accordingly, the electrode is used to estimate glycols.[549,550] Similarly, the periodate–α-amino-alcohol reactions have been followed.[551] It was found that $Mn^{II}$ ions catalysed strongly the periodate–triethanolamine reaction in the presence of trinitrilotriacetic acid. This fact has been utilized, and a kinetic procedure has been described to estimate micro-amounts of manganese. Similarly, ultramicro-quantities of $Mn^{II}$ and $Cr^{III}$ have been determined by following their catalytic effects on, respectively, periodate–acetylacetone[552] and periodate–arsenite[553] reactions. Periodate–diethylaniline is another indicator reaction by which catalytic titrations of EDTA and several metal ions may be followed with the help of the perchlorate electrode.[554] Also, the accurate estimation of glucose has been made by following its reaction with periodate.[555] The perchlorate electrode has been used to follow the comparative potentiometric estimation of perchlorate with tetraphenylarsonium chloride, tetraphenylphosphonium chloride, and tetra-n-pentylammonium bromide as titrants.[556] The tetraphenyl-onium salts were found to be equivalent in that they gave identical potentiometric breaks, whereas the tetra-n-pentylammonium bromide gave smaller breaks. Therefore the latter cannot be used as a titrant. Also, the lower limits of these titrations have been explored ($\sim 0.09$ mmol per 50 ml and $\sim 0.01$ mmol per 50 ml if Gran plots are used).

[546] A. C. Wilson and K. H. Pool, *Talanta*, 1976, **23**, 387.
[547] A. G. Fogg, A. S. Pathan, and D. T. Burns, *Anal. Chim. Acta*, 1974, **73**, 220.
[548] Yu. I. Urusov, V. V. Sergievskii, A. Ya. Syrachenkov, A. F. Zhukov, and A. V. Gordievskii, *Zh. Anal. Khim.*, 1975, **30**, 1757.
[549] C. E. Efstathiou and T. P. Hadjiioannou, *Anal. Chem.*, 1975, **47**, 864.
[550] C. E. Ffstathiou, T. P. Hadjiioannou, and E. McNelis, *Anal. Chem.*, 1977, **49**, 410.
[551] C. E. Efstathiou and T. P. Hadjiioannou, *Anal. Chem.*, 1977, **49**, 414.
[552] C. E. Efstathiou and T. P. Hadjiioannou, *Talanta*, 1977, **24**, 270.
[553] C. E. Efstathiou and T. P. Hadjiioannou, *Anal. Chim. Acta*, 1977, **89**, 391.
[554] T. P. Hadjiioannou, M. A. Koupparis, and C. E. Efstathiou, *Anal. Chim. Acta*, 1977, **88**, 281.
[555] C. E. Efstathiou and T. P. Hadjiioannou, *Anal. Chim. Acta*, 1977, **89**, 55.
[556] W. Selig, *Microchem. J.*, 1977, **22**, 1.

*Perrhenate- and tetrachloroaurate-selective membrane electrodes.* A solution of Brilliant Green perrhenate in *o*-dichlorobenzene, saturating a natural rubber membrane, has been used as a membrane electrode that is selective to perrhenate, exhibiting a Nernstian response in the concentration range of perrhenate of $10^{-5}$—$10^{-2}$ mol l$^{-1}$.[557] The response was independent of pH in the range 5—7.2. Similarly, a natural rubber membrane soaked overnight in a saturated solution of Safranine O tetrachloroaurate(III) in *o*-dichlorobenzene gave a Nernstian response in the pAuCl$_4^-$ range 3.6—6.6.[558]

*Phosphate-selective membrane electrode.* Several attempts to prepare membrane electrodes that are selective to phosphate have not been successful. The search for materials and methods to find a satisfactory electrode is continuing. Several polyphenyl-onium bases and other materials have been examined for the construction of both solid and liquid phosphate-selective electrodes.[559] Of the several electrodes tested, liquid membrane electrodes based on phosphonium and triphenyltin salts achieved good sensitivity but were deficient in selectivity for use in routine analysis. Similarly, silver phosphate has been tried as the electro-active material for the construction of a phosphate electrode which consisted of a Teflon-impregnated graphite rod, the end of which had been rubbed with the electro-active material.[560] The electrode showed Nernstian behaviour, with a response time of 2 min, the detection limit for total phosphate being $10^{-5}$ mol l$^{-1}$. No interferences were found from polyphosphates, nitrates, or sulphates, but chloride ion did interfere. A dual enzyme electrode that is sensitive to phosphate, probably the most useful electrode developed so far, has been described.[561] The immobilized enzymes alkaline phosphatase and glucose oxidase catalyse a reaction involving phosphate ion. With β-D-glucose-6-phosphate as the substrate, the reactions involved are

$$\text{glucose-6-phosphate} + H_2O \xrightarrow[\text{phosphatase}]{\text{alkaline}} \text{glucose} + HPO_4^{2-}$$

$$\text{glucose} + O_2 \xrightarrow[\text{oxidase}]{\text{glucose}} \text{gluconic acid} + H_2O_2$$

As glucose oxidase has sufficient activity to oxidize glucose faster than its rate of formation, the rate of the total reaction is determined by that of the first step. Consequently phosphate can be measured amperometrically by means of the oxidation of glucose. Two other enzyme electrodes utilizing enzymes, (i) phosphorylase *a*, phosphoglucomutase, and glucose-6-phosphate dehydrogenase, and (ii) glyceraldehyde–phosphate dehydrogenase, phosphoglycerate kinase, and hexokinase, for the specific assay of phosphate have been described.[562] In both cases, assay is carried out by amperometric monitoring of NADH or NADPH formed in the reaction, at an applied potential of +0.65 V.

[557] A. G. Fogg and A. A. Al-Sibaai, *Anal. Lett.*, 1976, **9**, 39.
[558] A. G. Fogg and A. A. Al-Sibaai, *Anal. Lett.*, 1976, **9**, 33.
[559] M. Nanjo, T. J. Rohm, and G. G. Guilbault, *Anal. Chim. Acta*, 1975, **77**, 19.
[560] I. Novozamsky and W. H. van Riemsdijk, *Anal. Chim. Acta*, 1976, **85**, 41.
[561] G. G. Guilbault and M. Nanjo, *Anal. Chim. Acta*, 1975, **78**, 69.
[562] G. G. Guilbault and T. Cserfalvi, *Anal. Lett.*, 1976, **9**, 277.

*Sulphate-selective membrane electrode.* Freshly precipitated $BaSO_4$ incorporated into a PVC matrix has been used as an electrode that responds to the activity of the sulphate ion in a Nernstian fashion in the concentration range $10^{-6}$—$2 \times 10^{-1}$ M-$Na_2SO_4$.[563] No interference was noted with nitrate, bromide, iodide, and chloride, but the other ions (acetate, phosphate, and carbonate) interfered with the response of the electrode to sulphate.

An enzyme electrode based on immobilized arylsulphatase for the selective assay of sulphate ion has been described.[564] The enzyme is chemically immobilized in a layer on a platinum electrode. The steady-state current arising from the reaction

$$\text{4-nitrocatechol sulphate} + H_2O \xrightarrow[\text{sulphatase}]{\text{aryl}} \text{4-nitrocatechol} + SO_4^{2-}$$

is measured at $+0.8$ V *vs.* SCE. The competitive inhibition of this reaction by added $SO_4^{2-}$ causes a decrease in the steady-state current, the decrease being linearly proportional to $pSO_4^{2-}$ in the range 2—4. By proper selection of the concentration of nitrocatechol sulphate, the electrode can be rendered specific for $SO_4^{2-}$ ions. Only molybdate ions have been found to interfere with the electrode response.

*Sulphide-selective membrane electrode.* An evaluation of the detection limits of Orion solid-state $Ag_2S$ membrane ion-selective electrode has been made, taking precautions against oxidation of sulphide ions.[565] The electrode showed a Nernstian response down to $10^{-7}$ mol l$^{-1}$ for $Ag^+$ ions and $2 \times 10^{-7}$ mol l$^{-1}$ for sulphide ions. The solubility of the membrane material determined the detection limits, and the defects in the crystal lattice had no significant role in practical measurements.

The electrode has been used in the following measurements: (*a*) to determine sulphur in sulphur compounds,[566] sulphides in waste waters[567] and in natural sea-water,[568] and in an automatic semi-continuous titration procedure in which a density gradient of mercuric nitrate is employed;[569] (*b*) in the determination of standard potentials of electrodes made from mixed metal sulphides of $Ag_2$–MS type;[570] (*c*) in the potentiometric titration of phenylthiourea and *NN*-diphenylthiourea in alkaline, neutral, and acidic media[571] and of 2-thiouracil, 6-methyl-2-thiouracil, and 2,4-dithiouracil;[572] (*d*) to determine dithio-oxamide by titration with silver nitrate;[573] and (*e*) in automatic determinations of protein[574] and in monitoring and analysing individual proteins and protein mixtures in serum[575] and in proteins involved in antibody and antigen reactions.[576]

[563] O. G. Taksishvili, E. P. Motsonelidze, Yu. M. Karachentseva, and P. I. Lavitaya, *Zh. Anal. Khim.*, 1975, **30**, 1629.
[564] T. Cserfalvi and G. G. Guilbault, *Anal. Chim. Acta*, 1976, **84**, 259.
[565] D. J. Crombie, G. J. Moody, and J. D. R. Thomas, *Anal. Chim. Acta*, 1975, **80**, 1.
[566] U. Clysters and F. Adams, *Anal. Chim. Acta*, 1977, **92**, 251.
[567] E. W. Baumann, *Anal. Chem.*, 1974, **46**, 1345.
[568] E. Mor, V. Scotto, G. Marcenaro, and G. Alabiso, *Anal. Chim. Acta*, 1975, **75**, 159.
[569] B. Fleet and A. Y. W. Ho, *Anal. Chem.*, 1974, **46**, 9.
[570] M. Koebel, *Anal. Chem.*, 1974, **46**, 1559.
[571] M. K. Papay, V. P. Izvekov, K. Toth, and E. Pungor, *Anal. Chim. Acta*, 1974, **69**, 173.
[572] N. T. Neshkova, V. P. Izvekov, M. K. Papay, K. Toth, and E. Pungor, *Anal. Chim. Acta*, 1975, **75**, 439.
[573] N. M. Sheina, V. P. Izvekov, M. K. Papay, K. Toth, and E. Pungor, *Anal. Chim. Acta*, 1977, **92**, 261.
[574] P. W. Alexander and G. A. Rechnitz, *Anal. Chem.*, 1974, **46**, 860.
[575] P. W. Alexander and G. A. Rechnitz, *Anal. Chem.*, 1974, **46**, 250.
[576] P. W. Alexander and G. A. Rechnitz, *Anal. Chem.*, 1974, **46**, 1253.

*Sulphonate-selective membrane electrode.* This electrode may be formed by incorporating an alkylbenzenesulphonate–ferroin complex into a PVC matrix.[577] The sulphonate electrode responded selectively to sulphonate in the concentration range $10^{-6}$—$10^{-2}$ mol $l^{-1}$ in the presence of sulphate, phosphate, nitrate, and chloride ions.

A nitrobenzene extract of the ion pair consisting of 8-quinolinol-5-sulphonate (Hqs$^-$) and the zephiramine cation (benzyldimethyltetradecylammonium) has been used as the ion-exchanging liquid membrane for the electrode responding to Hqs$^-$ ions.[578] The response of the electrode was Nernstian, and the electrode was applied to following the potentiometric titration of $Cu^{II}$ with $H_2qs$.

MISCELLANEOUS MEMBRANE ELECTRODES. A probe for hydrogen peroxide has been constructed, using a double membrane of catalase–collagen and Teflon, an alkaline electrolyte, and a pair of electrodes (Pt cathode and Pb anode).[579] Immobilization of catalase, which catalyses the reaction

$$H_2O_2 \xrightarrow{\text{catalase}} H_2O + \tfrac{1}{2}O_2,$$

has been carried out electrolytically.

A solution containing collagen fibrin (200 ml, 0.45%, pH 3.8) and 20 ml of aqueous 0.45% catalase was electrolysed by passing a constant current (3.2 mA) for 1 h between two Pt electrodes at 5 °C. The catalase–collagen membrane formed at the cathode was used with a Teflon membrane in an electrochemical cell (see Figure 2) to sense $H_2O_2$. The double membrane is in contact with the Pt cathode. The oxygen formed in the reaction may be determined amperometrically or galvanostatically. The current due to oxygen was measured both in the presence and the absence of $H_2O_2$ in a control solution that was saturated with dissolved oxygen. The measured current varied linearly with the concentration of $H_2O_2$ in the solution in the range 0—1.5 mmol $l^{-1}$ at the optimum pH of 6.2. The sensor took 1.5—2 min, at 20 °C, to reach a steady-state current.

Another peroxide-sensing device has been made by covering an oxygen electrode with a membrane that catalyses the breakdown of $H_2O_2$ to $O_2$. This is similar to the one described above except that, instead of catalase, an inorganic catalyst has been used.[580] Metal oxides and hydroxides, particularly those of Mn, Co, and Ru, were deposited in the dialysis membrane.

An electrode that is selectively responsive to saccharin has been made by dissolving an association complex of iron(II)–bathophenanthroline chelate and saccharin in nitrobenzene.[581] Saccharin in the presence of other sweetening substances such as saccharose, glucose, sodium cyclamate, and sorbitol has been estimated in the concentration range $10^{-1}$—$10^{-5}$ mol $l^{-1}$. Replacement of saccharin with sulfa drugs in the preparation of the electrode gave other electrodes that were selective to the sulfa drug used. In this way electrodes selective to sulfisomidine and sulfamera-

---

[577] T. Tanaka, K. Hiiro, and A. Kawahara, *Anal. Lett.*, 1974, **7**, 173.
[578] M. Sugawara, T. Nakajima, and T. Kambara, *J. Electroanal. Chem. Interfacial Electrochem.*, 1976, **67**, 315.
[579] M. Aizawa, I. Karube, and S. Suzuki, *Anal. Chim. Acta*, 1974, **69**, 431.
[580] S. J. Updike, M. C. Shults, J. K. Kosovich, I. Treichel, and P. M. Treichel, *Anal. Chem.*, 1975, **47**, 1457.
[581] N. Hazemoto, N. Kamo, and Y. Kobatake, *J. Assoc. Off. Anal. Chem.*, 1974, **57**, 1205.

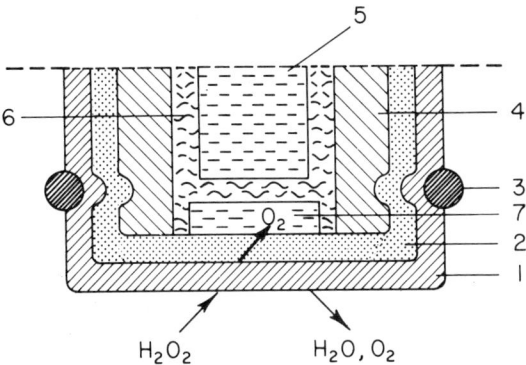

**Figure 2** Schematic representation of the sensor for hydrogen peroxide: 1, catalase–collagen membrane; 2, Teflon; 3, rubber O-ring; 4, insulator; 5, anode (platinum or lead); 6, electrolyte solution (potassium hydroxide); and 7, platinum cathode.
(Reproduced by permission from 'Membrane Electrodes', Academic Press, New York, 1976)

zine in the presence of urea, glycine, aminopyrine, or $p$-aminobenzoic acid have been prepared.[582]

A membrane electrode that is responsive to hapten trimethylphenylammonium ion has been prepared and used to study the binding of rabbit $\gamma$-globulin antibody to hapten.[583]

Anion- and cation-exchange membranes have been used to construct cells of the type

| reference electrode | reference solution (1) | $\|\|$ | $A_m$ | $\|\|$ | test solution (2) SDS | $\|\|$ | $C_m$ | $\|\|$ | reference solution (1) | reference electrode |

where test solution (2) contained sodium dodecylsulphate or cetylpyridinium bromide confined between an anion-exchange membrane $A_m$ and a cation-exchange membrane $C_m$. These cells have been used to determine the critical micellar concentration (CMC) of surfactants. The potential of the cell is given by the equation $E = (2RT/F) \ln (a_1/a_2)$. At concentrations below the CMC, a Nernstian response was observed. When concentrations exceeded the CMC level, $E$ deviated from Nernstian behaviour. These procedures could be used to estimate the activity of surfactants, provided that the membranes did not lose their permselectivity on prolonged soaking.[584] Powdered hexadecyltrimethylammonium dodecylsulphate has been incorporated into a silicone rubber membrane. This, when used as an electrode, responded to cationic and not to anionic detergents.[585] It showed a strong memory effect on going from a high to a low concentration.

A dodecylsulphate-selective electrode has been evaluated for its use in sufactant

---

[582] N. Hazemoto, N. Kamo, and Y. Kobatake, *J. Pharm. Sci.*, 1976, **65**, 435.
[583] M. Meyerhoff and G. A. Rechnitz, *Science*, 1977, **195**, 494.
[584] B. J. Birch and D. E. Clarke, *Anal. Chim. Acta*, 1974, **69**, 473.
[585] A. F. Fogg, A. S. Pathan, and D. T. Burns, *Anal. Chim. Acta*, 1974, **69**, 238.

solutions containing polymers [dextran, poly(vinyl alcohol), and poly(vinylpyrrolidone)] and a protein (bovine serum albumin).[586] Some of the membrane electrodes, such as precipitate-based silicone rubber ion-selective electrodes, have been used in voltammetry.[587]

ENZYME MEMBRANE ELECTRODES. Two methods have been described for the construction of enzyme electrodes.[588] In method one, a cation-selective electrode is dipped into a solution of enzyme, albumin, and glutaraldehyde. A thin layer of crosslinked protein coats the electrode bulb. In method two, two steps are used. The active membrane containing the enzyme is first made by crosslinking protein with glutaraldehyde on a glass plate. This membrane is next fitted with a silicone membrane on a gas electrode. Both methods have been used to construct urea and glucose electrodes.

Enzyme-coated electrodes sensing $NH_4^+$ ion or ammonia, with the help of the air-gap electrode,[303, 589] have been used to estimate urea. The reaction involved is

$$CO(NH_2)_2 + 2H_2O + H^+ \xrightarrow{urease} HCO_3^- + 2NH_4^+$$

This reaction can be followed with a carbon dioxide gas electrode.[588] Using a nonactin–silicone-rubber-based ammonium-selective electrode, the rate of dissolution of urea has been followed.[590] An improved electrode in which urease is chemically bound and attached to a Teflon membrane which is an integral part of the ammonia gas electrode has been used in the determination of urea.[591] Another urea-sensing electrode in which a small quantity of EDTA-stabilized urease solution is held between an external dialysis membrane and a gas-permeable membrane of a conventional ammonia-selective membrane electrode has been shown to function reliably for urea in aqueous as well as in blood and serum samples.[592] A modification of this urea electrode has been to place enzyme directly on a magnetic stirrer and hold it in place by a nylon net. The enzyme-coated stirrer stirs the solution and brings about an enzymatic transformation, permitting assay of a substrate such as urea.[593] An air-gap electrode can be used to sense ammonia produced in the reaction. Similarly, the air-gap electrode has been used in the kinetic chemical analysis of urease by monitoring the initial rate of release of ammonia.[594] This procedure has been applied to assay arginase activity also. The reactions involved are[594]

$$\text{L-arginine} \xrightarrow{arginase} \text{urea} + \text{L-ornithine} \xrightarrow{urease} CO_2 + 2NH_3$$

Excess urease is added to ensure that the reactions proceed. Earlier, an ammonium-selective electrode containing immobilized urease was used to estimate the activity

---

[586] B. J. Birch, D. E. Clarke, R. S. Lee, and J. Oakes, *Anal. Chim. Acta*, 1974, **70**, 417.
[587] F. Feher, G. Nagy, K. Toth, and E. Pungor, *Analyst (London)*, 1974, **99**, 699.
[588] C. Tran-Minh and G. Brown, *Anal. Chem.*, 1975, **47**, 1359.
[589] G. G. Guilbault and M. Tarp, *Anal. Chim. Acta*, 1974, **73**, 355.
[590] I. Fritz, G. Nagy, L. Fodor, and E. Pungor, *Analyst (London)*, 1976, **101**, 439.
[591] M. Mascini and G. G. Guilbault, *Anal. Chem.*, 1977, **49**, 795.
[592] D. S. Papastathopoulos and G. A. Rechnitz, *Anal. Chim. Acta*, 1975, **79**, 17.
[593] G. G. Guilbault and W. Stokbro, *Anal. Chim. Acta*, 1975, **76**, 237.
[594] N. R. Larsen, E. H. Hansen, and G. G. Guilbault, *Anal. Chim. Acta*, 1975, **79**, 9.

*Membrane Phenomena* 109

of arginase.[595] The assay procedure consisted in injecting 1 ml of enzyme solution (arginase) into buffered arginine solution and following the potential as a function of time with the urease electrode and SCE. The initial slope ($\Delta E/\Delta t$) or the lapsed-time slope (slope after 30 s), when plotted against enzyme concentration or substrate concentration, gave a straight line.

Glucose, on oxidation, gives $H_2O_2$ according to the reaction

$$\text{glucose} + O_2 \xrightarrow{\text{glucose oxidase}} \text{gluconic acid} + H_2O_2$$

$H_2O_2$ can be decomposed by catalase according to

$$H_2O_2 \xrightarrow{\text{catalase}} H_2O + \tfrac{1}{2}O_2$$

and oxygen can be sensed by using an oxygen electrode covered with the double enzyme membrane.[588]

Glucose can also be measured by following amperometrically the production of $H_2O_2$, using a single enzyme membrane covering the oxygen electrode. The electrode is placed in a solution of glucose and a potential of 0.6 V (*vs.* SCE) is impressed across it. The $H_2O_2$ formed by the enzyme reaction diffuses out of the enzyme layer toward the Pt sensor electrode, where it is oxidized thus:

$$H_2O_2 \rightleftharpoons O_2 + 2H^+ + 2e^-$$

Typical response curves of the glucose electrode showed that the current initially decreased for about 2 s because of diffusion of glucose through the cellophane and/or enzyme layers and then started to increase with time, reaching a steady-state value. Both the initial rate of increase of current and the final steady-state current can be used to analyse glucose concentrations. This method, although it gave accurate values for glucose, was not as sensitive as the method in which $p(O_2)$ in the solution is monitored.[596] In the latter procedure the enzyme and calomel electrodes are placed in a phosphate buffer solution and a potential of $-0.6$ V is applied. When the current is constant, the sample solution is added and the initial rate of change in the current due to oxygen is recorded for the assay of glucose. In this manner, amino-acids have also been analysed.[596] The general reaction involved is

$$\underset{NH_2}{\underset{|}{R-\overset{\overset{H}{|}}{C}-CO_2H}} + H^+ + H_2O + O_2 \xrightarrow[\text{oxidase}]{\text{L-amino-acid}} R\underset{\parallel}{\overset{}{C}}CO_2H + NH_4^+ + H_2O_2$$

Also, amperometric measurements at $+0.6$ V (*vs.* SCE) have been carried out to estimate certain L-amino-acids, such as the L-isomers of leucine, phenylalanine, methionine, tryptophan, tyrosine, and cysteine.[597] Other amino-acids, such as glycine and the L-isomers of alanine, serine, threonine, asparagine, valine, glutamine, and lysine, could not be estimated amperometrically using the L-amino-acid oxidase enzyme anode poised at 0·35 V (*vs.* SCE).

---

[595] H. E. Booker and J. L. Haslam, *Anal. Chem.*, 1974, **46**, 1054.
[596] M. Nanjo and G. G. Guilbault, *Anal. Chim. Acta*, 1974, **73**, 367.
[597] G. G. Guilbault and G. J. Lubrano, *Anal. Chim. Acta*, 1974, **69**, 183.

Alcohol oxidase reacts with primary aliphatic alcohols according to the reaction

$$RCH_2OH \xrightarrow{\text{alcohol oxidase}} RCHO + H_2O_2$$

This reaction can be followed amperometrically by monitoring $H_2O_2$. Two methods have been used. In method one,[598] a clean, bright platinum electrode and a SCE were used in a buffer solution containing the sample. A potential of 0.35 V was applied until the current decreased to a low value. Then alcohol oxidase was added and the initial rate was measured. This rate varied linearly with alcohol concentration. In this way alcohol and blood alcohol can be estimated. In method two, an immobilized enzyme (alcohol oxidase) electrode mounted on the surface of a platinum wire was used.[599] A potential of $-0.6$ V (vs. SCE) was applied to the electrode, placed in a stirred buffer solution. When the current reached a steady value, the sample was added and the initial rate of change in current and the final steady-state current were noted. The amount of substrate present was calculated from calibration curves.[599] The alcohol oxidase electrode responded also to aldehydes and carboxylic acids.

A platinum electrode covered with an immobilized uricase layer has been used to assay uric acid in human serum and urine.[600] The reaction is

$$\text{uric acid} + O_2 \xrightarrow{\text{uricase}} \text{allantoin} + CO_2 + H_2O_2$$

The initial rate of disappearance of dissolved oxygen during enzymatic reaction of uric acid was directly measured at $-0.6$ V (vs. SCE) and was proportional to the amount of uric acid present. The reaction has also been followed by using a carbon dioxide electrode covered with an immobilized layer of uricase.[601]

The amperometric response following periods of electrode inactivity of an immobilized glucose oxidase enzyme electrode has been modelled by digital simulation.[602] Average currents greater than the steady-state value observed experimentally have been predicted by simulation. The amount of enhancement was related to the time allowed for the build-up of the electro-active product in the membrane, and was independent of the velocity of the enzyme reaction and the substrate concentration. The transient behaviour of the enzyme electrode potential has been mathematically considered by using Fourier analysis.[603]

A standard-addition method for determination of kinetic substrate in enzymatic analysis has been given for urea in aqueous samples and in sera.[604] This was carried out by using the air-gap electrode. Also, an enzyme reactor electrode for urea determination has been described.[605] The reactor contained urease, immobilized with glutaraldehyde, on porous glass. A buffer mixed with the sample was pumped through the reactor and the effluent was mixed with NaOH; the ammonia evolved

---

[598] G. G. Guilbault and G. J. Lubrano, *Anal. Chim. Acta*, 1974, **69**, 189.
[599] M. Nanjo and G. G. Guilbault, *Anal. Chim. Acta*, 1975, **75**, 169.
[600] M. Nanjo and G. G. Guilbault, *Anal. Chem.*, 1974, **46**, 1769.
[601] T. Kawashima and G. A. Rechnitz, *Anal. Chim. Acta*, 1976, **83**, 9.
[602] L. D. Mell and J. T. Maloy, *Anal. Chem.*, 1976, **48**, 1597.
[603] P. W. Carr, *Anal. Chem.*, 1977, **49**, 799.
[604] N. R. Larsen and E. H. Hansen, *Anal. Chim. Acta*, 1976, **84**, 31.
[605] G. Johansson and L. Ogren, *Anal. Chim. Acta*, 1976, **84**, 23.

was quantitatively sensed by an ammonia electrode. A similar reactor has been described for the determination of amino-acids.[606] A reactor–separator membrane was formed by moulding a urease gel layer to an anion-exchange membrane. When this was used as a barrier between donor urea and acceptor buffer solution, ammonia was enriched in the acceptor solution. The rate of production of ammonia was linear with time and was proportional to the concentration of donor substrate. A quantitative description of this transport has been given.[607]

One of the problems involved in immobilizing enzymes is related to loss of enzyme activity. Consequently, attempts have been made to incorporate them into matrices to preserve as much activity as possible. Cellulose acetate membranes have been used to immobilize enzymes. Chymotrypsin incorporated into cellulose acetate membrane either by incubation or by pressure showed activities as high as 75% of that in free solution.[608] Similarly, urease has been incorporated into collagen and used to assay for urea by measurement of ammonia or ammonium ion produced in the reaction.[609]

By using the electrochemical technique used for the preparation of catalase–collagen membrane that has been used for sensing $H_2O_2$ (see $H_2O_2$ electrode, p. 106), collagen-based membranes containing uricase[610] and a mixture of enzymes, *i.e.* invertase, mutarotase, and glucose oxidase,[611] have been prepared. The latter membrane, when used in sucrose solution, brought about the following reactions:

$$\text{sucrose} + H_2O \xrightarrow{\text{invertase}} \alpha\text{-D-glucose} + \text{D-fructose}$$

$$\alpha\text{-D-glucose} \xrightarrow{\text{mutarotase}} \beta\text{-D-glucose}$$

$$\beta\text{-D-glucose} + O_2 + H_2O \xrightarrow{\text{glucose oxidase}} \text{D-glucose-}\delta\text{-lactose} + H_2O_2$$

By monitoring the decrease in oxygen concentration, sucrose has been determined.

Enzymes encapsulated in liquid surfactant membranes have been shown to retain their catalytic activity. So a liquid membrane system has been used to encapsulate a bacterial cell-free homogenate. This system reduced effectively the nitrate to nitrite and to other nitrogen compounds of lower oxidation state.[612]

Lactate dehydrogenase (LDH)–nicotinamide adenine dinucleotide (NAD; oxidized form $NAD^+$ and reduced form NADH), when incorporated into a membrane or bound in agarose and held close to a glassy carbon electrode, served as an electrochemical sensor for lactate.[613] The reaction that is involved in the passage of current is

$$CH_3CH(OH)COO^- + NAD^+ \xrightarrow{\text{LDH}} CH_3COCOO^- + NADH + H^+$$

[606] G. Johansson, K. Edström, and L. Ogren, *Anal. Chim. Acta,* 1976, **85**, 55.
[607] W. J. Blaedel and T. R. Kissel, *Anal. Chem.,* 1975, **47**, 1602.
[608] H. P. Gregor and P. W. Rauf, *Biotech. Bioeng.,* 1975, **17**, 445.
[609] Y. Nakamoto, I. Karube, and S. Suzuki, *Biotech. Bioeng.,* 1975, **17**, 1387.
[610] S. Suzuki, N. Sonobe, I. Karube, and M. Aizawa, *Chem. Lett.,* 1974, **1**, 9.
[611] I. Satoh, I. Karube, and S. Suzuki, *Biotech. Bioeng.,* 1976, **18**, 269.
[612] R. R. Mohan and N. N. Li, *Biotech. Bioeng.,* 1974, **16**, 513.
[613] W. J. Blaedel and R. A. Jenkins, *Anal. Chem.,* 1976, **48**, 1240.

Catalytic oxidation of lactate by $NAD^+$ is coupled with the electrochemical oxidation of NADH on the inert electrode. The bound $NAD^+$ is electrochemically recycled and re-used by LDH. Similarly, lactate has been quantitatively detected in blood within a minute in an electrode system in which LDH catalyses the oxidation of lactate by hexacyanoferrate(III), the latter being recycled by electrochemical oxidation of the hexacyanoferrate(II) formed on an inert electrode.[614] The reaction is

$$CH_3CH(OH)COO^- + 2Fe(CN)_6^{3-} \xrightarrow{LDH} CH_3COCOO^- + 2\,Fe(CN)_6^{4-} + 2H^+$$

A number of other enzyme electrode systems that have been developed are the following:

(i) Phenylalanine lyase cleaves L-phenylalanine according to the reaction

$$PhCH_2-\underset{H}{\overset{NH_2}{\underset{|}{\overset{|}{C}}}}-CO_2H \xrightarrow{\text{L-phenylalanine lyase}} NH_4^+ + PhC\underset{H}{\overset{H}{\underset{|}{\overset{|}{=}C}}}-COO^-$$

and subsequently

$$NH_4^+ + OH^- \rightarrow NH_3\uparrow + H_2O$$

An air-gap electrode has been used to measure the ammonia evolved.[615]

(ii) A layer of suspended 5'-adenylic acid deaminase in conjunction with an ammonia-gas-sensing membrane electrode (Orion Model 95-10) has been used to sense the nucleotide 5'-adenosine monophosphate (5'-AMP).[616] The reaction is

$$5'\text{-AMP} + H_2O \xrightarrow{\text{enzyme}} 5'\text{-IMP} + NH_3$$

(IMP is inosine 5'-monophosphate).

Similarly, the important enzyme adenosine deaminase has been determined in an automated system. The amount of ammonia released in the reaction

$$\text{adenosine} \xrightarrow[\substack{\text{deaminase} \\ (\text{pH } 7.5,\, 37\,^\circ\text{C})}]{\text{adenosine}} \text{inosine} + NH_3\uparrow$$

is related to the enzyme activity.[617]

(iii) Creatininase or tripolyphosphate-activated creatininase, which converts creatinine into ammonia and N-methylhydantoin, have been used to construct an electrode to sense creatinine in aqueous samples[618] and in urine and serum samples.[619]

(iv) Procedures to assay L-asparaginase continuously by coupling with the

---

[614] H. Durliat, M. Comtat, J. Mahenc, and A. Baudras, *Anal. Chim. Acta*, 1976, **85**, 31.
[615] C. P. Hsiung, S. S. Kuan, and G. G. Guilbault, *Anal. Chim. Acta*, 1977, **90**, 45.
[616] D. S. Papastathopoulos and G. A. Rechnitz, *Anal. Chem.*, 1976, **48**, 862.
[617] C. E. Hjemdahl-Monsen, D. S. Papastathopoulos, and G. A. Rechnitz, *Anal. Chim. Acta*, 1977, **88**, 253.
[618] H. Thompson and G. A. Rechnitz, *Anal. Chem.*, 1974, **46**, 246.
[619] M. Meyerhoff and G. A. Rechnitz, *Anal. Chim. Acta*, 1976, **85**, 277.

*Membrane Phenomena* 113

glutamic dehydrogenase (GDH) reaction and a certain glass membrane electrode have been given.[620] The reactions involved are

$$\text{L-asparagine} + H_2O \xrightarrow{\text{L-asparaginase}} \text{L-aspartic acid} + NH_4^+$$

$$NH_4^+ + \alpha\text{-ketoglutarate} + NADH \xrightarrow{\text{GDH}} \text{L-glutamic acid} + NAD^+ + H_2O$$

A direct assay of asparaginase can be effected by the first reaction by using the ammonium-sensing electrode. Conversely, immobilized asparaginase has been used to assay L-asparagine.[621]

(v) A carbonate-selective membrane electrode has been used to determine L-arginine and L-lysine by both manual and automated flow methods.[622] The decarboxylation of the amino-acids was brought about by specific enzymes (*l*-arginine and *l*-lysine decarboxylases).

(vi) Galactose oxidase, immobilized in a membrane and covering an amperometric hydrogen peroxide electrode, has been used in the microdetermination of glalactose in plasma and whole blood.[623]

(vii) Cholinesterase catalyses the reaction

$$RCOYCH_2CH_2\overset{+}{N}Me_3 + H_2O \rightarrow RCOO^- + H^+ + HYCH_2CH_2\overset{+}{N}Me_3$$

where Y is O or S. This enzymatic reaction results in a linear change of pH with time at pH 8.2—7.9. So, by following the initial rate of enzymatic reaction with the pH electrode, the enzyme has been assayed.[624] Similarly, by immobilizing the substrate (acetylcholine chloride) by allowing it to react with a cation-exchange resin and holding it in a nylon net, cholinesterase in an aqueous sample has been determined by following the pH change with time.[625]

(viii) A procedure in which two enzymes have been employed to determine cholesterol in an automated system has been described.[626] The reactions involved are

$$\text{cholesterol esters} + H_2O \xrightarrow[\text{hydrolase}]{\text{cholesterol ester}} \text{free cholesterol} + \text{fatty acids}$$

$$\text{free cholesterol} + O_2 \xrightarrow[\text{oxidase}]{\text{cholesterol}} \text{cholest-4-en-3-one} + H_2O_2$$

$$H_2O_2 + 2I^- + 2H^+ \xrightarrow{\text{molybdenum(IV)}} I_2 + H_2O$$

Iodide concentration has been monitored by using a flow-through iodide-selective membrane electrode.

---

[620] D. A. Ferguson, Jr., J. W. Boyd, and A. W. Phillips, *Anal. Biochem.*, 1974, **62**, 81.
[621] R. Wawro and G. A. Rechnitz, *J. Membr. Sci.*, 1976, **1**, 143.
[622] S. L. Tong and G. A. Rechnitz, *Anal. Lett.*, 1976, **9**, 1.
[623] P. J. Taylor, E. Kmetec, and J. M. Johnson, *Anal. Chem.*, 1977, **49**, 789.
[624] K. Gibson and G. G. Guilbault, *Anal. Chim. Acta*, 1975, **76**, 245.
[625] G. G. Guilbault and A. Iwase, *Anal. Chim. Acta*, 1976, **85**, 295.
[626] D. S. Papastathopoulos and G. A. Rechnitz, *Anal. Chem.*, 1975, **47**, 1792.

(ix) Nitrite reductase, extracted from spinach leaves and purified, catalyses the reduction of nitrite to ammonia in the presence of reduced methyl viologen as the electron donor. Thus

$$NO_2^- + 6e^- + 8H^+ \xrightarrow{\text{nitrite reductase}} NH_4^+ + 2H_2O$$

Nitrite has been assayed by measurement of the ammonia formed, using the air-gap electrode.[627]

(x) Penicillinase that is immobilized by adsorption on a fritted glass disk and fixed to the end of a pH electrode that has a flat surface has been found to be sensitive to changes in penicillin concentration and insensitive to cations.[628] The enzyme hydrolyses penicillin to penicilloic acid according to the reaction

penicillin → penicilloic acid

The increase in pH sensed by the glass electrode is proportional to the penicillin concentration.

(xi) A layer of β-glucosidase trapped between the sensitive surface of a cyanide-selective electrode and dialysis paper served as an electrode to assay amygdalin at pH 7.[629] Several factors (such as substrate concentration, pH, amount of trapped enzyme, thickness of dialysis paper, and stirring) which control the response of the electrode have been evaluated.

**Diffusion of Non-electrolyte.**—The influence of membrane tortuosity as a variable on the transport properties of hydrophobic polymer membranes has been considered[630] from the standpoint of two treatments due to Mackie and Meares[631] and Yasuda et al.[632] The former treatment related tortuosity ($\theta$) to volume fraction of the polymer ($V_r$) by the expression

$$\theta = \left(\frac{1+V_r}{1-V_r}\right)^2 \quad (51)$$

The diffusion coefficient, $D$, of the diffusant in the membrane that one measures was related to the diffusion coefficient in the solvent ($D_0$) by the relation

$$\bar{D} = (D_0/\theta^2) = D_0 \left(\frac{1-V_r}{1+V_r}\right) \quad (52)$$

According to the second treatment, the diffusivity in a membrane in which diffusant

---

[627] C. A. Kiang, S. S. Kuan, and G. G. Guilbault, *Anal. Chim. Acta*, 1975, **80**, 209.
[628] L. F. Cullen, J. F. Rushing, A. Schleifer, and G. J. Papariello, *Anal. Chem.*, 1974, **46**, 1955.
[629] M. Mascini and A. Liberti, *Anal. Chim. Acta*, 1974, **68**, 177.
[630] H. W. Osterhoudt, *J. Phys. Chem.*, 1974, **78**, 408.
[631] J. S. Mackie and P. Meares, *Proc. R. Soc. London, Ser. A*, 1955, **A232**, 498.
[632] H. Yasuda, C. E. Lamaze, and L. D. Ikenberry, *Makromol. Chem.*, 1968, **118**, 19.

*Membrane Phenomena* 115

is excluded from the polymer matrix (no diffusant–matrix interaction) is given by

$$\bar{D} = D_0 \exp\left[-k\left(\frac{1}{V_0} - 1\right)\right] \quad (53)$$

where $V_0$ is the volume fraction of solvent and $k$ is a constant related to volumes of diffusant and solvent in the membrane. The parameter usually measured in dialysis experiments is permeability, $P$; this is related to $\bar{D}$ by $P = \bar{D} K$, where $K$ is the partition coefficient between membrane and solution phases, which for an uncharged diffusant of low molecular weight is considered equal to $V_0$. Thus equation (53) can be written as

$$\ln(\bar{D}K) = \ln D_0 + \ln V_0 + k - (k/V_0) \quad (54)$$

For values of $V_0 > 0.5$, the term $(1/V_0)$ becomes dominant. Since $V_0 + V_r = 1$, equation (54) becomes

$$\log(\bar{D}K) = \log D_0 + \left(\frac{k}{2.303}\right) - \left(\frac{k}{2.303(1 - V_r)}\right) \quad (55)$$

Equation (52) in terms of the parameter $(\bar{D}K)$ becomes

$$\log(\bar{D}K) = \log D_0 + \log\left[\frac{(1 - V_r)^3}{(1 + V_r)^2}\right] \quad (56)$$

Using different membrane materials whose volume fractions $(V_r)$ covered a range from 0.08 to 0.59, and permeants of low molecular weight (ethylene glycol and ethanol), it has been shown[630] that both equations (55) and 56) agreed reasonably well with experimental data obtained for moderately to lightly water-swollen membranes. But only equation (56) gave better agreement, covering the whole range of $V_r$ examined.

Equation (52) has been further considered by Brown and co-workers, who in a series of three papers[633–635] described measurements of diffusion of homologous series of solutes in tightly and loosely crosslinked water-swollen gels of cellulose, hydroxyethyl-cellulose, and polyacrylamide. The diffusion coefficients in the gel phase were substantially lower than the values for free diffusion in bulk solvents in all cases. The influences of solvent, temperature, and polymer concentration on this reduction in the values of diffusion coefficients have been discussed.

When the permeant interacts with the polymer matrix, causing changes in the nature of the species involved, the equations so far considered become inapplicable. A study of diffusion of water in cellulose acetate membranes as a function of membrane porosity using a carrier-gas flow technique showed that the diffusion coefficients for water could not be determined reliably for membranes of very high porosity.[636] In the case of dense membranes, the diffusion coefficient varied with the concentration of water in the membrane. This was attributed to the formation of water clusters.

Studies of the permeation of water, 1,4-dioxan, and their mixtures through

---

[633] W. Brown and K. Chitumbo, *J. Chem. Soc., Faraday Trans. 1*, 1975, **71**, 1.
[634] W. Brown and K. Chitumbo, *J. Chem. Soc., Faraday Trans. 1*, 1975, **71**, 12.
[635] W. Brown, G. Kloow, K. Chitumbo, and T. Amu, *J. Chem. Soc., Faraday Trans. 1*, 1976, **72**, 485.
[636] M. J. Palin, G. J. Gittens, and G. B. Porter, *J. Appl. Polym. Sci.*, 1975, **19**, 1135.

Nylon-6 membranes used as multilayers enabled the determination of concentration profiles within the membrane matrix.[637] This permitted the measurement of concentration-dependent diffusivities. Since water plasticizes nylon, the behaviour of the mixture could not be predicted from the performance of the pure components. When equimolar feed mixtures were used, transfer of water was enhanced while that of 1,4-dioxan was depressed. Acid treatment of the membrane enhanced the transfer of both components. Measurement of concentration profiles in the laminates during the non-steady state suggested that the polymer matrix was undergoing structural changes. Similarly, acrylic–methacrylic ester polymers have been used to study the influence of film surface on diffusion of urea in aqueous solutions.[638] A greater urea permeation rate was observed when the lower surface, *i.e.* the surface in contact with substrate during casting, was the entry surface for urea than when the upper surface, *i.e.* the surface in contact with the atmosphere during casting, was the entry surface. Scanning electron microscopy revealed that the lower surface formed larger pores than the upper surface, which formed smaller pores during exposure to urea solution. Aromatic polyamide membranes rejected urea, creatinine, and uric acid very efficiently and so have been used in the development of a miniature artificial kidney.[639] Cuprammonium membranes are the membranes of choice used in artificial kidney machines. Cellophane membranes possessing permeability characteristics superior to those of cuprammonium membranes have been prepared.[640]

An experimental method for measuring permeability of small tubular (hollow fibre) membranes by means of an unsteady-state diffusion experiment has been described.[641] This method is more convenient than the conventional one for determining the permeability of tubular membranes, which involves the construction of a small-scale dialyser. This new method has been used to obtain permeability data for collagen and commercial cellulose acetate membranes, using urea, sucrose, and poly(ethylene glycol).

The transport of organic liquids through polyurethane film has been measured.[642] The diffusion coefficients evaluated from the relation $P = \bar{D}S$ ($P$ is permeability, $S$ is solubility in the membrane) are given in Table 18. In similar types of studies, permeation of dichlorobenzene through a Teflon–FEP [fluorinated copoly(ethylene/propylene)] membrane has been measured by both liquid/liquid dialysis, using hexadecane as the dialysis solvent, and pervaporation.[643] Values of the permeation rate determined by the two methods at 72 °C were $4.6 \times 10^{-3}$ and $2.7 \times 10^{-3}$ g h$^{-1}$ for areas of 11 cm$^2$ respectively, the respective activation energies being 11 and 15 kcal mol$^{-1}$. These data indicate that the dialysis solvent, hexadecane, plasticized the FEP membrane, thereby increasing the permeation rate and decreasing the activation energy. The dialysis method gave a value of 10 h for the time lag ($t$) involved in diffusion of chlorobenzene across the membrane, whereas a corresponding value calculated from the relation $t = d^2/6D$, where $d$ is the

[637] R. W. Tock and J. Y. Cheung, *Sep. Sci.*, 1974, **9**, 361.
[638] S. A. M. Abdel-Aziz, W. Anderson, and P. A. M. Armstrong, *J. Appl. Polym. Sci.*, 1975, **19**, 1181.
[639] M. A. Kraus, M. A. Frommer, M. Nemas, and R. Gutman, *J. Membr. Sci.*, 1976, **1**, 115.
[640] D. W. Monk and E. Wellisch, *J. Appl. Polym. Sci.*, 1974, **18**, 2875.
[641] J. F. Stevenson, M. A. Von Deak, M. Weinberg, and R. W. Schuette, *AIChE J.*, 1975, **21**, 1192.
[642] G. W. C. Hung, *Microchem. J.*, 1974, **19**, 130.
[643] C. H. Lee, *J. Appl. Polym. Sci.*, 1975, **19**, 3087.

# Membrane Phenomena 117

**Table 18** *Permeation of organic liquids through polyurethane film at 30 °C*

| Compound | Molar volume /ml molecule$^{-1}$ × 10$^{23}$ | Solubility in membrane /(mg of solvent) (mg of membrane)$^{-1}$ | Diffusion coefficient /cm$^2$ s$^{-1}$ × 10$^8$ |
| --- | --- | --- | --- |
| Octan-1-ol | 26.20 | 0.2984 | 0.497 |
| Heptan-1-ol | 23.47 | 0.3718 | 0.863 |
| Hexane | 21.16 | 0.0825 | 5.64 |
| Hexan-1-ol | 20.72 | 0.4097 | 1.42 |
| 3-Methylbutan-1-ol | 18.02 | 0.3561 | 1.40 |
| Toluene | 17.64 | 0.4874 | 14.3 |
| Chlorobenzene | 16.33 | 0.9016 | 13.7 |
| Benzoic acid | 16.02 | 0.2259 | 0.165 |
| Ethyl acetate | 15.70 | 0.4115 | 11.2 |
| Benzene | 14.75 | 0.5585 | 12.9 |
| Methyl acetate | 12.92 | 0.4378 | 11.7 |
| Acetone | 11.77 | 0.3851 | 14.3 |

membrane thickness and $D$ the diffusion coefficient derived from measurements of permeability $P$ and solubility $S$, was about 2 minutes. Time lag in diffusion was much longer than that in pervaporation. A study of permeation of pentylbutazone through a poly(dimethylsiloxane) membrane also gave a lower value for the diffusion coefficient measured by the time-lag method than that derived from the relation $D = P/S$.[644] Several solutes added to the outside solution had no effect on the permeability of phenylbutazone through poly(dimethylsiloxane) membrane.[645] Similarly, permeation studies have been carried out using a number of other permeant–membrane systems, such as (i) passage of oestrogen through polyethylene, silicone, or a composite of these two films,[646] (ii) passage of lipid-soluble substances through collodion membranes containing plant lecithin,[647] (iii) transport of benzene and cyclohexane through polymeric alloys of polyphosphonates and acetyl-cellulose,[648] (iv) passage of methoxychlor through crosslinked polyethyleneimine–toluene-2,4-di-isocyanate membranes,[649] and (v) of sugars through fibroin membrane.[650]

A method to predict the transport of steroids in polymers has been presented.[651] The approach is based on the application of Hildebrand's theory of solubility of microsolutes in ordinary solvents and of the Flory–Huggins theory to the solubility of steroids in polymers. Membrane permeability is correlated with steroid crystalline melting temperature, entropy of fusion of the steroid, and the computed solubility parameters of steroid and polymer.

Benzene–methanol permeation in asymmetric membranes has been studied to understand the effect of asymmetric membrane structure on transport.[652] These membrane systems play a useful role in the separation of liquid mixtures. Surfactant

[644] E. G. Lovering and D. B. Black, *J. Pharm. Sci.*, 1974, **63**, 1399.
[645] E. G. Lovering and D. B. Black, *J. Pharm. Sci.*, 1974, **63**, 671.
[646] R. Bloch, P. F. Kraicer, H. Binder, and E. Lobel, *J. Pharm. Sci.*, 1975, **64**, 832.
[647] M. Falk and W. Fuerst, *Pharmazie*, 1974, **29**, 715.
[648] I. Cabasso, J. Jagur-Grodzinski, and D. Vofsi, *J. Appl. Polym. Sci.*, 1974, **18**, 2117.
[649] R. Seevers and M. Deinzer, *J. Phys. Chem.*, 1976, **80**, 761.
[650] M. Sugiura and T. Shinbo, *Nippon Nogei Kagaku Kaishi*, 1974, **48**, 627.
[651] A. S. Michaels, P. S. L. Wong, R. Prather, and R. M. Gale, *AIChE J.*, 1975, **21**, 1073.
[652] H. P. Stormberg, *Ber. Bunsenges. Phys. Chem.*, 1975, **79**, 512.

liquid membranes have been proposed for the separation of hydrocarbon mixtures.[653] Selective permeation brought about the separation of the constituents of the mixture.

**Diffusion of Electrolyte and Non-electrolyte.**—Methacrylate gels have been used to follow the diffusion of antitumour drugs through them.[654] Transport of three *p*-aminobenzoate esters (ethyl, butyl, and hexyl) through a tubular poly(dimethylsiloxane) membrane into a flowing liquid has been studied.[655] The tubular configuration enabled the determination of the contribution that convective diffusion made to the total membrane transport. The observed transport behaviour ranged from complete convective diffusion control for the hexyl ester to complete membrane control for the ethyl ester. The butyl ester showed a change in the control with flow rate. In another study,[656] it was found that permeation of butamben through a dimethicone membrane was controlled by the diffusion layer.

Several strategies are employed to control the release of substances from certain depots or reservoirs. Membranes play a significant role in this process.[657] In the drug industry, several membrane matrices are used to hold the drug so that it can be released under controlled conditions. Several studies have been carried out in this direction. For example, it has been found that the solubility of the drug in the material of the matrix determines the rate of its release.[658] The kinetics of drug release from polymeric films of ethylcellulose doped with caffeine or salicylic acid have been followed.[659] The rates of release followed both the first-order equation [log (amount of drug retained) *versus* time] and diffusion-controlled models (drug release proportional to square root of time). In a number of cases, release of drug from the polymer matrix conformed to a diffusion-layer-control model. These kinetics can be changed to zero-order (rate is proportional to time) by laminating a second film that does not contain the drug to the releasing side of the film with the dispersed drug.[660] The composition of the vehicle, which determines the solubility of the drug, is very important in securing timed release of the drug. Consequently the influence of the polarity of the vehicle on drug release has been studied.[661]

Effects of pH of the medium and $pK_a$ of the drug (chlordiazepoxide) on its permeation through poly(dimethylsiloxane) membrane have been studied.[662] Similarly, the use of complexation in accelerating or retarding drug release has been explored. The effects of caffeine, $\beta$-cyclodextrin, and povidone on the permeation behaviour of butamben from saturated solutions in these complexing agents through a poly(dimethylsiloxane) membrane have been investigated.[663] In all the systems, the agents increased the rate of release over that of the plain saturated drug

---

[653] R. P. Cahn and N. N. Li, *J. Membr. Sci.*, 1976, **1**, 129.
[654] J. Drobnik, P. Spacek, and O. Wichterle, *J. Biomed. Mater. Res.*, 1974, **8**, 45.
[655] D. R. Flanagan and S. H. Yalkowsky, *J. Pharm. Sci.*, 1977, **66**, 337.
[656] K. G. Nelson and A. C. Shah, *J. Pharm. Sci.*, 1977, **66**, 137.
[657] R. W. Baker and H. K. Lonsdale, *Chemtech.*, 1975, **5**, 668.
[658] Y. W. Chen, H. J. Lambert, and T. K. Lin, *J. Pharm. Sci.*, 1975, **64**, 1643.
[659] M. Doubrow and M. Friedman, *J. Pharm. Sci.*, 1975, **64**, 76.
[660] S. Borodkin and F. E. Tucker, *J. Pharm. Sci.*, 1975, **64**, 1289.
[661] S. H. Yalkowsky and G. L. Flynn, *J. Pharm. Sci.*, 1974, **63**, 1277.
[662] E. G. Lovering, D. B. Black, and M. L. Rowe, *J. Pharm. Sci.*, 1974, **63**, 1224.
[663] M. Nakano, K. Juni, and T. Arita, *J. Pharm. Sci.*, 1976, **65**, 709.

solution. In order to obtain sustained release, systems must contain stable complexes. The effects of tablet excipients such as talc, polysorbate 80, fumed $SiO_2$, lactose, starch, *etc.*, on drug (diazepam) permeation through poly(dimethylsiloxane) membrane have been evaluated.[664] Talc, polysorbate, and $SiO_2$ decreased the permeation rate but increased turnover time, whereas the other compounds were without any effect.

In a number of studies, permeation of dyes has been discussed.[665] The role of organic solvent in the transport of a solute (an organic dye) through a swollen polymer membrane has been studied. The solute diffusion coefficient ($D$) was calculated from the measured permeability ($P$) and solubility ($S$). The main parameters of the solvent that controlled diffusion were viscosity of the solvent and the extent to which it swelled the membrane.[666] At a high degree of membrane swelling, the results conformed to equation (52). At a very low degree of swelling, the diffusion coefficient was independent of solvent viscosity. Tetrachloroethylene increased the diffusion of C.I. disperse dyes into poly(ethylene terephthalate) compared to diffusion from an aqueous dye bath.[667]

Membrane filters impregnated with glycerine esters [glycerine monoricinoleate (GMRO), glycerine dioleate (GDO), and glycerine trioleate (GTO)] have been used to determine the permeation coefficients of urea and ammonium chloride.[668] One side of the permeation apparatus contained equimolar urea and $NH_4Cl$ solution (each 1M) and the other side contained distilled water. The increase of urea and $NH_4Cl$ concentration in distilled water was followed enzymatically. The flow rate ($dn/dt$) was given by

$$dn/dt = AKD(\Delta C/d) \qquad (57)$$

The permeation coefficient $DK$ was evaluated as $(dn/dt) \cdot (d/A\Delta C)$.

The diffusion coefficient was evaluated by measuring the time lag ($t$) by extrapolation of the permeation *vs.* time plot to intersect the time axis. Thus

$$D = d^2/6t \qquad (58)$$

The results obtained are shown in Table 19. The values of $\sigma$ indicate that the selectivity increases with increase in the number of free hydroxy-groups of the glyceride.

A technique involving the use of a liquid surfactant membrane has been used to separate phenol from waste water.[669]

### 4 Phenomena due to an Applied Electric Field

**Electrical Conductance.**—Electromigration in membranes has been studied using several ion-exchange membranes.[670, 671] The experiments were done in the absence of external electrolyte from conditions of virtually dry to fully water-saturated

[664] E. G. Lovering, C. A. Mainville, and M. L. Rowe, *J. Pharm. Sci.*, 1976, **65**, 207.
[665] D. R. Paul and S. K. McSpadden, *J. Membr. Sci.*, 1976, **1**, 33.
[666] D. R. Paul, M. Garcin, and W. E. Garmon, *J. Appl. Polym. Sci.*, 1976, **20**, 609.
[667] Z. Morita, R. Kobayashi, K. Uchimura, and H. Motomura, *J. Appl. Polym. Sci.*, 1975, **19**, 1095.
[668] W. Volkel, C. Wandrey, and K. Schugerl, *Sep. Sci.*, 1977, **12**, 425.
[669] R. P. Cahn and N. N. Li, *Sep. Sci.*, 1974, **9**, 505.
[670] R. A. Wallace and J. P. Ampaya, *Desalination*, 1974, **14**, 121.
[671] R. A. Wallace, *J. Appl. Polym. Sci.*, 1974, **18**, 2855.

**Table 19** *Permeability and selectivity ($\sigma$) of urea and $NH_4Cl$ in glycerine ester membranes*

| | Membrane | | |
|---|---|---|---|
| | GMRO | GDO | GTO |
| Temperature/°C | 25 | 25 | 25 |
| $A$/cm$^2$ | 35.18 | 35.18 | 35.18 |
| $\Delta C$/mol cm$^{-3}$ | $10^{-3}$ | $10^{-3}$ | $10^{-3}$ |
| $d/\mu$m | 165 | 165 | 165 |
| $(\mathrm{d}n/\mathrm{d}t)$/mol s$^{-1}$ | | | |
| for: urea | $4.5 \times 10^{-9}$ | $7.5 \times 10^{-10}$ | $1.2 \times 10^{-10}$ |
| $NH_4Cl$ | $2.9 \times 10^{-10}$ | $1.1 \times 10^{-11}$ | $5.5 \times 10^{-11}$ |
| $DK$/cm$^2$ s$^{-1}$ | | | |
| for: urea | $2.1 \times 10^{-9}$ | $3.5 \times 10^{-10}$ | $5.4 \times 10^{-11}$ |
| $NH_4Cl$ | $1.4 \times 10^{-10}$ | $5.3 \times 10^{-11}$ | $2.6 \times 10^{-11}$ |
| $\sigma_a$ | 150 | 6.6 | 2.1 |
| $D$/cm$^2$ s$^{-1}$ | | | |
| for: urea[b] | $2.1 \times 10^{-8}$ | $0.4 \times 10^{-8}$ | $0.2 \times 10^{-8}$ |
| $NH_4Cl$[c] | $0.5 \times 10^{-8}$ | $0.2 \times 10^{-8}$ | $0.1 \times 10^{-8}$ |
| $K$ | | | |
| for: urea | $9.8 \times 10^{-2}$ | $9.0 \times 10^{-2}$ | $2.7 \times 10^{-2}$ |
| $NH_4Cl$ | $2.5 \times 10^{-2}$ | $3.1 \times 10^{-2}$ | $3.9 \times 10^{-2}$ |

(a) $\sigma$ is defined as $(DK$ for urea$)/(DK$ for $NH_4Cl)$; (b) Value for water is $1.4 \times 10^{-5}$; (c) Value for water is $1.8 \times 10^{-5}$

membrane. In both strong acid and strong base membranes, at low membrane water contents, the migration of $H^+$ and $OH^-$ ions was similar in nature to the movement of $Na^+$ and $Cl^-$ ions under an applied electric field. Above certain levels of membrane water absorption, $H^+$ and $OH^-$ ions were transferred by an additional fast transport mechanism.[670] The study of conductances and mobilities of the potassium ion in three different cation-exchange membranes (AMF C-60; Asahi DK-1; and Nepton CR-61, AZL 183) containing sulphonic acid groups revealed the influence of structural variables on the mobilities of $K^+$ ions.[671] Nepton CR-61 membrane, which is backed with Dynel, showed lower $K^+$ conductance compared to the other two membranes, which had no fabric support. At high water contents, the conductance of Asahi DK-1 membrane was about 1.5 times greater than that of AMF C-60 membrane. This was attributed to the presence of higher concentrations of $K^+$ ions in the membrane. Similarly, the study of ion movements in cellulose acetate membranes showed that, as the ion size increased, physical friction between the ion and the membrane pore wall became very important, particularly when the ionic radius came closer to intermolecular gaps between polymer chains.[672] Mobilities of $Na^+$ and $Cl^-$ ions $(2—3 \times 10^{-7}$ cm$^2$ s$^{-1}$ V$^{-1})$ were about 3—4 orders of magnitude less than those in aqueous solution. The electrochemical properties of mica membranes with pores of known geometry in contact with aqueous KCl solutions have been studied by conductance measurements.[673] The results have been discussed in terms of a model of the membrane with narrow pores.

[672] S. Horigome and Y. Taniguchi, *J. Appl. Polym. Sci.*, 1977, **21**, 343.
[673] B. Klump and D. Woermann, *Ber. Bunsenges. Phys. Chem.*, 1977, **81**, 92.

## Membrane Phenomena

Membranes based on poly-(2-hydroxyethyl methacrylate) crosslinked with ethylene dimethacrylate were modified by introducing weakly acidic (methacrylic acid) and/or weakly basic (diethylaminoethyl methacrylate) groups.[674] A study of the specific resistance of these membranes as a function of pH showed that maxima appeared in specific resistance $vs.$ pH curves, the maximum being at pH $\sim 3.5$ for the cation-exchange membrane, at pH $\sim 10$ for anion-exchange membrane, and at pH $\sim 6$ for electroneutral membranes.

A method of measurement of membrane resistance, suggested years ago by Lakshminarayanaiah,[675] in which electrodes on either side of the membrane can be moved so that the contribution of electrolyte outside the membrane may be eliminated by extrapolation, has been used to study the conductance of the membrane as a function of external electrolyte solution.[676] It has been shown that the membrane has a unique resistance ($R_m$) that is independent of the external electrolyte concentration below a critical concentration of electrolyte. The ratio of conductance of the membrane at a given electrolyte concentration to that at higher concentrations has been quantitatively used as a measure of the permselectivity of the membrane.

A cable-like model membrane system has been described and the cable parameters, particularly the space constant, $\lambda$, given by $(r_m/r_a)^{\frac{1}{2}}$, where $r_m$ is the membrane resistance ($\Omega$ cm) for radial current flow and $r_a$ is the resistance ($\Omega$ cm$^{-1}$) for axial current flow, has been evaluated by measurement of electrical potential as a function of distance from the site of injection of current.[677]

The complete matrix of phenomenological coefficients (*e.g.* membrane resistance, salt flow, electro-osmotic flow, *etc.*) appearing in the linear laws of thermodynamics of irreversible processes has been determined for a strong acid ion-exchange membrane in KCl solution.[678] A study of the polarization of hydrated films of lysozyme as a function of peak applied potential showed that, in the low-voltage region, where there was a small rise in polarization, the conduction was due to acidic water of crystallization in the film. The steep rise at higher voltages was attributed to protonic conduction arising from the dissociation of water.[679]

Polymeric membranes containing quaternary ammonium salts are used in the construction of ion-selective membrane electrodes (see the section on Membrane Electrodes). The mechanism of charge conduction in these membranes has been elucidated by measurements of their electrical conductance as a function of temperature and pressure.[680] Conduction activation volumes of 60—80 cm$^3$ mol$^{-1}$ and a linear dependence of log $\sigma'$ on $(1/\varepsilon')$, where $\sigma'$ is conductivity at 25 °C and $\varepsilon'$ is the effective dielectric constant, indicated an ionic mechanism of charge conduction. However, this mechanism, which usually requires low activation energy, was difficult to reconcile with the observed high values for conduction activation energy.

Polyacrylonitrile film containing perchlorate salts showed a glass transition temperature exhibiting low activation energy for conduction in the rubbery state

---

[674] J. Vacik and J. Kopecek, *J. Appl. Polym. Sci.*, 1975, **19**, 3029.
[675] N. Lakshminarayanaiah, *J. Polym. Sci.*, 1960, **46**, 529.
[676] F. L. Ramp, *Desalination*, 1975, **16**, 321.
[677] S. Chammas and D. Woermann, *Ber. Bunsenges. Phys. Chem.*, 1975, **79**, 622.
[678] G. Wiedner and D. Woermann, *Ber. Bunsenges. Phys. Chem.*, 1975, **79**, 868.
[679] R. H. Tredgold and R. C. Sproule, *J. Chem. Soc., Faraday Trans. 1*, 1976, **72**, 509.
[680] G. D. Carmack and H. Freiser, *Analyt. Chem.*, 1975, **47**, 2249.

and high activation energy for conduction in the glassy state.[681] Water-soluble and -insoluble films have been compared with respect to their water sorption, fine structure, and electrical properties.[682] Water-soluble films were prepared from cellulose nitrate–nitrite by dissolving bleached sulphite pulp in a solution of nitrogen dioxide in dimethylformamide (DMF) at room temperature and heating the solution at 70 °C for about 3 h. This film, in comparison to the other water-soluble film of poly(vinyl alcohol), showed higher crystallinity, lower moisture absorption in the region of 0—80% relative humidity, lower permittivity and a.c. conductance, and higher d.c. conductance. Of the two water-soluble films (cellophane and a DMF-based one prepared from a solution of cellulose in $NO_2$–DMF mixture and cast immediately after dissolution of cellulose), the DMF-cast film absorbed less moisture than the cellophane film.

The salt transport parameter ($L_e$), the water transport parameter ($\beta$), and the cation transference number ($\bar{t}_+$) in asymmetric cellulose membranes cured at 70, 85, and 93 °C have been measured, using solutions of alkali-metal chlorides.[683] The relative transport rates defined by the product $L_e \bar{t}_+$ followed the orders $K^+ > Rb^+ > Cs^+ > Na^+ > Li^+$ (for loose membranes, cured at 70 °C and for medium membranes cured at 85 °C) and $Rb^+ > K^+ > Cs^+ > Na^+ > Li^+$ (for tight membranes cured at 93 °C).

A cation-exchange membrane between two identical solutions of $Na^+$ and $Ca^{2+}$ ions that were originally equilibrated in them underwent a significant decrease in its equivalent fraction of calcium upon passage of an electric current.[684] The transport number of the two ions also showed a strong dependence on current density.

A synthetic calcium carrier or ionophore embedded in a PVC matrix gave a value of unity for the transport number of $Ca^{2+}$ when pure solutions of $CaCl_2$ ($10^{-3}M$) in the anode and $10^{-3}$M-KCl in the cathode were used and subject to electrodialysis. The value of the transport number did not change with the change in the dielectric constant of the membrane solvent.[685] However, when mixed solutions of $CaCl_2$ ($5 \times 10^{-4}M$) and NaCl ($5 \times 10^{-4}M$) in the anode and KCl ($10^{-3}M$) in the cathode were used, the transport number of $Ca^{2+}$ was reduced from 0.98 in the membrane solvent of high dielectric constant to 0.63 in that of low dielectric constant, showing strong preference to univalent ion.[685] Similarly, other synthetic carriers for the ions $K^+$, $Na^+$, and $Li^+$ and for an organic cation ($\alpha$-phenylethylammonium) trapped in PVC matrix membranes gave a value of unity for the transport number of the ion concerned.[686]

Cellophane-supported $BaSO_4$ membrane in the 'conditioned' state is impermeable to $Ba^{2+}$ and $SO_4^{2-}$ ions, while being permeable to $H^+$, $OH^-$, $K^+$, and $Cl^-$ ions. Impressed electric potentials 'decondition' the membrane, and removal of the electric field re-conditions the membrane.[687] These membranes showed rectification which was destroyed by the impressed electric field but restored on removal of the

---

[681] S. Reich and I. Michaeli, *J. Polym. Sci., Polym. Phys. Ed.*, 1975, **13**, 9.
[682] A. Venkateswaran and L. P. Clermont, *J. Appl. Polym. Sci.*, 1974, **18**, 133.
[683] W. G. Sunu and D. N. Bennion, *Ind. Eng. Chem., Fundam.*, 1977, **16**, 283.
[684] Y. Oren and A. Litan, *J. Phys. Chem.*, 1974, **78**, 1805.
[685] W. Simon, W. E. Morf, E. Pretsch, and P. Wuhrman, in 'Calcium Transport in Contraction and Secretion', ed. E. Carafoli, F. Clementi, W. Drawbikowski, and A. Margrech, North-Holland Publishing Co., Amsterdam, The Netherlands, 1975, p. 15.
[686] A. P. Thoma, A. Viviani-Nauer, S. Arvanitis, W. E. Morf, and W. Simon, *Anal. Chem.*, 1977, **49**, 1567.
[687] G. Bahr and P. Hirsch-Ayalon, *J. Membr. Biol.*, 1974, **15**, 405.

field. These effects were caused by the removal and re-adjustment of adsorbed ions. A composite membrane of two layers, one layer being selective ($\bar{t}_+$ or $\bar{t}_- = 1$) and the second layer being non-selective (a layer of unstirred electrolyte solution), has been treated as an experimental model to examine the effects of current-induced electrolyte accumulation and depletion on the electrical properties of such a two-layer membrane system.[688] At the limiting current density (i.e. the current density at which the electrolyte concentration at one surface of the selective layer reached zero), voltage fluctuations in the form of periodic spikes were observed. This phenomenon was not observed when current flow led to accumulation of electrolyte in the non-selective layer. Also, S-shaped and N-shaped current–voltage curves could be realized with such a composite membrane system. Other composite or asymmetric membranes in which cationic and anionic regions exist showed also asymmetric current–voltage characteristics.[689] The pH dependence of the resistance of an amphoteric membrane (cast sulphochloride polyethylene film treated with 1-amino-3-dimethylaminopropane) showed that there is loss of membrane conductance at pH 8.[690] Between pH 13 and 14, the resistance decreases dramatically, due to liberation of sulphonate groups. Similarly, between pH 2 and 5, the membrane conductance is high, due to primary amino-groups.

The impedance characteristics of some liquid membranes have been described. A bulk membrane of nitrobenzene formed between two supports showed two semicircles in the impedance plot at different frequencies.[691] This indicated two rate processes, the fast one characteristic of membrane resistance and capacitance and the slow one indicating interfacial transport. Similar characteristics were also observed in a nitrocellulose membrane impregnated with glyceryl α-mono-oleate in 100mM-KCl solution.[692]

The piezoelectric strain constant ($q$) for polymer films is usually defined as the polarization ($P$) produced by a unit magnitude of stress ($T$) under the condition that the electric field $E$ across the film is zero. That is

$$q = (\partial P/\partial T)_{E=0}$$

where $q$ is a complex quantity expressed as $q = q' - iq''$.[693]

An apparatus has been devised to measure the real and imaginary components for polymer films with a d.c. bias field. Electric-field-induced piezoelectricity has been observed in several polymers. The ratio of the piezoelectric constant to the d.c. bias field gave $(\varepsilon + k)/G$, where $\varepsilon$ is the dielectric constant, $k$ is the electrostriction constant, and $G$ is the elastic constant. The temperature dependence of the field-induced piezoelectricity gave combined information on the dielectric constant and the elastic properties of polymers. After heating to about 95 °C followed by cooling to room temperature under a constant d.c. bias on a poly(vinyl chloride) film, piezoelectricity was observed at zero electric field. This indicated production of a residual polarization in the film, and thus a charged film or electret is formed. Such electrets, which store energy, have been formed from membranes of polyelectro-

[688] R. C. Macdonald, *Biochim. Biophys. Acta*, 1976, **448**, 199.
[689] V. Enkelmann and G. Wegner, *J. Appl. Polym. Sci.*, 1977, **21**, 997.
[690] F. De Korosy, *Desalination*, 1975, **16**, 85.
[691] D. E. Mathis and R. P. Buck, *Anal. Chem.*, 1976, **48**, 2033.
[692] T. Yoshida, S. Ogura, and M. Okuyama, *Bull. Chem. Soc. Jpn.*, 1975, **48**, 2775.
[693] R. L. Zimmerman, C. Suchicital, and E. Fukada, *J. Appl. Polym. Sci.*, 1975, **19**, 1373.

lytes,[694] poly(vinylidene fluoride),[695,696] poly(methyl methacrylate),[697] and cellulose acetate.[698]

In the absence of isotope interaction, the flux ratio $f(=$ influx $\div$ outflux) has been shown to be given by[699]

$$f = \exp(X/RT) \tag{59}$$

where $X = -zF\Delta\psi$ and $z$ is the valency, $F$ the faraday constant, and $\Delta\psi$ the transmembrane electric potential.

The phenomenological relation $\bar{R} = X/J$, where $J$ is the net flow of anion in an anion-selective membrane, so that $\bar{t}_- = 1$. Electric current $I = zFJ$. The d.c. electrical resistance, $\bar{R}_{el}$, $(= -\Delta\psi/I)$ is related to $\bar{R}$ by

$$\bar{R} = -zF\Delta\psi/J = -(zF)^2\Delta\psi/I = (zF)^2\bar{R}_{el} \tag{60}$$

The exchange resistance, $\bar{R}_{ex}$, when both $\Delta C$ and $\Delta\psi$ are zero is given by

$$\bar{R}_{ex} = (-RT\Delta\rho/J^x)_{J=0} \tag{61}$$

where $J^x$ is tracer flux, $\Delta\rho$ is the difference in specific activity across the membrane, and $J$ is the net flux.

In the presence of isotope interaction, the flux ratio $f$ is given by[699]

$$f = \exp[(\bar{R}_{ex}/\bar{R})(X/RT)] \tag{62}$$

The relations shown in equations (59) and (62) have been tested by Essig and colleagues, using an ion-exchange membrane and several anionic species. The $\bar{R}_{el}$ of the anion-exchange membrane [porous collodion activated by poly(vinylbenzyltrimethylammonium chloride) and converted into the required anionic form] was measured by applying a potential of $\pm 10\,mV$ and measuring the current. From this measured value of $\bar{R}_{el}$, $\bar{R}$ was evaluated from equation (60). The exchange resistance $\bar{R}_{ex}$ was evaluated by measuring the isotope flux by using identical solutions ($\Delta C = 0$) in the absence of an electric field across the membrane according to equation (61). The labelled species used were $^{131}I^-$, $^{36}Cl^-$, $^{35}SO_4^{2-}$, and [$^{14}C$]-acetate.[700]

Influx was measured by keeping one side 'cold' and the other side 'hot', by using a radioactive solution, and applying a known electric field, $\Delta\psi$. Next, the outflux $\overrightarrow{J}$, as opposed to the influx $\overrightarrow{J}$, was measured by reversing the polarity of the electric field. When this was done, the cold side was emptied and refilled with the cold solution. Thus the flux ratio may be evaluated.[700] Some of the values realized for several solutions are given in Table 20. Theoretically, the values given in columns 4 and 5 should be unity. Only the values in column 5 are near unity, in accordance with equation (62), which contains the correction term for isotope interaction. The sulphate ion shows some disparity, which is more marked in the value given in

---

[694] I. F. Miller and J. Mayoral, *J. Phys. Chem.*, 1976, **80**, 1387.
[695] N. Murayama, *J. Polym. Sci., Polym. Phys. Ed.*, 1975, **13**, 929.
[696] N. Murayama, T. Oikawa, T. Katto, and K. Nakamura, *J. Polym. Sci., Polym. Phys. Ed.*, 1975, **13**, 1033.
[697] H. Solunov and T. Vassilev, *J. Polym. Sci., Polym. Phys. Ed.*, 1974, **12**, 1273.
[698] R. A. Wallace and R. J. Gable, *Polym. Eng. Sci.*, 1974, **14**, 92.
[699] O. Kedem and A. Essig, *J. Gen. Physiol.*, 1965, **48**, 1047.
[700] J. H. Li, R. C. DeSousa, and A. Essig, *J. Membr. Biol.*, 1974, **19**, 93.

Membrane Phenomena 125

**Table 20** *Flux ratios observed and calculated*

| Solutions | $(\bar{R}_{ex}/\bar{R})$ | $\Delta\psi/\text{mV}$ | $f/\exp(J\bar{R}/RT)$ | $f/\exp(J\bar{R}_{ex}/RT)$ |
|---|---|---|---|---|
| 0.03M-KCl | 0.71 | 41.8 | 0.72 | 1.05 |
| 0.01M-KCl | 0.76 | 25.0 | 0.88 | 1.00 |
| 0.01M-KI | 0.44 | 75.0 | 0.38 | 1.02 |
| 0.01M-KOAc | 0.56 | 50.0 | 0.78 | 0.98 |
| 0.01M-K$_2$SO$_4$ | 0.52 | 50.0 | 0.31 | 0.87 |

column 4 than that in column 5. The effects of $X$ on ion tracer flows and flux ratios were also studied, using heterogeneous membranes which contained parallel pathways of different intrinsic resistance.[701] The total resistance to net flow exceeded the tracer exchange resistance and the flux ratio was abnormal.

The passage of an electric current through a cation-exchange membrane containing 0.1M-NaCl solution on either side results generally in depletion of electrolyte at the membrane/solution interface on the anode side and accumulation on the cathode side. Laser interferometry has been applied to study the concentration gradients in these diffusion layers,[702, 703] which were non-linear. The thickness of these stagnant layers has been estimated to be 265 μm. In the case of the polarization induced by a concentration gradient, using a cellulose membrane, the thickness was 575 μm.[704]

A new technique for the observation and interpretation of concentration profiles in diffusion layers has been developed.[705] This consisted in the use of a wedge laser interferometer for the observation of concentration profiles near the membrane surface while an electric current (1.15 and 2.3 mA cm$^{-2}$) passed through the membrane/solution system. A concentration minimum developed when the current was reversed following pre-polarization periods of 5—60 s. The image of this minimum in relation to the surface of the membrane could be determined accurately, as it was not subject to optical deflection. From its trajectory, the salt concentration at the membrane surface at the moment of reversal of current could be determined. When 0.01M-KCl solution was used, this concentration was found to be of the order of $10^{-3}$—$10^{-4}$ mol l$^{-1}$.

A new technique of multiple-beam real-time holographic interferometry has been described.[706] By this technique, concentration gradients inside transparent or translucent ion-exchange membranes have been recorded by using a television optical system. The results showed that anion-exchange membranes in an electrodialysis set-up were completely polarized, while cation-exchange membranes appeared not to have reached their maximum polarization.

Noise originating from movement of ions through pores or channels existing in membranes has been considered by a number of investigators. Läuger[707] has presented a theoretical treatment for the discontinuous movement of ions using two

[701] J. H. Li and A. Essig, *Biochim. Biophys. Acta*, 1977, **465**, 421.
[702] D. Lerche and H. Wolf, *Z. Phys. Chem. (Leipzig)*, 1974, **255**, 126.
[703] D. Lerche, *J. Bioelectrochem. Bioenerg.*, 1975, **2**, 304.
[704] D. Lerche, *J. Membr. Biol.*, 1976, **27**, 193.
[705] C. Forgacs, J. Leibovitz, R. N. O'Brien, and K. S. Spiegler, *Electrochim. Acta*, 1975, **20**, 555.
[706] R. N. O'Brien, *Electrochim. Acta*, 1975, **20**, 447.
[707] P. Läuger, *Biochim. Biophys. Acta*, 1975, **413**, 1.

simple models; one is a pore with a single energy barrier and the other is also a pore but with a binding site separated from aqueous solutions by energy barriers on either side. White noise originates from the single-barrier pore with a current spectral density ($S_j$) that is proportional to the total rate of jumps over the barrier. Around the equilibrium potential of the permeable ion, the spectral density $S_j$ became identical with the spectral density of thermal (Johnson) noise. At higher voltages, $S_j$ approached the spectral density of a Schottky (shot) noise source. A similar behaviour was noted in the case of membrane pores with a double energy barrier at low frequency, $\omega$ ($\omega = 2\pi f$). At higher frequencies, a dispersion of spectral density near $\omega = 1/\tau$ occurred, where $\tau$ is the average lifetime of the occupied state of the pore. Similarly, mathematical treatments of $(1/f)$ ($f$ = frequency) noise generated by diffusion of current carriers (ions) in an electric field have been given by Neumcke[708] and Frehland.[709]

Experimentally, noise power spectra related to ion transport across an anion-exchange membrane in equilibrium with HCl or $H_3PO_4$ solutions have been established.[710] One type of spectrum gave a slope (plot of log power vs. log $f$) of $-1.5$ and the other type gave a slope of $-1$ at lower frequencies and nearly $-2$ at higher frequencies. The slope of $-1.5$ is considered characteristic of noise arising from diffusion[711] (also see Dorset and Fishman[712] and Frehland[713]).

**Electro-osmosis.**—Track-etched mica membranes have been used to study electrokinetic phenomena, particularly conduction and electro-osmosis, as a function of pore size and electrolyte (KCl) concentration.[714] Capillary pores of radius comparable to and even less than the solution Debye-length parameter at electrolyte concentrations $10^{-3}$—$10^{-4}$ mol $l^{-1}$ were used. To augment electrostatic effects, heparin, a negatively charged polyelectrolyte, was adsorbed onto the pore wall. Numerical solutions to steady-state double-layer and transport equations were obtained for both circular and elliptical pores. The observed variables (conductivity of pore liquid and electro-osmotic coefficient) were a function of $(r_0/L)$ and $r_0\sigma$, where $r_0$ is the radius of a circle of equivalent pore area, $L$ the Debye length, and $\sigma$ the charge density at the pore wall due to ion adsorption. The ratio of pore to bulk solution conductivity calculated from the classical diffuse double-layer model was the same for pores of both shapes, while the electro-osmotic coefficient was sensitive to pore shape, being nearly 20 to 30% larger for circular pores.

The charge on the pore wall was varied by adsorption of heparin, which increased the electrostatic effects. Wall charge, dependent on heparin adsorption, increased with electrolyte concentration and heparin concentration.

The values for charge density obtained from the data for conduction and electro-osmosis at corresponding physical conditions disagreed. This discrepancy was attributed to a basic defect in the classical theory for the double layer and related electrokinetic phenomena.

[708] B. Neumcke, *Biophys. Struct. Mechan.*, 1975, **1**, 295.
[709] E. Frehland, *Z. Naturforsch., Teil. B*, 1976, **31b**, 942.
[710] M. E. Green, *J. Phys. Chem.*, 1974, **78**, 761.
[711] M. E. Green, *J. Membr. Biol.*, 1976, **28**, 181; 1977, **32**, 197.
[712] D. L. Dorset and H. M. Fishman, *J. Membr. Biol.*, 1975, **21**, 291.
[713] E. Frehland, *J. Membr. Biol.*, 1977, **32**, 195.
[714] W. Koh and J. L. Anderson, *AIChE J.*, 1975, **21**, 1176.

## Membrane Phenomena

Nickel-hydroxide-supported cellulose membrane was found to be anion-selective in that the transport number of anions determined by measurements of the membrane potential was 0.98.[715] These membranes transported 4.5 moles of water per Faraday in chloride solutions (0.1N), 5.3 moles in perchlorate solution (0.1N), 4.9 moles in nitrate solution (0.1N), and 1.1 moles in sulphate solution (0.1N). When stronger (0.5N) sulphate solution was used, the character of the membrane changed, due to binding of sulphate in conferring negativity to the membrane. Accordingly, $\bar{t}_{Na}$ in 0.5N-$Na_2SO_4$ solution was 0.5, compared to a value of 0.12 in 0.5N-NaCl solution.

Water transport (electro-osmosis) through cation-exchange membranes in general takes place on application of an electric field from the anode to the cathode side, and the reverse occurs in anion-exchange membranes. In the case of a cationic montmorillonite membrane in equilibrium with a solution of NaCl in water on one side and dioxan on the other, it has been reported[716] that transport of water (6.5 mol per Faraday) occurred from cathode to anode.

In collodion membranes, little transport of water occurred when the applied electric field was 3.7 V or less; however, when this value was exceeded, mass flux occurred, following a non-linear exponential relation.[717] Similarly, the electro-osmosis of water, ethanol, methanol, water–ethanol, and water–methanol mixtures through Zeokarb 226 membrane was found to vary with potential difference in a non-linear manner.[718] Also, the electro-osmotic pressure showed a non-linear behaviour. The sign of these showed dependence on the composition of the mixture used in the measurements. The electro-osmotic flux was satisfactorily described by the non-linear phenomenological equation

$$J_v = L_{11}(\Delta P/T) + L_{12}(\Delta \phi/T) + \tfrac{1}{2}L_{122}(\Delta \phi/T)^2 \quad (63)$$

where $L$'s are the phenomenological coefficients and $\phi$ is the electric potential. When $\Delta \phi = 0$, equation (63) becomes

$$(J_v)_{\Delta \phi = 0} = L_{11}(\Delta P/T) \quad (64)$$

In all the cases studied, equation (64) was followed. When $\Delta P = 0$, equation (63) becomes

$$(J_v)_{\Delta P = 0} = L_{12}(\Delta \Phi/T) + \tfrac{1}{2}(L_{122}/T^2)\Delta \phi^2 \quad (65)$$

Accordingly, plots of $[(J_v)_{\Delta P = 0}/\Delta \phi]$ versus $\Delta \phi$ gave straight lines in all cases. Further, it has been shown that the non-linearity arose from changes in the characteristics of the electrical double layer. Similar types of studies have been reported for a Pyrex membrane in contact with water, methanol, and their mixtures.[719] Also, electrophoretic mobility measurements using suspensions of Pyrex glass in water–methanol mixtures have been reported.

The electro-osmotic flux in terms of the zeta potential, $\zeta$, can be written as

$$J_v = \left(\frac{nr^2 \zeta \varepsilon}{4\eta d}\right)\Delta \phi \quad (66)$$

---

[715] M. Takashita and N. Sato, *J. Electroanal. Chem. Interfacial Electrochem.*, 1975, **62**, 127.
[716] T. Mussini and P. Loughi, *Chim. Ind. (Milan)*, 1975, **57**, 97.
[717] R. C. Srivastava and A. K. Jain, *Indian J. Chem.*, 1975, **13**, 1306.
[718] R. P. Rastogi, K. Singh, and J. Singh, *J. Phys. Chem.*, 1975, **79**, 2574.
[719] K. Singh and J. Singh, *Colloid Polym. Sci.*, 1977, **255**, 379.

where $\varepsilon$ is the dielectric constant, $\eta$ is the viscosity coefficient, and $n$ is the number of pores of radius $r$ in the membrane of thickness $d$. Equating equation (66) to equation (65), assuming linearity of flux with $\Delta\phi$, gives

$$\frac{(J_v)_{\Delta P=0}}{\Delta\phi} = \frac{nr^2\zeta\varepsilon}{4\eta d} = L_{12}/T \qquad (67)$$

The electrophoretic mobility of particles that is induced by $\Delta\phi$, using electrodes positioned at distance $d'$, is given by

$$\left(\frac{V_e}{\Delta\phi}\right) = \frac{\varepsilon\zeta}{4\pi\eta d'} \qquad (68)$$

The assumption of equivalence of $\zeta$ in equation (67) and in equation (68) gives

$$\frac{L_{12}}{T} = \left(\frac{nr^2 V_e}{d\Delta\phi}\right)\pi d' \qquad (69)$$

Equating Poiseuille's law with equation (64) gives

$$(J_v)_{\Delta\phi=0} = \left(\frac{n\pi r^4}{8\eta d}\right)\Delta P = L_{11}(\Delta P/T) \qquad (70)$$

Dividing equation (70) by equation (69) gives

$$\frac{L_{11}}{L_{12}}\left(\frac{V_e}{\Delta\phi}\right) = \frac{r^2}{8\eta d'} \qquad (71)$$

The pore radius, $r$, can be determined provided the permeability data (hydraulic, electro-osmotic, and electrophoretic) are available. The data obtained for Pyrex glass are given in Table 21. In order to use equation (71), three independent measurements must be made. Some time ago[720] it was pointed out that only two measurements (hydraulic permeability and isotope permeability) are enough to derive a value for the pore radius of membranes. In addition, the limitations of equation (71) have been discussed.

Non-linear electro-osmotic flow in a Pyrex membrane system conformed to the equation

$$J_v = L_{11}(\Delta P/T) + L_{12}(\Delta\phi/T) + L_{122}(\Delta\phi/T)^2 + L_{121}(\Delta\phi\Delta P/T^2) + L_{111}(\Delta P/T)^2 \qquad (72)$$

**Table 21** *Pore radius, $r$, of Pyrex membrane calculated from equation (71)*

| % Water in water–methanol mixture | $(L_{11}/T) \times 10^3$/ cm$^5$ dyn$^{-1}$ s$^{-1}$ | $(L_{12}/T) \times 10^4$/ cm$^3$ s$^{-1}$ V$^{-1}$ | $(L_{122}/T^2) \times 10^7$/ cm$^3$ s$^{-1}$ V$^{-2}$ | $r \times 10^4$/cm |
|---|---|---|---|---|
| 0  | 6.5 | 2.0 | 2.1 | 12.3 |
| 5  | 5.3 | 2.8 | 0   | —    |
| 10 | 5.0 | 1.8 | 4.3 | 15.1 |
| 20 | 3.8 | 2.4 | 0.8 | 12.5 |
| 30 | 9.0 | 2.6 | 0.8 | 13.3 |
|    |     |     |     | Average: 13.3 |

---

[720] N. Lakshminarayanaiah, *J. Phys. Chem.*, 1969, **73**, 4428.

# Membrane Phenomena

When $\Delta P=0$, the slope of the plot of $(J_v/\Delta\phi)$ vs. $\Delta\phi$ gives $(L_{122}/T^2)$. These values are also shown in Table 21.

Under conditions when $J_v=0$, equation (72) becomes

$$L_{111}(\Delta P/\Delta\phi)^2 + L_{121}(\Delta P/\Delta\phi) + L_{122} = 0 \qquad (73)$$

since $-L_{11}\Delta P = +L_{12}\Delta\phi$.

As $L_{122}$ can be evaluated (see the foregoing), substitution of experimental values of $\Delta P$ and $\Delta\phi$ for the condition $J_v=0$ in the above equation (73) gives values for $L_{111}$ and $L_{121}$. The values so derived for the second-order coefficients for Pyrex glass membrane are given in Table 22.[721–723]

Electrokinetic parameters of cellulose acetate membranes have been measured by Srivastava and Jain.[724] Hydraulic permeability $L_{11}$ $[(J_v)_{\Delta\phi=0}=L_{11}\Delta P]$, electro-osmotic permeability $L_{12}$ $[(J_v)_{\Delta P=0}=L_{12}\Delta\phi]$, streaming potential $[(\Delta\phi/\Delta P)_{I=0}= -(L_{21}/L_{22})]$, and streaming current $[(I/\Delta P)_{\Delta\phi=0}=L_{21}]$ have been evaluated. The values for the phenomenological coefficients (see Table 23) derived were dependent on the direction of flow. This indicated that the membranes were anisotropic. In similar studies, membranes of sintered Pyrex glass have been used. The phenomenological coefficients for the several solvents used are given in Table 24. In all these cases, the coefficients $L_{11}$ and $L_{12}$ followed the linear relationships that are represented by

$$L_{11} = (L_{11})_M x_M + (L_{11})_W x_W$$

**Table 22** Values of second-order phenomenological coefficients for a Pyrex glass membrane

| Solvent | $L_{122}/$ cm$^3$ K$^2$ s$^{-1}$ V$^{-2}$ | $L_{121}/$ cm$^5$ K$^2$ dyn$^{-1}$ s$^{-1}$ V$^{-1}$ | $L_{111}/$ cm$^7$ K$^2$ dyn$^{-1}$ s$^{-1}$ | Ref |
|---|---|---|---|---|
| Water + methanol: | | | | |
| 25% Methanol | $8 \times 10^{-2}$ | $-3.3$ | 33 | 721 |
| 75% Methanol | $13.3 \times 10^{-2}$ | $-5.9$ | 65.8 | 721 |
| Acetonitrile | 0.21 | $-0.03$ | 0.001 | 722 |
| Methanol + acetonitrile: | | | | |
| 25% Methanol | 1.6 | $-0.004$ | 0.005 | 723 |
| 75% Methanol | 0.7 | $-0.002$ | 0.005 | 723 |

**Table 23** Dependence of phenomenological coefficients on the direction of flow in cellulose acetate membranes

| Coefficient | Units | Values for opposite directions of flow | |
|---|---|---|---|
| $L_{11}$ | cm$^5$ dyn$^{-1}$ s$^{-1}$ | $2.95 \times 10^{-10}$ | $1.69 \times 10^{-10}$ |
| $L_{12}$ | cm$^3$ A J$^{-1}$ | $1.3 \times 10^{-6}$ | $2.02 \times 10^{-6}$ |
| $L_{21}$ | cm$^3$ A J$^{-1}$ | $1.33 \times 10^{-6}$ | $2.14 \times 1^{-5}$ |
| $L_{22}$ | $\Omega^{-1}$ | $1.72 \times 10^{-5}$ | $1.72 \times 10^{-5}$ |

[721] R. L. Blokhra and T. C. Singhal, J. Electroanal. Chem. Interfacial Electrochem., 1974, 57, 19.
[722] R. L. Blokhra and M. L. Parmar, J. Colloid Interface Sci., 1975, 51, 214.
[723] R. L. Blokhra and M. L. Parmar, J. Electroanal. Chem. Interfacial Electrochem., 1975, 62, 373.
[724] R. C. Srivastava and A. P. Jain, J. Polym. Sci., Polym. Phys. Ed., 1975, 13, 1603.

**Table 24** *Phenomenological coefficients (for sintered Pyrex membrane)*

| Mass fractions, as appropriate[a] | $L_{11} \times 10^6/$ cm$^5$ dyn$^{-1}$ s$^{-1}$ | $L_{12} \times 10^4/$ cm$^3$ A J$^{-1}$ | $L_{21} \times 10^4/$ cm$^3$ A J$^{-1}$ | $L_{22} \times 10^6$ /$\Omega^{-1}$ | Ref. |
|---|---|---|---|---|---|
| Acetone–methanol mixture: | | | | | 725 |
| $x_M$ | | | | | |
| 0 | 9.3 | 3.1 | 3.3 | 7.1 | |
| 0.3 | 7.9 | 4.2 | 3.9 | 12.7 | |
| 0.7 | 6.3 | 5.6 | 5.2 | 25.1 | |
| 1.0 | 5.1 | 6.5 | 6.3 | 52.4 | |
| Acetone–water mixture: | | | | | 726 |
| $x_W$ | | | | | |
| 0 | 9.3 | 3.3 | 3.3 | 7.1 | |
| 0.3 | 6.9 | 5.2 | 5.3 | 9.5 | |
| 0.7 | 3.8 | 8.2 | 8.2 | 18.1 | |
| 1.0 | 1.4 | 10.2 | 10.4 | 50.0 | |
| Methanol–water mixture: | | | | | 727 |
| $x_M$ | | | | | |
| 0 | 1.4 | 10.5 | 11.0 | 55.0 | |
| 0.3 | 2.6 | 9.1 | 8.8 | 52.6 | |
| 0.7 | 4.1 | 7.7 | 7.8 | 40.0 | |
| 1.0 | 5.1 | 6.3 | 6.3 | 25.0 | |

(a) $x_M$ is the mass fraction of methanol and $x_W$ is the mass fraction of water.

where $x$'s are the mass fractions of the components. However, the dependence of $L_{22}$ on concentration was not linear. The data have been used to calculate the efficiencies ($\eta$) of electrokinetic energy conversion for both electro-osmosis ($\eta_e$) and streaming potential ($\eta_s$). These efficiencies are given by[725–727]

$$\eta_e = -(J\Delta P/I\Delta\phi) = -[(J\Delta P)/\{(\Delta\phi)^2/R\}] \tag{74}$$

$$\eta_s = -(I\Delta\phi/J\Delta P) = -[\{(\Delta\phi)^2/R\}/J\Delta P] \tag{75}$$

where $R$ is the electrical resistance of the system.

In an electro-osmosis experiment, when $\Delta P$ is equal to electro-osmotic pressure corresponding to a particular input $\Delta\phi$, the net water flux $J$ is zero. Thus it is seen from equation (74) that, for a fixed value of input force $\Delta\phi$, $\eta_e = 0$ when either $\Delta P = 0$ or $\Delta P$ equals electro-osmotic pressure. Consequently the $\eta_e$ vs. $\Delta P$ curve for a fixed value of $\Delta\phi$ would pass through a maximum as $\Delta P$ is varied from zero to the electro-osmotic pressure. Similar considerations apply to $\eta_s$. It has been shown that (i) $\eta_e$ would be maximum when $\Delta P$ is half the electro-osmotic pressure, (ii) values of $(\eta_e)_{max}$ were independent of input forces, and (iii) $(\eta_e)_{max} = (\eta_s)_{max}$, a consequence of the Onsager theorem.

The forces $\Delta P$ and $\Delta\phi$ in electro-osmotic phenomena are related to the flows $J_v$ and $I$ (employing the resistance formulation)[728] by the relations

$$\Delta P = R_{11}J_v + R_{12}I \tag{76}$$

$$\Delta\phi = R_{21}J_v + R_{22}I \tag{77}$$

[725] R. C. Srivastava and M. G. Abraham, *J. Colloid Interface Sci.*, 1976, **57**, 58.
[726] R. C. Srivastava and M. G. Abraham, *J. Chem. Soc., Faraday Trans. 1*, 1976, **72**, 2631.
[727] R. C. Srivastava, M. G. Abraham, and A. K. Jain, *J. Phys. Chem.*, 1977, **81**, 906.
[728] Ref. 68, p. 112.

where $R_{11}$ and $R_{22}$ are the straight coefficients, which are always positive, and $R_{12}$ and $R_{21}$ are the cross coefficients; the latter may be positive or negative, and obey the Onsager relation, and so $R_{12} = R_{21}$. These coefficients have been evaluated for the systems Pyrex sinter membrane plus acetone, methanol, or methyl ethyl ketone[729] by performing the appropriate experiments.

Measurement of $J_v$ as a function of $\Delta P$ when $I = 0$ gives

$$R_{11} = (\Delta P/J_v)_{I=0} \tag{78}$$

Similarly, $R_{22}$ may be evaluated by measuring the resistance of the membrane permeant system. Thus

$$R_{22} = (\Delta\phi/I)_{J_v=0} \tag{79}$$

When $J_v = 0$, equations (76) and (77) give

$$(\Delta P/\Delta\phi)_{J_v=0} = (R_{12}/R_{22}) \tag{80}$$

Similarly, when $I = 0$, one gets

$$(\Delta\phi/\Delta P) = (R_{21}/R_{11}) \tag{81}$$

Thus the coefficients $R_{12}$ and $R_{21}$ may be evaluated from the data on electro-osmotic pressure and streaming potential respectively. The values derived for the resistance coefficients for the three systems are given in Table 25, from which it is obvious that the cross coefficients obey the Onsager reciprocity relation.

Electro-osmotic properties of charged polymer membranes (AMF C-60, polystyrene-based and containing sulphonic acid groups; Nafion XR 170, perfluorosulphonic acid) have been described in terms of the binary electrolyte model.[730] Similarly, the transport properties of a quaternary ammonium anion-exchange membrane (AMF A-104) in chloride and iodide forms have been described. In these studies, both equilibrium and transport parameters of the membrane in chloride[731] and in iodide[732] forms have been measured. The data have been analysed from the standpoint of irreversible thermodynamics and values for the friction coefficients $(f_{ij})$ have been derived. These values, in 0.1M sodium halide solution, were as follows:

|  | $f_{14} \times 10^{-10}$ | $f_{13} \times 10^{-9}$ | $f_{31} \times 10^{-8}$ | $f_{34} \times 10^{-8}$ | $f_{43} \times 10^{-9}$ |
|---|---|---|---|---|---|
| Iodide form | 4.3 | 1.7 | 4.4 | 3.7 | 1.4 |
| Chloride form | 0.2 | 0.9 | 1.0 | 1.7 | 1.5 |

**Table 25** *Friction coefficients in the Pyrex membrane system*

| System | $R_{11} \times 10^{-6}/$ dyn s cm$^{-5}$ | $R_{22} \times 10^{-5}/\Omega$ | $R_{12} \times 10^{-7}/$ V s cm$^{-3}$ | $R_{21} \times 10^{-7}/$ V s cm$^{-3}$ |
|---|---|---|---|---|
| Pyrex/acetone | 1.0 | 5.0 | 5.3 | 5.4 |
| Pyrex/methanol | 1.4 | 1.3 | 2.5 | 2.6 |
| Pyrex/methyl ethyl ketone | 0.8 | 1.8 | 1.1 | 1.0 |

[729] M. L. Srivastava and R. Kumar, *Colloid Polym. Sci.*, 1977, **255**, 595.
[730] R. G. Cameron, I. G. Lyle, J. F. Walker, and R. Paterson, *Desalination*, 1975, **17**, 313.
[731] C. McCallum and R. Paterson, *J. Chem. Soc., Faraday Trans. 1*, 1974, **70**, 2113.
[732] C. McCallum and R. Paterson, *J. Chem. Soc., Faraday Trans. 1*, 1976, **72**, 323.

where 1 represents counter-ion, 3 water, and 4 the fixed groups in the membrane. The value of $f_{14}$ is high for the iodide form, indicating a large amount of friction between iodide and the fixed charge on the matrix. This is also reflected in the large selectivity that the membrane shows for the iodide over the chloride.[732] The specific conductance of the iodide form of the membrane was low ($0.24 \times 10^{-3} \, \Omega^{-1} \, cm^{-1}$) compared to the chloride form ($3.4 \times 10^{-3}$), and the water transport was 2.1 (iodide form) and 3.3 (chloride form) moles per Faraday.

**Electrodialysis.**—The preparation and the physicochemical properties of polyelectrolyte membranes for use in electrodialysis operation have been reviewed.[733-735]

Two serious problems related to efficient electrodialytic operation are membrane fouling and the effects related to polarization phenomena at the membrane interfaces. Considerable attention has been given to eliminating or rather minimizing the effects of these on the efficiency of the operation. A study of the effects of surfactants has shown that considerable adsorption of the surfactants on the membrane surface occurred. This slowed down the exchange process and finally the membrane was fouled.[736] It was found that cation-exchange membranes were more prone to fouling by surfactants than anion-exchange membranes. Several procedures have been suggested to eliminate or reduce this fouling. It was also found that an anion-exchange membrane that had been modified by sulphonating the surface of the base membrane composed of styrene–divinylbenzene (DVB) copolymer and poly(vinyl chloride) before its methylation and subsequent quaternization was resistant to fouling.[737] In pilot plant electrodialytic operation, the use of anionic membranes of low permselectivity in the place of conventional anion-exchangers eliminated the effects of fouling and scaling.[738] However, the use of these membranes required more electric power and higher flow rates. Consumption of power in electrodialytic experiments using interpolymer membranes of polyethylene, styrene, and DVB was found to depend on the salinity of waters, the power requirement being roughly one kilowatt hour for 1000 p.p.m. of salt removed from a litre of water.[739] The current efficiency was 75—95%.

A new method to reduce polarization in electrodialysis has been suggested.[740] This procedure required the use of ion-conducting spacers which led to improvements in the performance of electrodialysis stacks. Decreased resistance, increased current density, and reduced polarization were the advantages noted. Another procedure introduced to reduce polarization was to place ion-exchange resins between permselective membranes in an electrodialysis unit.[741] This procedure also reduced the electrical resistance of the stack and increased the current density.

Theoretical models to describe the steady-state transport of ions in an electrodialytic set-up have been presented. Grossman[742] has emphasized the effects of

---

[733] G. Richter, *Chem.-Ing.-Tech.*, 1975, **47**, 909.
[734] W. Pusch, *Chem.-Ing.-Tech.*, 1975, **47**, 914.
[735] K. Bunzl and B. Sansoni, *Chem.-Ing.-Tech.*, 1975, **47**, 925.
[736] C. McCallum and P. Meares, *Electrochim. Acta*, 1974, **19**, 537.
[737] K. Kusumoto and Y. Mizutani, *Desalination*, 1975, **17**, 111.
[738] E. Korngold, *Desalination*, 1974, **14**, 359.
[739] K. P. Govindan and P. K. Naryanan, *Indian J. Technol.*, 1975, **13**, 76.
[740] O. Kedem, *Desalination*, 1975, **16**, 105.
[741] E. Korngold, *Desalination*, 1975, **16**, 225.
[742] G. Grossman, *J. Phys. Chem.*, 1976, **80**, 1616.

water dissociation and related ion and electrolyte transport. Equations have been given to describe the distributions of potential and the concentrations of ionic species. Homogeneous anion- and cation-exchange membranes showed linear current–voltage characteristics, while bipolar membranes exhibited anisotropy with respect to current direction, showing current saturation in one direction. This saturation or plateau in the current–voltage curve may change to a region with rapid current increase with further increase in voltage. This is said to be due to a progressing participation of dissociated water in charge transfer. This splitting of water leads to considerable pH shifts in the layers near the membranes. A diffusion model for dissociation of water within an unstirred layer has been considered by Rubinstein,[743] who showed that the change in pH was dependent on voltage and independent of bulk concentration of salt, the latter being the determinant of current efficiency.

A bipolar membrane (a sandwich membrane composed of two distinct parts that are selective to ions of opposite charge), used in conjunction with other cation- and anion-selective membranes to form a cell assembly, has been shown to separate and concentrate hydrogen and hydroxyl ions from water molecules at high efficiency under the influence of an applied direct current.[744] The process can be represented by

$$MX + H_2O \rightleftharpoons HX + MOH$$
$$\text{salt} \qquad \text{acid} \quad \text{base}$$

and thus the electrically driven process generated acids and bases from their salts. An engineering and economic analysis of the process has been given.

Electrodialysis has been used in several separation processes. The separation of a mixture can be effectively planned if the transport characteristics of the constituents are known in advance. Accordingly, the transport numbers of several aliphatic and aromatic carboxylate ions across an anion-exchange membrane were determined on the basis of the transport numbers of co-ions across the membrane in electrodialysis.[745] Also, the specific conductivity and water content of the membrane in the particular carboxylate ionic form were measured. It was generally observed that lower aliphatic carboxylate ions, because they had high transport numbers even at high concentration, were able to separate from other ions with high current efficiency. However, ions with lower transport numbers, such as caprylate, caprate, and hydroxybenzoate, increased the resistance to current flow, and the system showed poor current efficiency. The role of several additives, such as complexing agents (*e.g.* thioxine[746] and valinomycin[747]), in increasing ion transport in dialysis and electrodialysis by reducing the surface resistance of liquid membranes has been described.

Electrodialytic procedures have been used: (i) to prepare acids and bases from salt solutions,[744, 748] (ii) to separate nickel and cobalt in the presence of EDTA[749] and

---

[743] I. Rubinstein, *J. Phys. Chem.*, 1977, **81**, 1431.
[744] K. Nagasubramanian, F. D. Chlanda, and K. Liu, *J. Membr. Sci.*, 1977, **2**, 109.
[745] R. Dohno, T. Azumi, and S. Takashima, *Desalination*, 1975, **16**, 55.
[746] V. N. Golubev and B. Purin, *Dokl. Akad. Nauk SSSR.*, 1975, **221**, 94.
[747] V. N. Golubev and B. Purin, *Biofizika*, 1975, **20**, 738.
[748] F. V. Rauzen, S. S. Dudnik, G. Z. Nefedova, M. N. Tereshchenko, and M. A. Zhukov, *Zh. Prikl. Khim. (Leningrad)*, 1974, **47**, 347.
[749] M. Labbe, J.-Cl. Fenyo, and E. Selegny, *Sep. Sci.*, 1975, **10**, 307.

rhenium from molybdenum, potassium, and calcium,[750] (iii) in the desalination of brackish waters having high silica content,[751] and (iv) to desalt brackish ground water by using a commercial multi-stage batch-system electrodialysis plant (the capacity was 200 m³ per day; raw water saline concentration was 6000 p.p.m.).[752] In order to make the electrodialytic operations efficient, effects of temperature have been explored. For these studies, new materials that can withstand high-temperature electrodialysis are required. New components, membranes, spacers, etc. which can withstand temperatures up to 80 °C have been developed and used in plant operations.[753]

**Oscillatory Phenomena.**—An algorithm for the computer simulation of transient concentration profiles and other electrical properties of charged membranes has been developed and described.[754] The algorithm is based on the use of Nernst–Planck and Poisson equations. Simulations of transient and steady-state properties and a.c. impedances are compared with analytical solutions. The stability condition is that no reduced mobility should exceed 0.5, otherwise oscillations would occur on simulation of concentrations.

Polycations, e.g. polylysine, and polyanions, e.g. poly(glutamic acid), may interact at an interface to give a structure in which a cationic surface could be separated from an anionic surface by a neutral polyampholyte zone. Such a structure has been shown to be capable of exhibiting electrical oscillations in an electric field.[755] Under voltage-clamp conditions, the current–voltage curves showed regions of negative resistance, typical of the initial inward current of the $i$–$V$ curve of the squid axon membrane. This region existed at the polarization conditions required for the appearance of electrical oscillations.

Studies of electric relaxation across the cell 0.1N-NaCl|membrane (cellulose acetate cured at 110 °C for 2 h)|0.1N-NH$_4$Cl, under voltage- and current-clamp conditions, gave oscillating response curves with more than one maximum or minimum.[756] A simple model of time-dependent changes has been used to explain the results.

## 5 Phenomena due to an Applied Pressure Gradient

The effect of pressure on the electrical conductivity of Dowex-1 in its Cl$^-$, Br$^-$, I$^-$, and NO$_3^-$ forms has been determined.[757] Pressures ranging up to 2000 atmospheres were employed. Assuming invariance of membrane geometry with pressure, the activation volume for conduction, $\Delta V^*$, is given by

$$\Delta V^* = (\partial \Delta G^*/\partial P)_T = RT(\partial \ln\sigma/\partial P)_T \tag{82}$$

---

[750] V. A. Zarinskii, L. V. Borisova, O. D. Prasolova, and A. N. Ermakov, *Zh. Prikl. Khim. (Leningrad)*, 1975, **48**, 646.
[751] G. Boari, C. Merli, R. Passino, and G. Tiravanti, *Desalination*, 1974, **15**, 3.
[752] K. Kusakari, F. Kawamata, N. Matsumoto, H. Saeki, and Y. Terada, *Desalination*, 1977, **21**, 45.
[753] F. B. Leitz, M. A. Accomazzo, and W. A. McRae, *Desalination*, 1974, **14**, 33.
[754] J. R. Sandifer and R. P. Buck, *J. Phys. Chem.*, 1975, **79**, 384.
[755] V. E. Shashoua, *Faraday Symp. Chem. Soc.*, 1975, **9**, 174.
[756] J. E. Anderson and W. Pusch, *J. Membr. Sci.*, 1977, **2**, 101.
[757] G. Carmack and H. Freiser, *Anal. Chem.*, 1977, **49**, 767.

# Membrane Phenomena

where

$$\sigma = \sigma_0 \exp(\Delta G^*/RT) \qquad (83)$$

$\Delta G^*$ is the molar free energy of activation for the conduction process, $\sigma$ is the observed volume conductivity, and $\sigma_0$ is dependent on geometric factors and the initial concentration of charged species in the membrane. Data from measurements of conductivity at high pressure gave linear plots of $\ln \sigma$ vs. $P$, with a negative slope, from which the values for $\Delta V^*$ given in Table 26 were calculated. Molar volumes, $V_m$, for the ions, calculated from $V_m = \frac{4}{3}\pi r^3 N$, where $r$ is the Pauling ionic radius and $N$ is Avogadro's number, are also given in Table 26. These values approximately follow the values derived from measurements of conductance. This is considered to show that the mechanism of conduction is ionic and not electronic, since the latter mechanism should yield negative values for $\Delta V^*$ due to increased orbital overlap between adjacent negative molecules during compression.

The theoretical treatments by Gardner et al.[758] (volume–salt–current flow formulation) and by Leitz[759] (water–salt flow formulation) of piezodialysis (which is the reverse of reverse osmosis) have been compared in a recent paper.[760] The critical requirement for successful piezodialysis is the availability of suitable membranes. The membranes generally used are mosaic ones containing patches of cation-passing and anion-passing resins, each resin having a high degree of coupling between mobile counter-ions and water. A latex polyelectrolyte membrane, cross-linked synthetic rubber film, has an anion-exchange resin containing a relatively continuous network of very loose cation-exchange resin. The preparation of such a membrane by three methods has been described.[761] Similarly, the method of preparation of a piezodialysis membrane based on styrene–butadiene copolymer has been described.[762]

The other pressure-induced process is pervaporation, which is being used increasingly for the separation of liquid mixtures. The mechanism of the pervaporation process involves dissolution of the liquid in the membrane, transport through the membrane, and evaporation of the permeant molecules from the downstream side of the membrane.

The effects of several factors such as temperature, pressure, and film thickness on

**Table 26** *Activation volume* ($\Delta V^*$) *for electrical conduction of Dowex-1 ion-exchange membrane in several anionic forms*

| Form of membrane | $\Delta V^*/\text{cm}^3 \text{ mol}^{-1}$ | Ionic radius/Å | Ionic molar volume, $V_m/\text{cm}^3 \text{ ml}^{-1}$ |
|---|---|---|---|
| $\text{Cl}^-$ | 19 | 1.81 | 15.0 |
| $\text{Br}^-$ | 22 | 1.95 | 18.7 |
| $\text{NO}_3^-$ | 23 | 1.93 | 18.1 |
| $\text{I}^-$ | 30 | 2.16 | 25.4 |

[758] C. R. Gardner, J. W. Weinstein, and S. R. Caplan, *Desalination*, 1973, **12**, 19.
[759] F. B. Leitz, *Desalination*, 1973, **13**, 373.
[760] J. N. Weinstein and F. B. Leitz, *Desalination*, 1975, **16**, 245.
[761] J. Shorr and F. B. Leitz, *Desalination*, 1974, **14**, 11.
[762] T. Yamabe, K. Umezawa, Sh. Yoshida, and N. Takai, *Desalination*, 1974, **15**, 127.

the fractionation of binary liquid mixtures by pervaporation through poly(tetrafluoroethylene) films grafted with $N$-vinylpyrrolidone or 4-vinylpyridine have been investigated,[763] and the results analysed.[764] Pervaporation rate was increased at constant selectivity when the liquid charge was at a higher temperature or the downstream vapour pressure or the film thickness was reduced. The rate showed little change when the liquid was subject to pressures higher than one atmosphere.[763] Liquid permeation occurred by a simple diffusion mechanism. The selectivity of the membrane to a particular liquid was not confined to the interfaces, but occurred along the diffusion paths through the membrane.[764]

Membranes of polymeric alloys of polyphosphonates and acetyl-cellulose are highly permeable to benzene and to cyclohexene but impermeable to the aliphatic hydrocarbons, cyclohexane, and decalin.[765] The pervaporation technique, when applied to the separation of the mixture benzene–cyclohexane, showed that the flux of the permeate increased with increase in (i) temperature, (ii) concentration of benzene in the feed solution, and (iii) fraction of polyphosphonate in the membrane. Similarly, poly(tetrafluoroethylene) films grafted with $N$-vinylpyrrolidone have been used to fractionate azeotropic mixtures.[766] The mass flux was brought about by maintaining the downstream side of the membrane at zero vapour pressure by using a vacuum pump.

Experiments have been performed to show that the important factor controlling transport of liquid through swollen polymeric membranes is the concentration gradient. The external force establishing this concentration gradient across the membrane had no role in controlling the rate of transport. So long as the activity gradient was kept constant across the membrane, hydraulic flux and pervaporation flux were equal.[767]

**Streaming Potential.**—In the measurement of electrical potentials arising across membranes when subject to a pressure gradient, two reversible electrodes, one on each side of the membrane, are used. When chloride solutions are used with the membrane, Ag/AgCl electrodes may be used. The potential responses of these electrodes have to be corrected for the effect of pressure.[768] The correction has been shown to be given by $(1/F)(\bar{V}_{AgCl} - \bar{V}_{Ag})\Delta P$, where the $\bar{V}$'s are partial molar volumes. This, for $\Delta P = 5$ decabar, gives a value of 0.8 mV {i.e. $[(25.8 - 10.28)/(9.65 \times 10^4)] \times (50 \times 10^6/10^7)$} which, although small, is not negligible.

Linear phenomenological equations can be written for the dependence of flow on forces when the same solution exists in the two compartments separated by a membrane. Thus

$$J_v = L_p \Delta P + L_{pe} \Delta E \quad (84)$$

$$I = L_{ep} \Delta P + L_e \Delta E \quad (85)$$

---

[763] P. Aptel, J. Cuny, J. Jozefonvicz, G. Morel, and J. Neel, *J. Appl. Polym. Sci.*, 1974, **18**, 351.
[764] P. Aptel, J. Cuny, J. Jozefonvicz, G. Morel, and J. Neel, *J. Appl. Polym. Sci.*, 1974, **18**, 365.
[765] I. Cabasso, J. Jagur-Grodzinski, and D. Vofsi, *J. Appl. Polym. Sci.*, 1974, **18**, 2137.
[766] P. Aptel, N. Challard, J. Cuny, and J. Neel, *J. Membr. Sci.*, 1976, **1**, 271.
[767] D. R. Paul and J. D. Paciotti, *J. Polym. Sci., Polym. Phys. Ed.*, 1975, **13**, 1201.
[768] K. S. Spiegler, *Desalination*, 1974, **15**, 135.

When $I=0$,

$$-(\Delta E/\Delta P)=(L_{ep}/L_e) \qquad (86)$$

Dividing equation (84) by equation (85) gives, for the condition when $\Delta P=0$, the relation

$$(J_v/I)_{\Delta P=0}=(L_{pe}/L_e) \qquad (87)$$

Since $L_{pe}=L_{ep}$, equations (86) and (87) become identical. Electro-osmotic permeability ($\beta$), however, is given by[769]

$$\beta=F\bar{X}L_pR \qquad (88)$$

where $\bar{X}$ is the concentration of charges in the membrane and $R$ is the membrane resistance, and is equal to $L_e$. Consequently

$$(\Delta E/\Delta P)_{I=0}=F\bar{X}L_pR \qquad (89)$$

Thus, by measuring streaming potential ($\Delta E/\Delta P$), $L_p$, and $R$, the value of $\bar{X}$ on the membrane may be evaluated. Such a procedure was used to derive values for $\bar{X}$ for asymmetric cellulose acetate membranes[770] cured at 83 °C and in 0.2M-NaCl solution. The measured values were $(\Delta E/\Delta P)=1.85$ mV atm$^{-1}$, $L_p=2.77\times 10^{-5}$ cm s$^{-1}$ atm$^{-1}$, and $R=325\Omega$ cm$^2$. Substitution of these values in equation (89) gave a value of $2.3\times 10^{-3}$ equivalents per litre of pore volume for $\bar{X}$. The ion-exchange capacity of the same membrane, as determined by titration, was $5\times 10^{-3}$ equivalents per litre of pore volume.

Bone shavings, packed into acrylic disks, were used in measurements of streaming potential, which was related to zeta potential by the equation[771]

$$\zeta=5.7\times 10^5(\Delta E/\Delta P)k$$

where $\Delta E$ is the streaming potential (mV), $k$ is the electrolyte conductance ($\Omega^{-1}$ cm$^{-1}$), and $\Delta P$ is the pressure (inches of water). Polyvalent cations such as Al$^{3+}$ and Fe$^{3+}$, when added to the buffered neutral electrolytes, had no effect on the bone zeta potential. Similarly, the streaming-potential method has been used to study the influence of reaction between a dye molecule and cotton cellulose on the zeta potential of cellulosic fibres.[772] Anionic dyes increased the negative zeta potential of poly(formaldehyde) fibres.[773] The increase in $\zeta$ was a function of concentration and basicity of the dye in the streaming solution. The addition of cationic surfactant brought about a decrease in the negative $\zeta$. Surface charge density ($\sigma$) on the fibres also increased with increase in the concentration of the dye.

**Hyperfiltration and Reverse Osmosis.**—These separation processes have assumed great technical importance, as evidenced by the several references already given in the Introduction.[4,5,7,24—28] In addition, two volumes of *Desalination Journal*[774,775]

---

[769] Ref. 68, p. 245.
[770] H. U. Demisch and W. Pusch, *J. Electrochem. Soc.*, 1976, **123**, 370.
[771] C. N. Wilson, A. D. Miller, and J. L. Nilles, *J. Biomed. Mater. Res.*, 1975, **9**, 265.
[772] H. T. Lokhande and A. S. Salvi, *Colloid Polym. Sci.*, 1976, **254**, 1030.
[773] H. T. Lokhande and A. S. Salvi, *J. Appl. Polym. Sci.*, 1977, **21**, 277.
[774] *Desalination*, 1976, Vol. 19.
[775] *Desalination*, 1977, Vol. 20.

are devoted to the publication of the Proceedings of the first desalination congress of the American continent, held in Mexico City in 1976. They contain articles related to hyperfiltration, reverse osmosis, and desalination on both experimental and pilot-plant scales. Volume 20[775] contains reviews of activities related to desalting on the American continent, in Europe and the Mediterranean, North Africa and the Gulf States, in the Pacific, and in Asia. Channabasappa[776] has reviewed the status of reverse osmosis as a desalination process used in the U.S.A. He has also pointed out the areas in which improvements have to be made to make the process of reverse osmosis cheap and useful.[777] There seems to be a great need for new and better membranes, despite all the research and efforts put in so far. A review of the development of selective membranes for use in reverse osmosis to desalt sea-water and to separate the constituents of a liquid mixture has been presented.[778] Similarly, reverse osmosis membranes for use in the separation of toxic compounds have been evaluated.[779] The principles and technology related to both reverse osmosis and hyperfiltration have been reviewed.[780] The results of a number of plant tests involving reverse osmosis have been published.[781]

Membranes used in several pressure-driven separation processes retain particles of different sizes, depending on the membrane structure (particularly its porosity). The pore size of the membrane and its ability to retain particles of several sizes are indicated in Figure 3.[782]

As the success of reverse osmosis depends upon the availability of useful membranes, considerable effort has been put into improving the characteristics of existing membranes and into developing new membranes. Several different compositions of casting solution, involving cellulose acetate, acetone, and aqueous magnesium perchlorate, have been used under different temperature conditions to prepare membranes suitable for hyperfiltration and reverse osmosis.[783,784] The data related to their separation characteristics have been presented. An apparatus for casting and testing of tubular cellulose acetate membranes has been described.[785] Also, an improved arrangement for the preparation of tubular cellulose acetate membranes for reverse osmosis and hyperfiltration has been given.[786] The application of light-scattering measurements to follow the formation of porous structure at the surface of cellulose acetate membranes has been described.[787]

Besides cellulose acetate, other polymeric materials have been tested for their suitability for use in the formation of membranes for reverse osmosis. Non-cellulosic asymmetric membranes (details not disclosed) with performance characteristics intermediate between that of a conventional reverse-osmosis membrane and an ultrafilter have been prepared.[788] These membranes gave high fluxes at low

[776] K. C. Channabasappa, *Desalination*, 1975, **17**, 31.
[777] K. C. Channabasappa, *Desalination*, 1976, **18**, 15.
[778] E. R. Becerro and A. Bellido, *Energ. Nucl. (Madrid)*, 1974, **18**, 223.
[779] E. S. K. Chian and H. H. P. Fang, *AIChE J.*, 1974, **70**, 497.
[780] R. Rautenbach and K. Rauch, *Chem.-Ing.-Tech.*, 1977, **49**, 223.
[781] *Desalination*, 1974, Vol. 14.
[782] M. C. Porter, *Chem. Eng. Prog.*, 1975, **71**, 55.
[783] B. Kunst and S. Sourirajan, *J. Appl. Polym. Sci.*, 1974, **18**, 3423.
[784] O. Kutowy and S. Sourirajan, *J. Appl. Polym. Sci.*, 1975, **19**, 1449.
[785] W. L. Thayer, L. Pageau, and S. Sourirajan, *J. Appl. Polym. Sci.*, 1974, **18**, 1891.
[786] W. L. Thayer, L. Pageau, and S. Sourirajan, *Desalination*, 1977, **21**, 209.
[787] B. Kunst, D. Skevin, G. Dezelic, and J. J. Petres, *J. Appl. Polym. Sci.*, 1976, **20**, 1339.
[788] S. B. Sachs, E. Zisner, and G. Herscovici, *Desalination*, 1976, **18**, 199.

# Membrane Phenomena 139

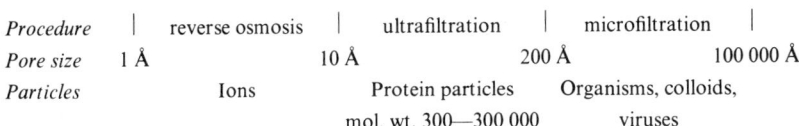

**Figure 3** *The range of sizes of pore membranes as related to the procedure for which membranes are used and the sizes of the entities that are retained or separated*

pressures and moderate rejection of various salts. Aromatic polyamide (Nomex, from DuPont) asymmetric membranes have been obtained by the Loeb–Sourirajan technique of membrane casting.[789] The rate of precipitation controls the porosity of membranes, low precipitation rates giving membranes with sponge-like structures and high precipitation rates giving membranes with larger pores. The former had high salt rejection and low water fluxes, whereas the latter had low salt rejection and high water fluxes. Similarly, asymmetric membranes from poly(*m*-phenyleneisophthalamide),[790] poly(piperazinamides),[791,792] polyacrylonitrile, polyamide, or polyimide,[793] aromatic polyamides,[794] polytetrafluoroethylene,[795] crosslinked polyethyleneimine,[796] polyurethane and polystyrene–polyisoprene,[797] and an aromatic polymer containing a benzimidazolone ring[798] have been prepared and tested. In addition, ion-exchange membranes from several materials [*e.g.* sulphonated aromatic polysulphone,[799] poly(styrenesulphonic acid)–poly(vinylidene fluoride),[800] membranes of polystyrene–polyisoprene copolymer containing sulphonate groups introduced by *N*-chlorosulphonyl isocyanate,[801] polypropylene cloth, coated with a mixture of styrene–DVB–polybutadiene–t-amyl alcohol and polymerized, subsequently chloromethylated and quaternized,[802] and other anion-exchange membranes that had been rendered microporous[803]] have been prepared and tested for the water-flux and salt-rejection properties. Also, the plasma polymerization method has been used to prepare composite reverse-osmosis membranes.[804,805]

The theory of reverse osmosis and other membrane-permeation processes has been considered by several investigators. The general permeation equations for various transport processes in membranes have been correlated by using the solution diffusion theory.[806] Under certain conditions and assumptions, the

---

[789] H. Strathmann, K. Kock, P. Amar, and R. W. Baker, *Desalination*, 1975, **16**, 179.
[790] H. D. Saier, H. Strathmann, and U. Von Mylius, *Angew. Makromol. Chem.*, 1974, **40–41**, 391.
[791] L. Credali, A. Chiolle, and P. Parrini, *Desalination*, 1974, **14**, 137.
[792] L. Credali, P. Parrini, E. Leonelli, and A. Chiolle, *Chim. Ind. (Milan)*, 1974, **56**, 19.
[793] H. D. Saier and H. Strathmann, *Chem.-Ing.-Tech.*, 1974, **46**, 109.
[794] R. Endoh, T. Tanaka, M. Kurihara, and K. Ikeda, *Desalination*, 1977, **21**, 35.
[795] S. Munari, F. Vigo, M. Nicchia, and P. Canepa, *J. Appl. Polym. Sci.*, 1976, **20**, 243.
[796] H. H. P. Fang and E. S. K. Chian, *J. Appl. Polym. Sci.*, 1976, **20**, 303.
[797] D. M. Koenhen, M. H. V. Mulder, and C. A. Smolders, *J. Appl. Polym. Sci.*, 1977, **21**, 199.
[798] S. Hara, K. Mori, Y. Taketani, T. Noma, and M. Seno, *Desalination*, 1977, **21**, 183.
[799] Cl. Brousse, R. Chapurlat, and J. P. Quentin, *Desalination*, 1976, **18**, 137.
[800] C. C. Gryte and H. P. Gregor, *J. Polym. Sci., Polym. Phys. Ed.*, 1976, **14**, 1839, 1855.
[801] P. M. Van der Velden and C. A. Smolders, *J. Appl. Polym. Sci.*, 1977, **21**, 1445.
[802] K. Kusumoto, H. Ihara, and Y. Mizutani, *J. Appl. Polym. Sci.*, 1976, **20**, 3207.
[803] Y. Mizutani, K. Kusumoto, M. Nishimura, and T. Nishimura, *J. Appl. Polym. Sci.*, 1975, **19**, 2537.
[804] H. Yasuda and H. C. Marsh, *J. Appl. Polym. Sci.*, 1975, **19**, 2981.
[805] H. Yasuda, H. C. Marsh, E. S. Brandt, and C. N. Reilley, *J. Appl. Polym. Sci.*, 1976, **20**, 543.
[806] C. H. Lee, *J. Appl. Polym. Sci.*, 1975, **19**, 83.

permeation properties for reverse osmosis can be obtained from those of pervaporation.

The relationships between concentration, solute mass, and membrane rejection have been derived.[807] Salt rejection, $R$, is given by

$$R = 1 - (C''/C') \quad (90)$$

where $C'$ is the concentration of salt in retentate and $C''$ is the concentration of salt in permeate. At any instant

$$M = CV \quad (91)$$

where $M$ is solute mass in retentate, $C$ is the solute concentration, and $V$ is the volume of retentate. Differentiation of equation (91) gives

$$d(M) = d(CV) = C_p \, dV \quad (92)$$

where $C_p$ is the solute concentration of permeate increment at any instant. Thus, one has

$$V \, dC + C \, dV = C_p \, dV$$

and on rearrangement this gives

$$-V \, dC = C \, R \, dV \quad (93)$$

Integration and rearrangement gives

$$R = [\ln(C'/C_0)/\ln(V_0/V')]$$
$$\text{i.e. } (C'/C_0) = (V'/V_0)^{-R} \quad (94)$$

As $V_0 = V' + V''$ ($V_0$ is the initial feed volume, $V'$ the final retentate volume, and $V''$ is final permeate volume), equation (94) becomes

$$(C'/C_0) = [1 - (V''/V_0)]^{-R} \quad (95)$$

Substituting from equation (91) gives

$$(M'/M_0) = (V'/V_0)[1 - (V''/V_0)]^{-R} = [1 - (V''/V_0)]^{1-R} \quad (96)$$

The behaviour of reverse osmosis and hyperfiltration membranes is generally described in terms of a pore model. A theory of hyperfiltration by charged micropores has been presented.[808] Assumptions of co-ion exclusion and flat concentration profiles in the membrane are made. Salt rejection by both charged and uncharged porous membranes has been considered. The effect of electrostatic screening on salt rejection has been explored.[809] A theory of salt rejection based on the distribution of electrolyte between cellulose acetate membrane (dielectric constant $\varepsilon_m$) and aqueous solution (dielectric constant $\varepsilon_0$) has been developed.[810] The

[807] M. Weintraub and S. Loeb, *Desalination*, 1975, **17**, 131.
[808] L. Dresner, *Desalination*, 1974, **15**, 109.
[809] L. Dresner, *Desalination*, 1974, **15**, 39.
[810] E. Glueckauf, *Desalination*, 1976, **18**, 155.

distribution of salt between membrane and external solution ($K^*$) has been shown to be given by[811]

$$K^* = \exp\left[-\left(\frac{e^2z^2(1-\alpha)Q}{2\varepsilon_0 kT(r+\alpha bQ)}\right)\right] \quad (97)$$

where $e$ is electronic charge, $z$ is valency, $Q=(\varepsilon_0-\varepsilon_m)/\varepsilon_0$, $\alpha=1-(1+L^2r^2)^{-\frac{1}{2}}$, $r$ is the membrane pore radius, $b$ is the ionic radius, $k$ is the Boltzmann constant, and $L$ is defined by

$$L=\left(\frac{4\pi e^2}{\varepsilon_0 kT}\sum n_i z_i^2\right)^{\frac{1}{2}} \quad (98)$$

Distribution coefficient data obtained for cellulose acetate membranes (35 μm thick, 12.5% water content) and the electrolytes NaCl, KCl, and $MgCl_2$ were found to agree with the predictions of equation (97).[812] The values used for the several parameters were: $\varepsilon_0 = 78.3$; $\varepsilon_m = 5.0$; $b = 0.95$ (Na), 1.33 (K), and 0.65 Å (Mg); $K^*$(experimental) = 0.25 (NaCl), 0.33 (KCl), and 0.036 ($MgCl_2$). The values for $r$ calculated from equation (98) were 2.0 (NaCl), 2.3 (KCl), and 2.4 Å ($MgCl_2$). Other investigators have used the pore model[813] and a modified diffusion model[814] to describe the selectivity of cellulose acetate membrane and the transport of salt and water through the membrane.

A friction model has been used by some investigators to describe the transport properties of the reverse-osmosis and hyperfiltration membranes.[815–817] A relation between salt rejection and the reflection coefficient has been described.[818] An algorithm to optimize the performance of cellulose acetate membranes has been presented.[819] For this study, three variables, i.e. (i) concentration of formamide in the casting solution, (ii) time of evaporation of membrane prior to gelling, and (iii) annealing temperature of the membrane, were chosen. The maximum flux predicted was 14.52 gallons per day per ft$^2$, at a level of 97% rejection of 0.5% NaCl solution at 600 p.s.i.g. These predictions were confirmed by experiment.

Several criteria of ion separation by reverse osmosis have been proposed. Fang and Chiang[820] found enthalpy of ion hydration to be the determining factor for effective ion separation. With high enthalpy of hydration, an increase in ion bulk prevented the ion from crossing the NS-100 (crosslinked polyethyleneimine) membrane. Similarly, it has been found that ion–water rather than ion–membrane interactions are the controlling factors in rejection of cations by cellulose acetate membranes.[821] The presence of a membrane-impermeable ion has been found to affect the rejection of membrane-permeable co-ions in neutral and cation- and anion-exchange membranes.[822] Membrane porosity is shown to affect concentra-

[811] E. Glueckauf, *Proc. Int. Symp. Water Desalination, Washington, D. C.*, 1967, **1**, 143.
[812] Y. Taniguchi and S. Harigome, *Desalination*, 1975, **16**, 395.
[813] A. Yiyan, *Desalination*, 1975, **17**, 367.
[814] P. Bo and V. Stannett, *Desalination*, 1976, **18**, 113.
[815] G. Jonsson and C. E. Boesen, *Desalination*, 1975, **17**, 145.
[816] G. Belfort and J. Scherfig, *Desalination*, 1976, **18**, 43.
[817] G. Belfort, *Desalination*, 1976, **18**, 259.
[818] W. Pusch and R. Riley, *Desalination*, 1974, **14**, 389.
[819] E. S. K. Chian and H. H. P. Fang, *J. Appl. Polym. Sci.*, 1975, **19**, 251.
[820] H. H. P. Fang and E. S. K. Chian, *J. Appl. Polym. Sci.*, 1975, **19**, 2889.
[821] H. K. Johnston, *Desalination*, 1975, **16**, 205.
[822] H. K. Lonsdale, W. Pusch, and A. Walch, *J. Chem. Soc., Faraday Trans. 1*, 1975, **71**, 501.

tion polarization on the high-pressure side of the membrane.[823] In the case of glycol chitosan–chondroitin sulphate C and glycol chitosan–heparin complex membranes, both salt rejection and flux were dependent on the composition of membranes and the pH of the feed solutions.[824] Steric resistance for the solute flow through the membrane and relative non-polarity and anion-exchanger characteristics of the NS-100 membrane seem to control separation of alcohols, amines, and aliphatic acids in aqueous solution.[825] Rejection of solute (inorganic ions) by ion-exchange membranes has been explained on the basis of the Donnan ion-exclusion principle.[826] Desalination efficiency by reverse osmosis has been shown to depend on the homogeneity of the distribution of water in polymer membranes.[827] In cellulose acetate membranes, extensive data indicate that salt rejection is strongly influenced by the distribution of solute between the membrane and the solution phases.[828] The distribution of solute is, in turn, influenced by temperature, pH, and ion pairing. Since cellulose acetate membranes exhibit swelling and shrinkage on exposure to certain aqueous solutions, these factors also influence salt rejection. At low concentrations, charges on the membrane also affect salt rejection.[829]

The effects of particle deposition[830] and of mineral fillers[831] to improve the resistance of membranes to compaction have been investigated. Also, chemical interactions in membrane separation experiments have been studied by using surfactant solutions.[832] Ultrafiltration measurements with uncharged polyacrylonitrile membranes showed that there was a difference in the dependence of flux and salt rejection on the concentration of sodium dodecyl sulphate (SDS) both below and above the critical micellar concentration (CMC). The difference was attributed to the formation of micelles above the CMC, when rejection was increased. Cation-exchange membranes also showed increased rejection at concentrations of SDS below the CMC. This was also attributed to the formation of micelles at the interface due to the participation of ionic groups in the membrane which acted like an inert electrolyte, lowering the CMC.

In swollen ionic membranes, fluxes show an initial decrease, while rejections show an increase. These phenomena are referred to as compaction. A model is proposed in which the loosely bound water of hydration is initially removed under the influence of pressure and salt concentration.[833] The removal of water has been described by an exponential decrease of the thickness of the wet membrane.

The solute transport parameter, $(D_m/k\delta)/\text{cm s}^{-1}$, in reverse osmosis experiments using aromatic polyamide membranes and NaCl feed solutions (0.0013—1.051 mol $\text{kg}^{-1}$) has been determined at several operating pressures.[834] The transport parameter increased with increasing operating pressure and solute concentration. The

---

[823] B. Kunst, G. Arneri, and Z. Vajnaht, *Desalination*, 1975, **16**, 169.
[824] A. Nakajima and K. Shinoda, *J. Appl. Polym. Sci.*, 1977, **21**, 1249.
[825] H. H. P. Fang and E. S. K. Chian, *J. Appl. Polym. Sci.*, 1975, **19**, 1347.
[826] D. Bhattacharyya, J. M. McCarlhy, and R. B. Grieves, *AIChE J.*, 1974, **20**, 1206.
[827] H. Strathmann and A. S. Michads, *Desalination*, 1977, **21**, 195.
[828] M. E. Heyde, C. R. Peters, and J. E. Anderson, *J. Colloid Interface Sci.*, 1975, **50**, 467.
[829] M. Bender, J. K. Moon, J. Stine, A. Fried, R. Klein, and R. Bonjouklian, *J. Chem. Soc., Faraday Trans. 1*, 1975, **71**, 491.
[830] C. C. Hung and C. Tien, *Desalination*, 1976, **18**, 173.
[831] I. Goossens and A. Van Haute, *Desalination*, 1976, **18**, 203.
[832] P. M. Van der Velden and C. A. Smolders, *J. Colloid Interface Sci.*, 1977, **61**, 446.
[833] P. M. Van der Velden and C. A. Smolders, *J. Appl. Polym. Sci.*, 1976, **20**, 1153.
[834] J. M. Dickson, T. Matsuura, P. Blais, and S. Sourirajan, *J. Appl. Polym. Sci.*, 1976, **20**, 1491.

## Membrane Phenomena

solute transport parameter ($D_m/k\delta$) for sodium, determined for an American standard type tubular module, was found to be $1.38 \times 10^{-5}$ cm s$^{-1}$.[835] Elimination of concentration polarization in ultrafiltration of a latex suspension by use of Kenics static mixer in a tubular membrane gave significant improvement in the flux of pure water.[836] Polarization effects, *i.e.* salt concentration profiles, in a reverse osmosis system have been measured with a Mach–Zehnder interferometer.[837]

Plasma-polymerized membranes of allylamine showed high rejection of NaCl and a high water flux.[838] The mobilities (defined as salt diffusion parameter × the transference number) of several ions determined during reverse osmosis experiments using cellulose acetate membranes cured at 85 °C followed the order $K^+ > Rb^+ > Cs^+ > Na^+ > Li^+$ and $Be^{2+} > Sr^{2+} > Ca^{2+} > Ba^{2+} > Mg^{2+}$.[839] The mobility of $Be^{2+}$ was exceptionally large, nearly the same as for the alkali-metal ions, while all other doubly charged ions had mobilities considerably smaller than for singly charged ions. In several reverse osmosis studies, flux and rejection data have been given for several membranes, including dynamically formed membranes and salt solutions.[840–846]

Sourirajan and co-workers, in a series of studies, have directed their attention to formulate and interrelate parameters characteristic of the membrane and the chemical and physical properties of the permeating species, with the help of which predictions regarding the separation of the constituents of an aqueous solution by reverse osmosis or ultrafiltration experiments could be made. To do this, extensive data must be compiled, and so the following systems have been studied:

(i) Reverse osmosis experiments using porous cellulose acetate membranes to separate single and mixed solutions of alcohols.[847]

(ii) Separation of amino-acids in aqueous solutions.[848]

(iii) Reverse osmosis separations of some alcohols and phenols[849] and of organic and inorganic solutes[850] in aqueous solutions by using aromatic polyamide membranes.

(iv) Separation of several inorganic ions in aqueous solution using porous cellulose acetate membranes.[851, 852]

(v) Separation of aldehydes, ketones, and ethers in aqueous solution.[853]

---

[835] H. Ohya and T. Moriyama, *Desalination*, 1975, **16**, 235.
[836] A. L. Copas and S. Middleman, *Ind. Eng. Chem., Process Des. Dev.*, 1974, **13**, 143.
[837] A. R. Johnson, *AIChE J.*, 1974, **20**, 966.
[838] A. T. Bell, T. Wydeven, and C. C. Johnson, *J. Appl. Polym. Sci.*, 1975, **19**, 1911.
[839] K. M. Choi and D. N. Bennion, *Ind. Eng. Chem., Fundam.*, 1975, **14**, 296.
[840] R. McKinney, Jr., *Sep. Purif. Methods*, 1974, **3**, 87.
[841] P. Canepa, S. Munari, M. Nicchi, and F. Vigo, *Chim. Ind. (Milan)*, 1974, **56**, 535.
[842] F. Alfani and E. Drioli, *Chem. Eng. Sci.*, 1974, **29**, 2197.
[843] R. E. Minturn, J. S. Johnson, Jr., W. M. Schofield, and D. K. Todd, *Water Res.*, 1974, **8**, 921.
[844] I. K. Bansal and A. J. Wiley, *Tappi*, 1974, **57**, 129.
[845] J. W. Van Heuven and R. K. Bloebaum, *Desalination*, 1974, **14**, 229.
[846] E. Drioli, H. K. Lonsdale, and W. Pusch, *J. Colloid Interface Sci.*, 1975, **51**, 355.
[847] T. Matsuura, M. E. Bednas, and S. Sourirajan, *J. Appl. Polym. Sci.*, 1974, **18**, 567.
[848] T. Matsuura and S. Sourirajan, *J. Appl. Polym. Sci.*, 1974, **18**, 3593.
[849] T. Matsuura, P. Blais, J. M. Dickson, and S. Sourirajan, *J. Appl. Polym. Sci.*, 1974, **18**, 3671.
[850] J. M. Dickson, T. Matsuura, P. Blais, and S. Sourirajan, *J. Appl. Polym. Sci.*, 1975, **19**, 801.
[851] T. Matsuura, L. Pageau, and A. Sourirajan, *J. Appl. Polym. Sci.*, 1975, **19**, 179.
[852] R. Rangarajan, T. Matsuura, E. C. Goodhue, and S. Sourirajan, *Ind. Eng. Chem., Process Des. Dev.*, 1976, **15**, 529.
[853] T. Matsuura, M. E. Bednas, J. M. Dickson, and S. Sourirajan, *J. Appl. Polym. Sci.*, 1975, **19**, 2473.

(vi) Separation of higher alcohols in aqueous solution by using cellulose acetate membranes.[854]

Polar and steric effects in the separation by reverse osmosis of alcohols, aldehydes, ketones, and non-cyclic ethers in aqueous solution, involving single-solute systems and porous cellulose acetate membranes, have been discussed.[855]

A variation in the procedure to desalt sea water has been proposed. Instead of applying mechanical pressure directly, a hypertonic glucose solution is used as the energy source, with hollow fibres.[856]

Reverse osmosis and/or ultrafiltration have been used in a number of separation or purification processes, listed below.

(a) Kraft pulp-mill and bleach-plant effluents have been purified by using a dual-layer zirconium dioxide–poly(acrylic acid) membrane.[857]

(b) Pollution has been reduced by treatment of waste water from the pulp and paper industry.[858]

(c) $D_2O$ has been separated from water.[859]

(d) Water has been separated from 1,4-dioxan by selective permeation through a Nylon-6 membrane.[860]

(e) Amino-acids in a mixture have been separated.[861]

(f) The concentration of pure electrolyte solutions and the separation of salts have been effected by using strong- and weak-acid ion-exchange membranes.[862]

(g) Ultrafiltration of non-ionic surfactants and inorganic salts from complex aqueous suspensions has been carried out, using non-cellulosic membranes.[863]

(h) Calcium-base sulphite liquor has been fractionated.[864]

(i) The separation of heavy-metal salts and the treatment of electroplating rinse waters for recovery of metals and water re-use have been carried out, using a negatively charged membrane.[865]

(j) Zinc[866] and uranium[867] have been separated from dilute solutions.

(k) Ultrafiltration of protein–saline solution (removal of toxic substances) and whole blood have been carried out.[868]

**Hydrodynamic Flow.**—The hydraulic permeation of toluene–cyclohexanone and iso-octane–carbon tetrachloride mixtures through a rubber membrane gave no separation of components. Treating the mixture as a single component, values for diffusion coefficient were derived from the solution diffusion theory.[869] The diffu-

[854] T. Matsuura, A. G. Baxter, and S. Sourirajan, *Ind. Eng. Chem., Process Des. Dev.*, 1977, **16**, 827.
[855] T. Matsuura, M. E. Bednas, J. M. Dickson, and S. Sourirajan, *J. Appl. Polym. Sci.*, 1974, **18**, 2829.
[856] R. E. Kravath and J. A. Davis, *Desalination*, 1975, **16**, 151.
[857] J. S. Johnson, Jr., R. E. Minturn, and G. E. Moore, *Tappi*, 1974, **57**, 134.
[858] M. Pichon, E. Muratore, and P. Monzie, *Atip*, 1974, **28**, 9.
[859] N. Kaneko and Y. Yamamoto, *Kagaku Kogaku*, 1974, **38**, 756.
[860] R. W. Tock, J. Y. Choung, and R. L. Cook, *Sep. Sci.*, 1974, **9**, 361.
[861] C. Dennison, *Anal. Biochem.*, 1974, **58**, 646.
[862] F. Wolf and O. Kononowa, *Acta Hydrochim. Hydrobiol.*, 1975, **3**, 289, 295.
[863] D. Bhattacharyya, K. A. Garrison, A. B. Jumawan, Jr., and R. B. Grieves, *AIChE J.*, 1975, **21**, 1057.
[864] I. K. Bansal and A. J. Wiley, *Tappi*, 1975, **58**, 125.
[865] D. Bhattacharyya, D. P. Schaaf, and R. B. Grieves, *Can. J. Chem. Eng.*, 1976, **54**, 185.
[866] V. S. Sastri, *Sep. Sci.*, 1977, **12**, 257.
[867] V. S. Sastri and A. W. Ashbrook, *Sep. Sci.*, 1976, **11**, 361.
[868] W. J. Huffman, R. M. Ward, and R. C. Harshman, *Ind. Eng. Chem., Process Des Dev.*, 1975, **14**, 166.
[869] D. R. Paul, J. P. Paciotti, and O. M. Ebra-Lima, *J. Appl. Polym. Sci.*, 1975, **19**, 1837.

# Membrane Phenomena

sion coefficients were governed by frictional forces, which had a hydrodynamic origin.

A membrane wedge interferometer has been developed for use in studies to obtain steady-state transport data for liquid–membrane–liquid systems.[870] Several criteria are given for the successful use of the Raleigh interferometer in quantitative transport studies.[871]

Liquid permeation, induced under mechanical pressure, has been used in hydrocarbon separations using membranes of nitrile rubber.[872] The vapour-permeation technique has been investigated in attempts to separate benzene and cyclohexane by using solvent-modified vinylidene fluoride films.[873]

An apparatus for determining the hydraulic permeability ($L_p$) and the reflection coefficient ($\sigma$) of synthetic membranes, using pressures of 1—100 atmospheres, has been described.[874] Similarly, another apparatus for non-steady-state characterization of membrane transport has been constructed.[875] With the help of this equipment, the data required have been obtained in a single experiment for the cellophane–NaCl solution system to derive values for $L_p$, $\sigma$, and $\omega$.

The transport coefficients $L_p$, $L_\pi$ (osmotic permeability), $L_{\pi p}$ (coupling coefficient), and $\sigma$ have been measured for cellulose acetate membranes by using several solutes.[876, 877] With heteroporous cellulose acetate and other commercially available membranes, it has been demonstrated[878] that an asymmetry of transmembrane tracer solute flow can result from the interaction or drag that the solute (called the driver) exerts on the tracer solute (driven). The same coupling of solute flux in a homopore membrane (polycarbonate membranes that have been irradiated and track-etched) has also been demonstrated.[879] The characteristics of the track-etched homopore membranes are given in Table 27.

The basic flux equation for two electrically neutral solutes can be written as[879]

$$J_i = C_{i(\text{av})}(1 - \sigma_i)J_v + P_{ii}\Delta C_i + P_{ij}\Delta C_j \qquad (99)$$

where $J_i$ is the flux of solute $i$, and is written as the sum of volume flow ($J_v$), diffusional flow, and solute interaction; $C_{i(\text{av})}$ is the mean concentration of solute, given by $C_{i(\text{av})} = (C_{i(1)} + C_{i(2)})/2$, where (1) and (2) represent the two sides; $\sigma_i$ is the

**Table 27** *Characteristics of track-etched homopore membranes*

| | |
|---|---|
| Pore radius/Å | 150 |
| Pore density/cm$^{-2}$ | $6.2 \times 10^8$ (from electron microscopy) |
| Pore length/cm | $6 \times 10^{-4}$ |
| $L_p$/cm$^3$ dyn$^{-1}$ s$^{-1}$ | $0.39 \times 10^{-9}$ |
| $\sigma_{\text{sucrose}}$ | 0.0016 |

[870] S. Min, J. L. Duda, R. H. Notter, and J. S. Vrentas, *AIChE J.*, 1976, **22**, 175.
[871] P. H. Bollenbeck and W. F. Ramirez, *Ind. Eng. Chem., Fundam.*, 1974, **13**, 385.
[872] J. T. Brun, G. Bulvestre, A. Kergreis, and M. Guillou, *J. Appl. Polym. Sci.*, 1974, **18**, 1663.
[873] F. P. McCandless, D. P. Alzheimer, and R. B. Hartman, *Ind. Eng. Chem., Process Des. Dev.*, 1974, **13**, 310.
[874] W. Pusch and H. J. Wolff, *Rev. Sci. Instrum.*, 1974, **45**, 1403.
[875] A. Zelman, R. Tankersley, A. Ford, H. Wayt, and A. Schindler, *J. Electrochem. Soc.*, 1976, **123**, 1015.
[876] W. Pusch, *Desalination*, 1975, **16**, 65.
[877] G. Boari, C. Merti, G. Mossa, and R. Passino, *Desalination*, 1975, **16**, 271.
[878] W. R. Galey and J. T. Van Bruggen, *J. Gen. Physiol.*, 1970, **55**, 220.
[879] J. T. Van Bruggen, J. D. Boyett, A. L. Van Bueren, and W. R. Galey, *J. Gen. Physiol.*, 1974, **63**, 639.

reflection coefficient for solute $i$, $P_{ii}$ is the self-permeability coefficient, and $\Delta C_i$ is the concentration gradient of solute $i$ across the membrane; $P_{ij}$ is the cross coefficient between solutes $i$ and $j$ and $\Delta C_j$ is the concentration gradient of solute $j$ across the membrane.

Writing equation (99) for the two solutes, d (driver) and t (tracer, driven), for the flow in the direction $2 \rightarrow 1$,

$$J_d^{2 \rightarrow 1} = C_{d(av)}(1 - \sigma_d)J_v^{2 \rightarrow 1} + RT\omega_d(C_{d(2)} - C_{d(1)}) + f_{td}C_{d(av)}J_t^{2 \rightarrow 1} \qquad (100)$$

$$J_t^{2 \rightarrow 1} = C_{t(av)}(1 - \sigma_t)J_v^{2 \rightarrow 1} + RT\omega_t(C_{t(2)} - C_{t(1)}) + f_{dt}C_{t(av)}J_d^{2 \rightarrow 1} \qquad (101)$$

where $f$ is the interaction coefficient.

The experiments performed were such that the osmotic flow due to $\Delta C_d$ was balanced by a counter flow induced by the application of hydrostatic pressure, so that $J_v = 0$. So the first term in equations (100) and (101) drops out. Because of the use of low concentration of t on side 2, the last term in equation (100) becomes negligible. Consequently, equation (100) becomes

$$J_d^{2 \rightarrow 1} = \omega_d RT(C_{d(2)} - C_{d(1)}) \qquad (102)$$

and equation (101) for the tracer flow becomes

$$J_t^{2 \rightarrow 1} = \omega_t RT(C_{t(2)} - C_{t(1)}) + f_{dt}C_{t(av)}J_d^{2 \rightarrow 1} \qquad (103)$$

For the condition when solutes d and t are present on side 2 only (*i.e.* $C_{d(1)} = 0$, $C_{t(1)} = 0$, and dropping the subscript 2), substitution of equation (102) in equation (103) gives the unidirectional flow. Thus

$$J_t^{2 \rightarrow 1} = \omega_t RT\, C_t + f_{dt}C_{t(av)}\omega_d RT\, C_d \qquad (104)$$

For the flow in the opposite direction, the relation is

$$J_t^{1 \rightarrow 2} = \omega_t RT\, C_t - f_{dt}C_{t(av)}\omega_d RT\, C_d \qquad (105)$$

Evaluation of the net flow, $J_t = J_t^{2 \rightarrow 1} - J_t^{1 \rightarrow 2}$, on rearrangement, gives the value for the solute interaction coefficient. Thus

$$f_{dt} = \frac{J_t^{2 \rightarrow 1} - J_t^{1 \rightarrow 2}}{C_t \omega_d\, RT\, C_d} \qquad (106)$$

Expressing equation (106) in permeability coefficients, one gets

$$f_{dt} = \frac{P_t^{2 \rightarrow 1} - P_t^{1 \rightarrow 2}}{P_d\, C_d} \qquad (107)$$

where $J_t = P_t C_t$ and $P_d = \omega_d RT$.

Some of the values determined for the permeability coefficient for several solutes under different conditions are given in Table 28, from which it is seen that an increase in sucrose concentration causes a decrease in the value of $P$. The permeability values of the tracer in the two directions given in Table 29 indicate solute–solute interaction at a sucrose concentration of 1.5 mol l$^{-1}$ in the case of sucrose tracer, at 1.0 and 1.5 mol l$^{-1}$ in the case of trisaccharide tracer, and at 0.5, 1.0, and 1.5 mol l$^{-1}$ in the other two cases. The quantitative values for the interaction coefficient, $f_{dt}$, derived from equation (107) are given in Table 30. All the data indicate that

# Membrane Phenomena

**Table 28** *Effect of sucrose concentration, with sucrose on both sides of a homopore membrane, on the permeability of a tracer solute*

| Concentration of sucrose/mol l$^{-1}$ | Permeability coefficient × 10$^5$/cm s$^{-1}$ | | | |
|---|---|---|---|---|
| | Sucrose | Maltotriose | Inulin | Dextran |
| 0 | 3.1 | 2.3 | 0.5 | 0.6 |
| 0.5 | 2.1 | 1.5 | 0.37 | 0.41 |
| 1.0 | 1.1 | 0.7 | 0.23 | 0.26 |
| 1.5 | 0.6 | 0.4 | 0.1 | 0.08 |

**Table 29** *Values of the permeability of tracer, $P_t \times 10^5$/cm s$^{-1}$, in the two directions*

| Concentration of sucrose/mol l$^{-1}$ | | 0 | 0.5 | 1.0 | 1.5 |
|---|---|---|---|---|---|
| Sucrose: | $P_t^{2\to1}$ | 2.94 | 1.06 | 1.11 | 1.10 |
| | $P_t^{1\to2}$ | 3.27 | 1.83 | 0.95 | 0.63 |
| Maltotriose: | $P_t^{2\to1}$ | 2.19 | 1.25 | 1.06 | 1.42 |
| | $P_t^{1\to2}$ | 2.39 | 1.23 | 0.80 | 0.53 |
| Inulin: | $P_t^{2\to1}$ | 0.45 | 0.41 | 0.47 | 0.61 |
| | $P_t^{1\to2}$ | 0.45 | 0.13 | 0.05 | 0.06 |
| Dextran: | $P_t^{2\to1}$ | 0.55 | 0.43 | 0.53 | 0.55 |
| | $P_t^{1\to2}$ | 0.55 | 0.24 | 0.13 | 0.09 |

**Table 30** *Values of solute–solute interaction coefficient, $f_{dt}$/cm$^3$ mol$^{-4}$, for sucrose driver in homopore membranes*

| Concentration of sucrose/mol l$^{-1}$ | Sucrose | Maltotriose | Inulin | Dextran |
|---|---|---|---|---|
| 0.5 | <25 | <25 | 520 | 360 |
| 1.0 | <100 | 235 | 375 | 360 |
| 1.5 | 281 | 533 | 330 | 275 |

solute–solute interaction is determined by: (i) transmembrane flow of both the driver and the driven solutes, (ii) concentration of the driver solute, (iii) size of the driven solute, and (iv) membrane pore size.

The several interactions that occur when a non-electrolyte solute flows through a porous glass membrane have been described quantitatively by Smit *et al.*,[880] using a friction model.

The resistance coefficients $R_{11}$, $R_{12}$, and $R_{22}$ (where 1 refers to solvent and 2 to solute) have been shown to be related to the experimental quantities $L_p$, $\omega$, and $\sigma$ by the equations

$$R_{12} = \left(\frac{V_1 V_2}{d L_p}\right) - \left(\frac{(1-\sigma)\{1-(1-\sigma)C_{2(av)} V_2\} V_1}{d\omega}\right) \quad (108)$$

$$R_{11} = \left(\frac{V_1^2}{d L_p}\right) + \left(\frac{(1-\sigma)^2 C_{2(av)} V_1^2}{d\omega}\right) \quad (109)$$

$$R_{22} = \left(\frac{V_2^2}{d L_p}\right) + \left(\frac{\{(1-(1-\sigma)C_{2(av)} V_1\}^2}{d\omega C_{2(av)}}\right) \quad (110)$$

[880] J. A. M. Smit, J. C. Eijsermans, and A. J. Staverman, *J. Phys. Chem.*, 1975, **79**, 2168.

where $V$'s are partial molar volumes of species 1 and 2, $d$ is membrane thickness, and $C_{2(av)}$ is the concentration of solute averaged according to

$$C_{2(av)} = 1/[V_2 - (\Delta\mu_2^c/\Delta\mu_1^c)V_1]$$

where $\mu^c$ is the concentration-dependent part of the chemical potential.

The determination of $R_{ij}$ coefficients involved measurements of $L_p$, $\omega$, $\sigma$, partial molar volumes of solute and solvent, and the geometry of the membrane, Such measurements, under both dynamic and steady-state conditions, have been carried out using a porous glass membrane (Vycor, made by the Corning Glass Company) and several non-electrolyte solutes.[880]

The resistance coefficients have been related to friction coefficients by making the following assumptions: (i) the concentration profile of the solute in the membrane was linear, (ii) a single partition coefficient related the internal solute concentration to the external one, and (iii) the volume fraction of the membrane $\phi_m$ was independent of the $x$-co-ordinate, *i.e.* the membrane is uniform. These relationships are

$$R_{12} = -r_{12} \qquad (111)$$

$$R_{11} = V_1\left(r_{12} + \frac{f_{1m}V_2}{1-\phi_m}\right)K\left(\frac{C_2' + C_2''}{2}\right) + \left(\frac{f_{1m}V_1}{1-\phi_m}\right) \qquad (112)$$

$$R_{22} = \frac{1}{K}\left(\frac{r_{12}}{V_1} + \frac{f_{2m}}{1-\phi_m}\right)\left(\frac{\ln(C_2'/C_2'')}{C_2' - C_2''}\right) - r_{12}\left(\frac{V_2}{V_1}\right) \qquad (113)$$

where $r_{ij}$'s are the friction coefficients (related to $f_{ij}$'s by the relation $r_{ij}C_j = f_{ij}$), $K$ is the partition coefficient, and $'$ and $''$ represent the two membrane (m) faces at $x=0$ and $x=d$. Thus, from the measured values of $R_{11}$, $R_{22}$, $R_{12}$ and $(1-\phi_m)$, the coefficients $r_{12}$, $f_{1m}$, $f_{2m}$, and $K$ can be evaluated. These values for the glass membrane and some non-electrolytes are given in Table 31, from which the following information about friction coefficients may be noted: (i) $f_{21} > f_{2m}$; the solute experiences more friction from the solvent than from the membrane, and the friction increases with increase in the molecular weight of the solute. (ii) $f_{2m} \gg f_{1m}$; the membrane exerts more friction on the solute than on the solvent. Again the effect increases with increase in the molecular weight of the solute (pentaerythritol seems to be an exception). (iii) $f_{1m} > \simeq f_{12}$. (iv) $K \approx 1$. (v) The values of tortuosity factor, $\theta$, remain practically constant at 0.33. Facts (iv) and (v) indicate inertness of the membrane, whose chemical and/or physical properties exert little influence on the friction between solute and solvent.

**Nature of Flow through Membranes.**—In porous membranes, it is generally observed that water flux under applied pressure exceeds the flux calculated from the diffusion coefficient of tritiated water. It is assumed that both bulk (or viscous) flow and molecular diffusional flow occur in the membrane. For a discussion, see Peterlin and Yasuda[881] and Paul.[882] In order to clarify the relation in more detail, permeation of water through crosslinked cellulose membranes of different density of crosslinking, cellulose diacetate membranes, and cellulose triacetate membranes

---

[881] A. Peterlin and H. Yasuda, *J. Polym. Sci., Polym. Phys. Ed.*, 1974, **12**, 1215.
[882] D. R. Paul, *J. Polym. Sci., Polym. Phys. Ed.*, 1974, **12**, 1221.

**Table 31** Friction coefficients, $f_{i,j}/\text{N s m}^{-1}$, and partition coefficient, $K$, in a glass membrane at 25 °C

| Solute | $(1-\phi_m)$ | $r_{12}\times 10^{-9}$ | $r_{12}^f \times 10^{-7}$ | $\theta^2$ | $f_{21}\times 10^{-13}$ | $f_{2m}\times 10^{-13}$ | $f_{1m}\times 10^{-11}$ | $f_{12}\times 10^{-11}$ | $K$ |
|---|---|---|---|---|---|---|---|---|---|
| Pentaerythritol | 0.352 | 1.39 | 5.89 | 0.12 | 2.71 | 0.36 | 1.22 | $0.14^a$   $0.70^b$ | 1.03 |
| Mannitol | 0.309 | 2.11 | 6.72 | 0.10 | 3.61 | 0.26 | 1.44 | $0.21^a$   $1.06^b$ | 0.92 |
| Sucrose | 0.324 | 3.01 | 8.57 | 0.09 | 5.40 | 1.50 | 1.33 | $0.30^a$   $1.51^b$ | 1.06 |
| Raffinose | 0.288 | 3.67 | 10.28 | 0.10 | 5.85 | 2.67 | 1.25 | $0.37^a$   $1.85^b$ | 1.02 |

(a) Concentration $C_2$ is 10 mol m$^{-3}$; (b) Concentration $C_2$ is 50 mol m$^{-3}$.
$r_{12}^f$ is the friction coefficient in free solution; $\theta$ is the tortuosity factor, given by $\theta^2 = r_{12}^f/r_{12}(1-\phi_m)$. From values of $r_{12}$, two friction coefficients, $f_{21}$ and $f_{12}$, have been derived from $f_{21} = r_{21}C_1$ and $f_{12} = r_{12}C_2$.

has been studied.[883] In addition, permeation of water vapour through dry membranes of the same kind has been examined. The permeation mechanism has been discussed by comparing the results of the two types of measurements. Correlations between types of adsorbed water molecules and water permeability in swollen polymer membranes have been discussed.[884] Measurements of both diffusion coefficient and hydraulic permeability of collagen membranes showed that hydraulic permeability was three orders of magnitude larger than the salt diffusion coefficient.[885] An approximate equation has been derived to allow calculation of the solvent tracer diffusion coefficient in a homogeneous swollen membrane from measurements of the hydraulic permeability coefficient.[886] The hydraulic permeability of water through a crosslinked poly(vinyl alcohol) membrane has been measured as a function of pressure for temperatures ranging from 18 to 35.8 °C.[887] The data, when analysed according to solution diffusion theory, gave a value of 6.5 kcal mol$^{-1}$ for the activation energy for diffusion.

Schindler and Iberall[888] showed by model calculations that, in the time required for a water molecule to flow through a capillary pore of diameter 80 Å and length 0.1 $\mu$m (time = 5 ms), its diffusional displacement ($\bar{x}$) would far exceed the diameter of the pore. Consequently, they concluded that the interaction of water molecules with the wall of the pore was involved, rather than forces imposed by Poiseuille flow. This argument (*i.e.* the inapplicability of the Poiseuille law) has been questioned by Davis and Renkin,[889] who consider that the proper way to judge the applicability of Poiseuille's law to transport through small pores is to use the ratio of pore diameter to $\bar{x}$.

Poiseuille's law has been used by Van Bruggen *et al.*[879] to derive values for the sizes of pores present in track-etched polycarbonate membranes. The pore diameter has been also determined by electron microscopy and by studying the diffusion of tritiated water. The values for pore radius determined by these methods were 155 (electron microscopy), 145 ($^3$H$^1$HO measurement), and 178 Å ($L_p$ measurement). In view of these values, ordinary solutes such as mannitol, sucrose, and maltotriose would experience little hindrance to diffusion through the membrane. But an increase in solute size would increase hindrance to diffusion. This is confirmed by the data given in Table 32. Hindrance to diffusion is given by ($D_M/D_0$), the ratio of solute diffusivity in membrane ($D_M$) to that in free solution ($D_0$).

An interesting study in which homogeneity criteria are developed has been presented by Tanny and Kedem.[890] The streaming potential arising across an anion-exchange membrane (collodion containing polyvinylamine) under hyperfiltration conditions has been measured as a function of pressure and also at elevated temperature. According to the equation

$$E = -\beta \Delta P - 2t_1 \left(\frac{RT}{F}\right) \ln \left(\frac{1}{1-R_r}\right) \tag{114}$$

[883] M. Kawaguchi, T. Taniguchi, K. Tochigi, and A. Takizawa, *J. Appl. Polym. Sci.*, 1975, **19**, 2515.
[884] Y. J. Chang, C. T. Chen, and A. V. Tobolsky, *J. Polym. Sci., Polym. Phys. Ed.*, 1974, **12**, 1.
[885] G. Wiedner and G. Wilhelm, *Biochim. Biophys. Acta*, 1974, **367**, 349.
[886] D. R. Paul and O. M. Ebra-Lima, *J. Appl. Polym. Sci.*, 1975, **19**, 2759.
[887] O. M. Ebra-Lima and D. R. Paul, *J. Appl. Polym. Sci.*, 1975, **19**, 1381.
[888] A. M. Schindler and A. S. Iberall, *Biophys. J.*, 1973, **13**, 804.
[889] D. G. Davis and E. M. Renkin, *Biophys. J.*, 1974, **14**, 514.
[890] G. Tanny and O. Kedem, *J. Colloid Interface Sci.*, 1975, **51**, 177.

## Table 32 Diffusion in a track-etched polycarbonate membrane

| Solute | $(R_s/R_p)$ | Permeability /cm s$^{-1}$ | $(D_M/D_0)$ Experimental | Theoretical[a] |
|---|---|---|---|---|
| Mannitol | 0.030 | $3.66 \times 10^{-5}$ | 0.88 | 0.94 |
| Sucrose | 0.037 | $3.12 \times 10^{-5}$ | 0.90 | 0.92 |
| Maltotriose | 0.042 | $2.29 \times 10^{-5}$ | 0.81 | 0.91 |
| Inulin | 0.103 | $0.45 \times 10^{-5}$ | 0.41 | 0.78 |

(a) $(D_M/D_0)$ was calculated from the equation:
$(D_M/D_0) = [1-(R_s/R_p)]^2[1-2.104(R_s/R_p)+2.09(R_s/R_p)^3-0.95(R_s/R_p)^5]$,
where $R_s$ is the radius of the solute and $R_p$ is the pore radius

where $t_1$ is the transport number of the cation and $R_r$ is salt rejection, the slope of a plot of streaming potential $E$ against $\Delta P$ should give $\beta$, the electro-osmotic coefficient, defined by

$$\beta = (J_v/I)_{\Delta P, \Delta \mu_s} \qquad (115)$$

$\beta$ could be determined as a function of concentration, $C$, of the solute. From a plot of $(1/\beta)$ against $C$, a value for $\beta_0$ at $C = 0$ can be derived. This has been shown to be related to the membrane porosity ($r_\beta$ is the pore radius) by the relation

$$\left(\frac{1}{\beta_0 F}\right) - \left(\frac{\bar{X}}{\phi_w}\right) = \frac{8\eta}{f_{1w} r_\beta^2} \qquad (116)$$

But $f_{1w} = f_{1w}^0/\theta$, where $f_{1w}^0$ is the friction coefficient in free solution and $\theta$ is the tortuosity. As these values and others ($\bar{X}$, $\eta$, $\phi_w$) were known for the membrane system used, a value for $r_\beta$ could be derived. Also, by measurement of hydraulic permeability, $L_p$, the pore radius, $[r(L_p)]$, for the membrane may be determined, since $L_p$ is given by

$$[r(L_p)]^2 = 8\eta L_p d/\phi_w \theta \qquad (117)$$

Determination of $r_\beta$ and $[r(L_p)]$ as indicated above, using collodion-based polyvinylamine (PVA) membranes both at ordinary and elevated temperatures, gave the following values:

| | Ordinary temperature | | Elevated temperature | |
|---|---|---|---|---|
| | $r_\beta$/Å | $[r(L_p)]$/Å | $r_\beta$/Å | $[r(L_p)]$/Å |
| 1 atm | 26 | >420 | 19 | — |
| 20 atm | 43 | 36 | 28 | 22 |

These data indicate that membrane homoporosity is reflected by $r_p \approx [r(L_p)]$ and heteroporosity by $r_\beta < [r(L_p)]$.

Macrocyclic polyether–polyamide (PC-6) and its mixture with poly(vinylpyrrolidone) (PVP) may be easily cast into membranes. These are interesting membranes in that they selectively bind alkali-metal ions. The permeabilities of these membranes to salts and water have been measured.[891] PC-6 membranes strongly absorb

---

[891] E. Shchori and J. Jagur-Grodzinski, *J. Appl. Polym. Sci.*, 1976, **20**, 773.

sodium salts, and the salts thus have low mobility in the membrane ($D_{Na} = 5 \times 10^{-12}$ cm² s⁻¹). On the other hand, PC-6 and PVP polymeric alloy membranes (30% PVP) are highly permeable to salt and water ($D_{Na} = 1.7 \times 10^{-9}$ cm² s⁻¹; $D_w = 2$—$5 \times 10^{-7}$ cm² s⁻¹). The membrane porosity of some alloy membranes designated A(20) and A(30) was also determined by comparing total osmotic flow with tracer flow of tritiated water. The $g$ value, which describes membrane porosity, was evaluated from

$$g = \frac{\ln\{(J_t/C_t)/[(J_t/C_t) - J_v]\}}{2\sigma\phi\Delta C\bar{V}_w} \quad (118)$$

where $J_t$, $C_t$ are tracer flux and concentration, $\phi$ is the salt osmotic coefficient, and $\Delta C$ is the salt gradient: (3M-NaCl on one side and ³H¹HO on the other side).

Pore diameter, $2r$, is related to the $g$ value by

$$2r = \left[\frac{32\eta \bar{V}_w D_w^0}{RT}(g-1)\right]^{\frac{1}{2}}$$
$$= 7.3 (g-1)^{\frac{1}{2}} \text{ Å} \quad (119)$$

The values obtained for the several parameters were as follows:

| Membrane | $(J_t/C_t) \times 10^5$ /cm s⁻¹ | $J_v \times 10^5$ /cm s⁻¹ | $\sigma$ | $g$ | $2r$/Å |
|---|---|---|---|---|---|
| A(20) | 2.59 | 0.60 | 0.98 | 2.36 | 8.5 |
| A(30) | 6.45 | 2.15 | 0.95 | 3.88 | 12.4 |

Two Eastman 39·8% acetylated cellulose acetate membranes (UF 10 and RO97), having different porosity and structure (UF 10 is porous and RO97 is a composite membrane of a thicker porous layer and a thinner dense layer), have been used in studies of water transport.[892] Transport of water in UF 10 membrane followed Poiseuille's law, giving a value of 76 Å for the effective pore radius. It was estimated that the dense layer in the composite membrane was 730 Å thick and that 92% of the pressure drop occurred across this layer. Water flow through this layer was considered to arise from molecular diffusion.

A study of the distribution of pore size in ultra-thin membranes has been presented by Ohya et al.[893] Diffusion of inulin and p-aminohippuric acid (PAH) in combined aqueous solution through artificial membranes of pore diameters 26, 50, 100, 200, 250, 350, 510, or 990 Å has been measured.[894] Diffusion of PAH was only restricted by membranes of pore size 26 Å, but the diffusion of inulin was restricted at 100 Å. The diffusion of both solutes remained unrestricted by membranes with pores of diameter 200 Å or greater. In such a case, PAH diffused four times faster than inulin. In diffusion-restricted cases, this ratio was even greater. Similar behaviour, with the passage ratio lowered from 4 to 2, was noted during the filtration of solutes under slowly flowing conditions, even at elevated temperature and pressure.

A new transport equation developed by Verniory et al.[895] to describe the passage

---
[892] A. A. Blumberg, E. S. Haddadin, M. E. Chmielewski, and D. G. Storz, *J. Colloid Interface Sci.*, 1974, **49**, 24.
[893] H. Ohya, Y. Imura, T. Moriyama, and M. Kitaoka, *J. Appl. Polym. Sci.*, 1974, **18**, 1855.
[894] E. Middleton, *J. Membr. Biol.*, 1975, **20**, 347.
[895] A. Verniory, R. DuBois, P. Decoodt, J. P. Gassee, and P. P. Lambert, *J. Gen. Physiol.*, 1973, **62**, 489.

of uncharged macromolecules through porous membranes has been tested on several types of artificial membranes (Amicon PM-30, and XM-50 and XM-100 Diaflo ultrafilters), using a pluridisperse solution of poly(vinylpyrrolidone). From the sieving data collected, a mean pore radius and the width of the distribution function for pore radii have been calculated.[896] Another paper of interest, in which a model of cylindrical pores is proposed to explain electro-mechanical transduction in collagen and other polyelectrolyte membranes, has been published, by Grodzinsky and Melcher.[897]

## 6 Phenomena due to a Gradient of Temperature

Thermo-osmosis (osmosis or mass flow of water induced by a gradient of temperature) has been measured by several investigators. The maintenance of a temperature gradient across a cation-exchange membrane (Ionac MC 3470; a copolymer of styrene and DVB with sulphonic acid groups) in contact with the same solution on either side (solutes used were NaCl, KCl, glucose, and tetramethyl-, tetraethyl-, and tetrapropyl-ammonium chlorides) generated fluxes of solvent (thermo-osmosis) and of solute and an electrochemical potential difference (thermal membrane potential) across the membrane.[898] The thermo-osmotic flow was found always to occur in a direction opposed to the thermal gradient (from the cold side to the hot side). Similar results were obtained with oxidized collodion and collodion–sulphonated polystyrene membranes[899,900] when they were in contact with KCl solutions of different concentration. On the other hand, Dariel and Kedem[901] found that, with pure water on either side of a cellulose acetate membrane subject to a temperature gradient, the thermo-osmotic flow was in the same direction as the temperature gradient (hot side to cold side). However, the addition of salt (NaCl) to the hot side (cold side contained only water) decreased thermo-osmosis, and it became zero when the salt concentration was 0.27 mol l$^{-1}$. With further increase in salt concentration, the flow was from the cold to the hot side.

The problem with measurement of thermo-osmotic coefficient is the difficulty associated with an exact determination of the temperature gradient ($\Delta T_m$) across the membrane. Due to the presence of stagnant layers near the membrane surfaces, the temperature difference, $\Delta T_b$, measured in bulk solutions will be given by the relation

$$\Delta T_b = \Delta T_m + \Delta T_s \qquad (120)$$

where $\Delta T_s$ is the temperature drop across the unstirred layers. Dariel and Kedem[901] used two cellulose membranes, of thickness $d_1$ and $d_2$, and measured the thermo-osmotic flows $J_{v(1)}$ and $J_{v(2)}$ through them. A value for $\Delta T_m$ was evaluated from the relation

$$\Delta T_m = \left( \frac{1 - (J_{v(1)}/J_{v(2)})}{1 - (d_2/d_1)} \right) \Delta T_b \qquad (121)$$

[896] R. DuBois and E. Stoupel, *Biophys. J.*, 1976, **16**, 1427.
[897] A. J. Grodzinsky and J. R. Melcher, *IEEE Trans. Biomed. Eng.*, 1976, **BME 23**, 421.
[898] W. E. Goldstein and F. H. Verhoff, *AIChE J.*, 1975, **21**, 229.
[899] M. Tasaka and M. Nagasawa, *J. Polym. Sci., Polym. Symp.*, 1975, **49**, 31.
[900] M. Tasaka, S. Abe, S. Sugiura, and M. Nagasawa, *Biophys. Chem.*, 1977, **6**, 271.
[901] M. S. Dariel and O. Kedem, *J. Phys. Chem.*, 1975, **79**, 336.

On the other hand, Tasaka et al.[899,900] used thermal membrane potential data to evaluate $\Delta T_m$. It has been shown[902,903] that the thermal membrane potential is given by

$$-(\Delta\psi/\Delta T)=(2\bar{t}_+ - 1)(R/F)a_\pm + (t_+\alpha_+ + t_-\alpha_-) \quad (122)$$

where $\bar{t}_+$ is the transport number of the counter-ion in the membrane, and was determined by measurement of membrane potential under isothermal conditions. Thus

$$\Delta\psi' = \left(\frac{RT}{F}\right)(2\bar{t}_- - 1)\ln\left(\frac{a''_\pm}{a'_\pm}\right)$$

and $\bar{t}_+ + \bar{t}_- = 1$. The other symbols of equation (122) represent

$$\alpha_+ = \eta - (s^0_+/F) - t_w s_w; \qquad \alpha_- = \eta + (s^0_-/F) - t_w s_w$$

where $\eta$ is the differential thermoelectric potential and $s_w = s^0_w - R\ln a_w$, $s^0_i = (\partial\mu^0_i/\partial T)$ is partial molal entropy, $\mu^0_i$ is standard chemical potential, and $t_w$ is the reduced transport number of water, w.

With the help of equation (122), i.e. from known values of $a_\pm$, $\bar{t}_+$, $\alpha_+$, or $\alpha_-$ (see later for these values), values for $(\Delta\psi/\Delta T)$ can be calculated. By inserting the measured value for $\Delta\psi$, that of $\Delta T$ (i.e. $\Delta T_m$) can be derived. It has been further shown[900] that $\Delta T_m$ was linearly related to $\Delta T_b$ and that $(\Delta T_m/\Delta T_b)$ was independent of electrolyte concentration. Thus $\Delta T_m$ during thermo-osmosis can be estimated from bulk temperature gradient.

Tasaka et al.[899,900] found that the thermo-osmotic coefficient $D$ (cm s$^{-1}$ K$^{-1}$) evaluated from the linearity of flow ($J_v$) to $\Delta T$, i.e. $-J_v = D\Delta T$ (the sign is + when the flow is from the hot to the cold side), increased with increase in electrolyte concentration. As the charge on the membrane was increased, values for $D$ were reduced. This was attributed to interaction between fixed charges and counter-ions.

The thermo-osmotic pressures developed in cellulose acetate membranes have also been measured.[901] These were high, being of between $0.94 \times 10^6$ and $3.5 \times 10^6$ dyn cm$^{-2}$ K$^{-1}$.

The volume flow, $J_v$, induced by gradients of pressure and temperature is given by the equation[901]

$$J_v = L_p[(\Delta P - \Delta\pi) + (Q^*/V_w T)\Delta T] \quad (123)$$

where $\Delta\pi$ is the osmotic pressure of the solution and $Q^* = Q_m + \Delta H^s$, the total heat of transport being the sum of two parts, i.e. heat of transport in the membrane ($Q_m$) and heat of transfer between aqueous and membrane phases. $V_w$ is the partial molar volume of water.

When $J_v$ is zero, equation (123) becomes

$$[(\Delta P - \Delta\pi)/\Delta T]_{J_v=0} = -(Q^*/V_w T) \quad (124)$$

Thus by measuring total hydrostatic and osmotic pressure in the steady state when

---

[902] M. Tasaka, S. Morita, and M. Nagasawa, *J. Phys. Chem.*, 1965, **69**, 4191.
[903] M. Tasaka, K. Hanaoka, Y. Kurosawa, and C. Wada, *Biophys. Chem.*, 1975, **3**, 331.

# Membrane Phenomena

$J_v$ is zero, $Q^*$ can be evaluated. Also, by separate measurements of $J_v$ when $(\Delta P - \Delta \pi) = 0$ and when $\Delta T = 0$, $Q^*$ may be evaluated, since

$$\frac{(J_v/\Delta T)_{\Delta P - \Delta \pi = 0}}{[J_v/(\Delta P - \Delta \pi)]_{\Delta T = 0}} = \frac{Q^*}{V_w T} \tag{125}$$

The values determined for $Q^*$ using several cellulose acetate membranes were of the order of 0.1 kcal mol$^{-1}$. These values were dependent on the mean temperature. This dependence was not very marked compared to the dependence of thermo-osmosis on mean temperature.

If the membrane used in the measurement of thermal membrane potential is a permselective one, i.e. $\bar{t}_+ = 1$ (permselective for cations) or $\bar{t}_- = 1$ (permselective for anions), then equation (122) becomes

$$-(\Delta \psi / \Delta T) = (R/F) \ln a_\pm + \alpha_+ \tag{126}$$

$$-(\Delta \psi / \Delta T) = (R/F) \ln a_\pm + \alpha_- \tag{127}$$

According to equations (126) and (127), $\Delta \psi$ increases linearly with $\Delta T$, and the slope is equal to $2.303(R/F)$; the intercept at $\log a_\pm = 0$ gives $\alpha_+$ or $\alpha_-$.

**Table 33** *Characteristics of a charged membrane in potassium chloride solutions maintained at two different temperatures*

| Membrane type | Ion-exchange capacity$^a$ | Water content$^b$ | $t_w$ $^c$/mol Faraday$^{-1}$ | $\alpha_+$ or $\alpha_-$ | $\eta$/ mV K$^{-1}$ |
|---|---|---|---|---|---|
| Cation-exchange membranes | | | | | |
| 1. Heterogeneous: PVC binder (40%) + polystyrene sulphonate (Amberlite XE-69) | 1.89 | 0.37 | 6.05 | −0.24 | 5.20 |
| 2. Homogeneous: polystyrene sulphonate | 1.03 | 0.59 | 6.63 | −0.14 | 5.72 |
| 3. Homogeneous: polystyrene sulphonate (Nepton CR-61) | 1.21 | 0.43 | 22.6 | −0.07 | 17.36 |
| 4. Homogeneous: phenol sulphonate (Nepton CR-51) | 0.95 | 0.43 | 9.61 | −0.03 | 7.72 |
| 5. Interpolymer: collodion + 57% polystyrene sulphonate | 1.05 | 0.81 | 60 | 0.10 | 44.6 |
| 6. Liquid polyelectrolyte: 14.2% polystyrene sulphonate | 3.80 | 6.0 | 150 | 0.05 | 109.7 |
| 7. Liquid polyelectrolyte: 4.4% polystyrene sulphonate | 3.80 | 21.7 | — | 0.21 | — |
| Anion-exchange membranes | | | | | |
| 1. Heterogeneous: PVC binder (40%) + quaternized polystyrene (Amberlite XE-119) | 1.53 | 0.47 | −5.15 | 0.28 | −4.02 |
| 2. Homogeneous: quaternized polystyrene (Nepton AR-111) | 0.79 | 0.36 | −13.1 | 0.08 | −9.95 |

(a) The units are milliequivalents per gram of dry membrane; (b) Measured as grams of water per gram of dry membrane; (c) In 0·1M-KCl

For several ion-exchange membranes and different salt solutions, the slopes and intercepts have been determined.[899,903] By measuring transport numbers for water and using the literature values for $s^0_+$, $s^0_-$, and $s_w$ ($\approx 16.7$ cal K$^{-1}$ mol$^{-1}$), values for the thermoelectric potential coefficient, $\eta$, can be derived. Some of these values are given in Table 33. The values of $\eta$ derived are only relative, as the $s^0_+$ and $s^0_-$ are referred to the molar entropy of H$^+$ ion.

This type of study of thermal membrane potential has been extended to the bi-ionic case, using aqueous solutions of two 1:1 electrolytes with a common anion (e.g. NaCl+NH$_4$Cl; LiCl+NH$_4$Cl).[904] Here also, linear relationships between $\Delta\psi$ and $\Delta T$ and between ($\Delta\psi/\Delta T$) and log (activity of one cation) were observed.

Liquid membranes formed by alkylammonium salts in suitable organic solvents have also been used in the measurement of non-isothermal membrane potentials, using solutions of 1:1 electrolyte (mono-ionic)[905,906] and a mixture of two 1:1 electrolytes with a common cation (bi-ionic).[906]

For the mono-ionic system, the thermal membrane potential was dependent on the nature of the counter-ion, on its activity in the aqueous phase, and on the nature of the solvent used in the formation of the liquid membrane. In the case of the bi-ionic system, the measured thermal membrane potentials closely followed the equation developed from the application of the principles of irreversible thermodynamics.[906]

*Acknowledgements*: Thanks are due to Professor C. Paul Bianchi for his interest.

---

[904] M. Tasaka, N. Ichijo, S. Kobayashi, and H. Kobayashi, *Biophys. Chem.*, 1976, **4**, 269.
[905] G. Scibona, M. Magini, B. Scuppa, A. Castagnola, and C. Fabiani, *Anal. Chem.*, 1977, **49**, 212.
[906] G. Scibona, C. Fabiani, B. Scuppa, and P. R. Danesi, *Biophys. J.*, 1976, **16**, 691.

# 3
# The Application of A.C. Impedance Methods to Solid Electrolytes

BY W. I. ARCHER & R. D. ARMSTRONG

## 1 Introduction

In recent years, a great deal of attention has been given to the study of superionic conductors (solid electrolytes), particularly since the discovery of silver rubidium iodide[1,2] and the development of the sodium sulphur cell employing sodium $\beta$-alumina.[3]

The use of the complex impedance plane for the analysis of electrochemical reactions was introduced by Sluyters.[4] This method has subsequently found considerable application. In this review we shall discuss the information which can be obtained by measuring the a.c. impedance of cells containing superionic conductors. In particular we shall be concerned to show how the d.c. conductivity of such solids can be obtained, along with information concerning the rates of ion transfer across the interface between a metal electrode and a superionic conductor, and across grain boundaries in polycrystalline materials.

We shall first of all discuss some general concepts involved in a.c. impedance measurements, followed by models appropriate to solid electrolyte systems. We shall conclude with a discussion of some of the experimental results obtained recently.

## 2 Definitions and Derivation of Simple A.C. Impedance Theory

When a potential which is time-dependent in a sinusoidal manner, *i.e* $\Delta E \sin \omega t$, where $\Delta E < 5\,\text{mV}$, is applied across an electrochemical cell (see Figure 1), a sinusoidal current $\Delta i \sin(\omega t + \theta)$ will flow as a consequence. In addition, currents will also flow at angular frequencies $2\omega$, $3\omega$, *etc*. It should be noted that $\omega$ equals $2\pi f$, where $f$ is the sinusoid frequency/Hz. Thus, one can define an impedance, $z$, as having a magnitude given by

$$Z = \Delta E/\Delta i \qquad (1)$$

and a phase angle, $\theta$, which corresponds to the phase difference between the applied sinusoidal potential and the resultant sinusoidal current. An impedance is, therefore, a vector quantity, since it has both a magnitude and a direction. It is convenient to represent such two-component vectors as a point in a plane (Figure

---
[1] J. N. Bradley and P. P. Greene, *Trans. Faraday Soc.*, 1967, **63**, 2516.
[2] S. Geller, *Science*, 1967, **157**, 310.
[3] J. T. Kummer, *Progr. Solid State Chem.*, 1972, **7**, 141.
[4] J. H. Sluyters, *Recl. Trav. Chim. Pays-Bas*, 1960, **79**, 1092.

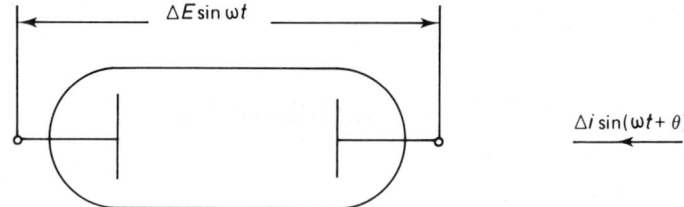

**Figure 1** *Schematic representation of a two-electrode electrochemical cell.*

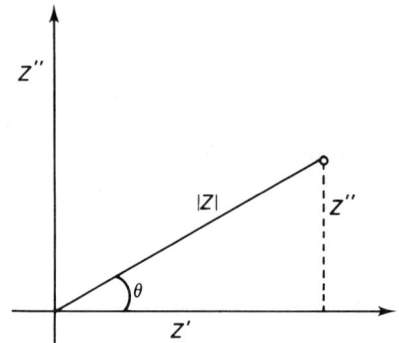

**Figure 2** *Impedance plotted in the complex plane.*

2). The vector is then characterized by $|Z|$ and $\theta$ or by its real and imaginary components, $Z'$ and $Z''$, projected on the $x$ and $y$ axes.

When the electrochemical cell of Figure 1 is replaced by a pure resistance of magnitude $R/\Omega$, we find that $Z = R$ and $\theta = 0$. Thus, a pure resistance is represented by a point on the $x$ axis for any frequency [Figure 3(a)]. When a pure capacitance, $C/F$, is substituted for the cell, the situation becomes somewhat more complicated. In this case, $\theta = 90°$ but $Z$ is frequency-dependent through the relationship $Z = 1/\omega C$. Therefore, as the frequency is varied, the representative point also varies, as in Figure 3(b). Figures 3(a) and 3(b) are the simplest forms of complex-plane impedance spectra, which are representations of the impedance being measured as a function of frequency.

In the majority of cases, the electrochemical cell is better represented by a more complicated network of resistances and capacitances, the so-called equivalent circuit. These show a more complex behaviour in the complex impedance plane. For example, a resistance and capacitance in series gives the impedance spectrum shown in Figure 3(c). In this case, $Z' = R_s$ and $Z'' = 1/\omega C_s$ and since

$$Z^2 = Z'^2 + Z''^2$$

then

$$Z^2 = R_s^2 + (1/\omega C_s)^2 \qquad (2)$$

# The Application of A.C. Impedance Methods to Solid Electrolytes

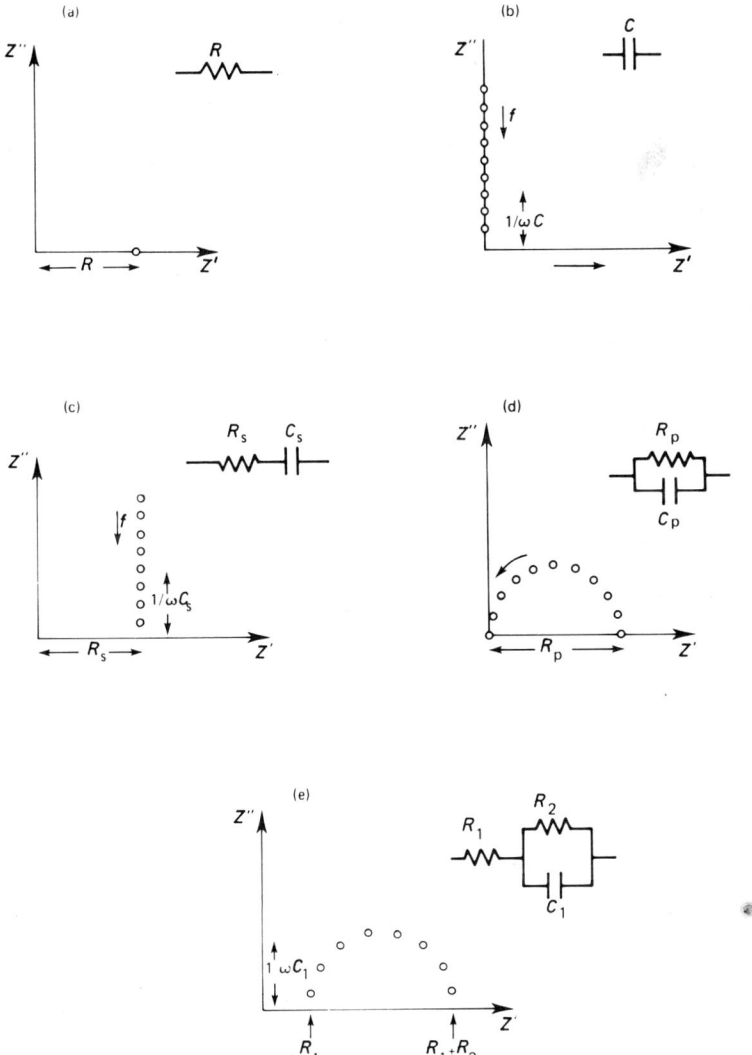

**Figure 3** *Complex-plane impedance spectra with their associated equivalent circuits.*

and $\theta$ can take all values between $0°$ and $90°$, depending upon the measurement frequency. From this, one can derive an equation for the impedance

$$Z = R_s - j/\omega C_s \qquad (3)$$

where $j = \sqrt{-1}$.

If, instead of a series combination of resistance and capacitance, we have a parallel circuit, the impedance spectrum is quite different [Figure 3(d)], and in this case the impedance is given by

$$Z=[(1/R_p)+j\omega C_p]^{-1} \quad (4)$$

When using a parallel combination of elements, the capacitance is additive, which can be advantageous. It is, therefore, sometimes more useful to plot the admittance, $Y$, given by

$$Y=\frac{1}{Z}=\left(\frac{1}{R_p}\right)+j\omega C_p \quad (5)$$

in the same manner described above.

It is thus useful to derive a set of equations to interconvert admittance and impedance.

Equation (4) gives

$$Z=[(1/R_p)+j\omega C_p]^{-1}$$

$$Z=\left(\frac{R_p}{1+\omega^2 C_p^2 R_p^2}\right)-\left(\frac{j\omega C_p R_p}{1+\omega^2 C_p^2 R_p^2}\right) \quad (6)$$

also

$$Z=R_s-\left(\frac{j}{\omega C_s}\right) \quad (3)$$

Thus, by comparison of equations (3) and (6), it is obvious that $R_s$ and $C_s$ are given by

$$R_s=R_p\left(\frac{1}{1+\omega^2 C_p^2 R_p^2}\right) \quad (7)$$

$$C_s=C_p\left(\frac{1+\omega^2 C_p^2 R_p^2}{\omega^2 C_p^2 R_p^2}\right) \quad (8)$$

Similarly, $R_p$ and $C_p$ are given by

$$R_p=R_s\left(\frac{1+\omega^2 C_s^2 R_s^2}{\omega^2 C_s^2 R_s^2}\right) \quad (9)$$

$$C_p=C_s\left(\frac{1}{1+\omega^2 C_s^2 R_s^2}\right) \quad (10)$$

It is clear, from what has been said previously, that, given an impedance (or admittance) spectrum, one can calculate the components of an equivalent circuit of resistances and capacitances responsible for it [Figure 3(e)]. With measurements on electrochemical cells, it is usual for the investigator to measure the impedance of the cell and subsequently to try to find the equivalent circuit and the significance of the different components. This is usually carried out by comparing the results with a theoretical model. Some theoretical models which are applicable to solid electrolytes are discussed in the following section.

### 3 Theoretical Models for the Metal/Superionic Conductor Interphase

**The Point-charge Model.**—Macdonald, in a series of papers,[5–11] has developed theoretical models to describe the a.c. impedance of a system in a number of different situations, some of which are applicable to the study of solid electrolytes. In the derivation of this theory, precursors of which are due to Chang and Jaffé[12, 13] and to Friauf,[14] Macdonald made a number of assumptions, given below:

(1) All the mobile ions in the electrolyte are point charges.
(2) The electrolyte consists of a continuum dielectric and these point charges.
(3) The results obtained apply only to samples in the form of a parallelepiped.
(4) Acceleration effects are neglected. In doing this, Macdonald assumes that the ions reach their terminal velocity spontaneously when subjected to the applied field. This assumption probably breaks down at high frequency ($> 10^8$ Hz).
(5) The distribution of counter-ions and co-ions around the ion under consideration is neglected, *i.e.* Debye–Hückel effects are neglected.
(6) Any compact double-layer effects are neglected. This means that the centre of any ion can be located exactly at the electrode surface. This is not the case if ionic size is taken into account, since the distance of closest approach of the ion to the electrode surface is then controlled by the ionic radius.

Macdonald has considered a number of cases, using these assumptions. The results obtained in some cases are shown below. In the ensuing discussion the terms blocking, non-blocking, and partial blocking will be used. It is thus appropriate, at this point, to define these terms.

These terms are applied to electrodes and they describe the degree to which the mobile ions in the electrolyte can penetrate the electrode material. A completely blocking electrode is one into which no penetration of the mobile ion can occur. An example of a completely blocking situation is that of gold or platinum electrodes on Na–$\beta$-alumina. In this case the mobile sodium ions in the electrolyte cannot cross the interface between the electrode and electrolyte to penetrate the metal of the electrode. The opposite situation is that involving completely non-blocking electrodes, an example of which is a sodium electrode on Na–$\beta$-alumina. In this case, the mobile sodium ions from the electrolyte can discharge at a very high (theoretically infinite) rate at the electrode by crossing the electrolyte/electrode interface and penetrating the electrode. In other cases, the rate of discharge of mobile ions at the electrode may be measurable. These electrodes are then said to be partially blocking.

Macdonald employs a number of parameters in his theories, some of which will

---
[5] J. R. Macdonald, *J. Chem. Phys.*, 1973, **58**, 4982.
[6] J. R. Macdonald, in 'Electrode Processes in Solid State Ionics', ed. M. Kleitz and J. Dupuy, Reidel Publishing Co., Dordrecht-Holland, 1976, p. 149.
[7] J. R. Macdonald, *J. Electroanal. Chem. Interfacial Electrochem.*, 1974, **53**, 1.
[8] J. R. Macdonald, *J. Chem. Phys.*, 1974, **61**, 3977.
[9] J. R. Macdonald, *J. Electroanal. Chem. Interfacial Electrochem.*, 1976, **70**, 17.
[10] J. R. Macdonald, *J. Chem. Phys.*, 1958, **29**, 1346.
[11] J. R. Macdonald, *J. Phys. C*, 1974, **7**, L327; 1975, **8**, L63.
[12] G. Jaffé, *Phys Rev.*, 1952, **85**, 354.
[13] H. Chang and G. Jaffé, *J. Chem. Phys.*, 1952, **20**, 1071.
[14] R. J. Friauf, *J. Chem. Phys.*, 1954, **22**, 1329.

be defined here in order that the reader can readily identify the different cases. Macdonald uses dimensionless boundary parameters, $r_p$ and $r_n$, for positive and negative species respectively, as a measure of the degree of blocking of the electrode to the species in question. Mobile charges may react at an electrode with a heterogeneous rate constant, related to the dimensionless boundary parameter, $r$, by

$$k = \left(\frac{D}{l}\right) \times r \qquad (11)$$

where $D$ is the diffusion coefficient of the species concerned and $l$ is the electrode separation. When $r_n = 0$, the negatively charged species are completely blocked, while $r_p = 0$ signifies complete blocking of the positive species. Thus, when both positive and negative species are completely blocked we have an $(r_p, r_n) = (0,0)$ situation. Similarly, if $r_p = 0$ and $r_n = \infty$, a $(0, \infty)$ situation exists, indicating that the electrode is completely blocking to the positive species but completely non-blocking or open to the negative species. When $r_n \neq 0$, the electrode is only partially blocking to the negative species. For negative species in the presence of adsorption, $r_n$ becomes complex and is denoted $r_n^*$. In such cases, $r_{n0}$, the low-frequency limiting value of $r_n^*$, and $r_{n\infty}$, the high-frequency limiting value of $r_n^*$, must be introduced, as well as the internal adsorption relaxation time, $\tau(A_n)$. This is also true for specifically adsorbed positive species. Macdonald defines the mobility ratio of the species as $\pi_m = \mu_n/\mu_p$, where $\mu_n$ and $\mu_p$ are the mobilities of the negative and positive species respectively. Another parameter, the definition of which may be useful to the reader, is $M = l/2L_D$, where $l$ is the electrode separation and $L_D$ is the Debye length of the electrolyte. Thus, a large value of $M$ means that a large number of Debye lengths are contained between the electrodes.

It should be noted here that all ensuing theory, and the equivalent circuits, applies only to one half cell of any cell considered, while the circuits given in Macdonald's original papers, in general, pertain to complete, two-electrode cells. We have chosen to do this because only certain cells can be set up in a two-electrode symmetrical manner. In certain cases, a particular electrode/electrolyte interphase can only be studied using a three-electrode cell in which the third electrode is a reference electrode, often a metal wire, and the conditions of the outermost electrodes are different. The two outermost electrodes may also be of different materials.

Thus, for example, we may be concerned with the $C/Ag_4RbI_5$ interface, which would be investigated in the cell $C|Ag_4RbI_5|Ag$ with a silver wire reference electrode, inserted at some point into the electrolyte. Since the total impedance of such a cell is extremely complex, it greatly simplifies our review to consider only the impedance of that part of the cell between the carbon electrode and the silver wire reference electrode.

*Completely Blocking Electrodes (0,0): No Specific Adsorption.* The equivalent circuit and complex-plane impedance spectrum predicted by Macdonald for the completely blocking situation are given in Figure 4. The circuit and spectrum apply for any number of mobile charged species, provided that there is no recombination of charged species. An example of such a case is the $C/Ag_4RbI_5$ interphase at potentials between 50 mV and 600 mV anodic to $Ag|Ag^+$.

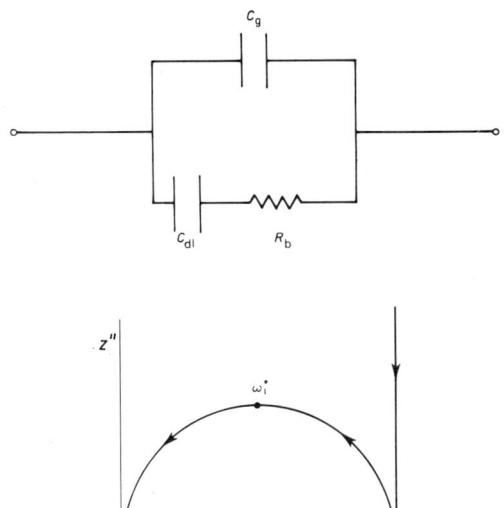

**Figure 4** *Equivalent circuit and predicted complex-plane impedance spectrum for the case of complete blocking, (0,0), in the absence of specific adsorption. (The arrows indicate the direction of increasing frequency.)*

$C_g$ is the geometric capacitance of the cell and is given by

$$C_g = \varepsilon_0 \varepsilon_r A / l \tag{12}$$

where $A$ is the electrode area, $l$ is electrolyte thickness under consideration, $\varepsilon_0$ is the permittivity of a vacuum, $\varepsilon_r$ is the dielectric constant of the electrolyte (low-frequency limit), and $R_b$ is the bulk resistance and is given by

$$R_b = l / A\sigma \tag{13}$$

where $\sigma$ is the electrolyte conductivity.

The parameter $\omega_1^*$, the frequency at the maximum of the semi-circle in Figure 4, is related to $C_g$ and $R_b$ by

$$\omega_1^* = (R_b C_g)^{-1} \tag{14}$$

In the limiting high-frequency case as $\omega \to \infty$, the $C_{dl}$ term in the equivalent circuit disappears, leaving only $C_g$ and $R_b$ in parallel.

The value of $C_{dl}$ depends upon the conditions of the system. For example, Macdonald gives, for one blocking electrode

$$C_{dl} = \frac{A \varepsilon_0 \varepsilon_r}{L_D} \cosh\left(\frac{V_a^*}{2}\right) \tag{15}$$

where $L_D$ is again the Debye length and $V_a^* = V_a/(kT/e)$, where $V_a$ is the applied potential difference, $k$ the Boltzmann constant, and $T$ the absolute temperature.

Equation (15) holds only when:

(i) The electrode separation is much greater than the Debye length.
(ii) The frequency of the applied potential is much lower than the frequency of relaxation of the diffuse double layer.
(iii) Positive and negative species do not recombine.
(iv) $Z_n$, the charge on the negative species, and $Z_p$, the charge on the positive species, are both equal to one.
(v) There is no specific adsorption at the electrodes.
(vi) $V_D$ (the diffusion potential) $= 0$.

If any of the above conditions are not met, then equation (15) no longer holds, and the reader is referred to Macdonald's original papers.[5-11]

*One Mobile Species: Blocked but Specifically Adsorbed (0).* Macdonald considers the case of two mobile charged species, the positively charged one being completely blocked at all frequencies. The other mobile species has a boundary parameter, $r_n^*$, which is complex. Thus, in Macdonald's notation this is a $(0, r_n^*)$ situation. For the specific case where $\pi_m \gg 1$, and $r_{n\infty} < 1$, and $r_{n0} = 0$, then in the low-frequency limit both species will be completely blocked. At the high-frequency limit, since $\pi_m \gg 1$, the positive species can be considered to be stationary since, as $\omega \to \infty$, the frequency of the applied signal will become so great that the positive species will not have time to move before the signal alternates. Thus, this case can be considered as that of one mobile charged species, blocked, but specifically adsorbed. We have rearranged Macdonald's equivalent circuit, for this situation, to conform with our conventions. However, in rearrangement, the circuit has remained equivalent to Macdonald's original. The resulting circuit and Macdonald's predicted complex-plane impedance spectrum are given in Figure 5, where $R_{ct}$ is the charge-transfer resistance which arises at the electrode/electrolyte interface because the electrode is partially blocking to the mobile charged species. At low frequency, where $r_{n0} = 0$, $R_{ct}$ will be infinitely large; $C_A$ is a capacitance term which arises from the specific adsorption of the mobile species. The terms $C_g$ and $R_b$ are again given by equations (12) and (13) respectively, and are related to $\omega_1^*$ by equation (14); $\omega_2^*$ is similarly given by

$$\omega_2^* = (R_{ct} C_{dl})^{-1} \tag{16}$$

An example of a completely blocking electrode with specific adsorption is the platinum electrode on $Ag_4RbI_5$ at potentials between 0 and 600 mV anodic to $Ag|Ag^+$.

*Non-blocking Electrodes with One Mobile, Charged Species ($\infty$).* The completely non-blocking situation involves infinitely rapid discharge of mobile, charged species at the electrodes. The case considered is that of one mobile charged species, non-blocked. The equivalent circuit and complex-plane impedance spectrum as predicted by Macdonald for this situation are given in Figure 6. As for the case of the blocking electrode, $C_g$ and $R_b$ are given by equations (12) and (13) respectively. The double-layer capacitance, $C_{dl}$, is zero in this case because of the infinitely rapid rate of discharge of the charged species across the electrolyte/electrode interface.

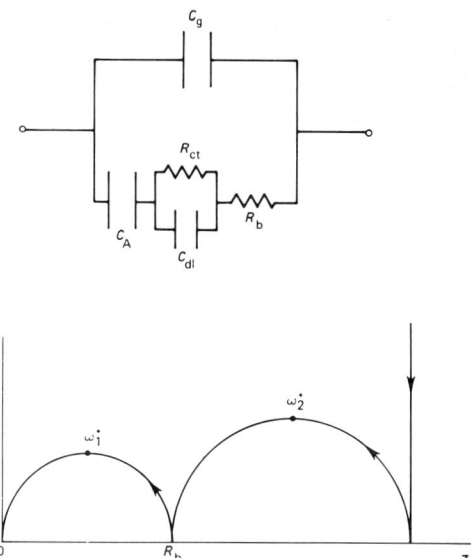

**Figure 5** *Equivalent circuit and predicted complex-plane impedance spectrum for the case of one mobile charged species, blocked but specifically adsorbed, (0). (The arrows indicate the direction of increasing frequency.)*

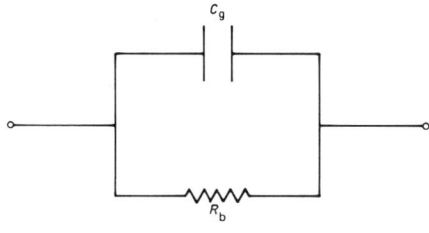

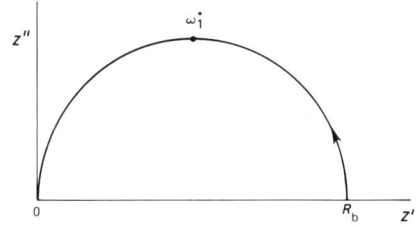

**Figure 6** *Equivalent circuit and predicted complex-plane impedance spectrum for the non-blocking case with one mobile species, ($\infty$). (The arrow indicates the direction of increasing frequency.)*

*One Mobile, Charged Species: partially Blocked* ($r_n$). Macdonald considers the case of two mobile charged species, one of which is completely blocked, while the other is only partially blocked, at the electrode. Thus, in Macdonald's notation this is a $(0, r_n)$ situation. For the specific case where the discharging species has a much greater mobility than the blocked species, *i.e.* $\pi_m \gg 1$, then, as $\omega \to \infty$, the frequency of the applied signal will become so great that the blocked species, which is of low mobility, can essentially be considered to be non-mobile. The situation at high frequency can thus be considered to approximate to that with only one mobile, charged species, which is partially blocked. This is the ($r_n$) case. Again we have rearranged Macdonald's equivalent circuit for this situation, to conform with our conventions. The resulting rearranged circuit and Macdonald's predicted complex-plane impedance spectrum are given in Figure 7. Once more, $C_g$ and $R_b$ are given by equations (12) and (13). In this case there is no simple relation describing $C_{dl}$, *i.e.* the double-layer capacitance, since it depends upon the degree of blocking of the electrodes, among other factors. In this case, $\omega_2^*$ is related to $C_{dl}$ and $R_{ct}$ through the relationship

$$\omega_2^* = (C_{dl} R_{ct})^{-1} \qquad (17)$$

*Two Mobile, Charged Species: One Non-blocked* $(0, \infty)$. In this case, the electrolyte contains two charged species of approximately equal mobilities, one of which is completely blocked at the electrode, while the other is completely non-blocked. Once again we have rearranged Macdonald's equivalent circuit. The resulting circuit and complex-plane impedance spectrum are shown in Figure 8. In the particular case given here, diffusion is the limiting step, and this gives rise to a Warburg impedance, $W$. In this case, the double-layer capacitance is zero because any charge which may build up on the electrode will be discharged immediately by the non-blocked mobile species. An example of such a case might be Na–$\beta$-alumina, with a small fraction of the sodium ions replaced by silver ions, using silver electrodes. Again, $C_g$ and $R_b$ are given by equations (12) and (13) and related to $\omega_1^*$ by equation (14).

The above examples are given merely as a general guide to Macdonald's theoretical models for some specific situations. More recently,[15] Franceschetti and Macdonald have explicitly taken account of compact double-layer effects. For other, more complex cases, the reader is referred to the original publications.[5–11]

**The Finite-ion-size Model.**—Armstrong has attempted to adapt some of the theoretical models for aqueous electrolyte systems in order to obtain a simple model for the metal/superionic conductor interphase which can explain some of the experimental observations that have been made on these systems.[16] Armstrong takes superionic conductors to be, typically, Na–$\beta$-alumina and Ag$_4$RbI$_5$, *i.e.* solids which have very high ionic conductivity (up to $10^{-1} \Omega^{-1} \text{cm}^{-1}$), negligible electronic conductivities, and very high concentrations of mobile species ($\approx 10 \text{mol l}^{-1}$). Initially, Armstrong assumes that the solid consists of immobile anions, and cations that are all equivalent and all mobile.

[15] D. R. Franceschetti and J. R. Macdonald, *J. Electroanal. Chem. Interfacial Electrochem.*, 1978, **87**, 419.
[16] R. D. Armstrong, *J. Electroanal. Chem. Interfacial Electrochem.*, 1974, **52**, 413.

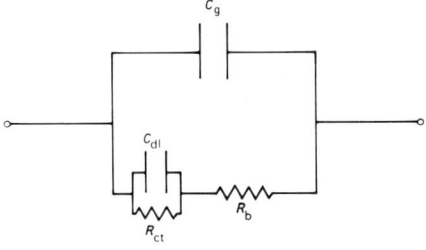

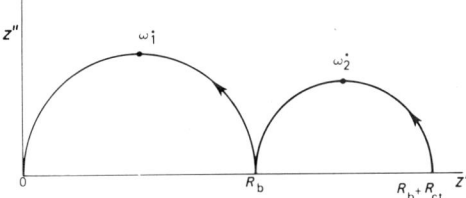

**Figure 7** Equivalent circuit and predicted complex-plane impedance spectrum for the case of two mobile species, one of which is partially blocked and the other completely blocked $(0, r_n)$. At high frequency, if the mobility of the partially blocked species is much greater than that of the blocked species, the situation then approximates to a one-mobile-species, partially blocked one, $(r_n)$. (The arrows indicate the direction of increasing frequency.)

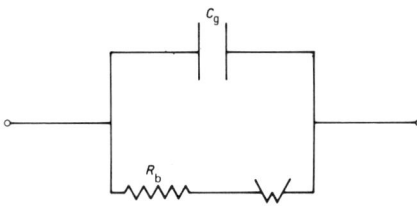

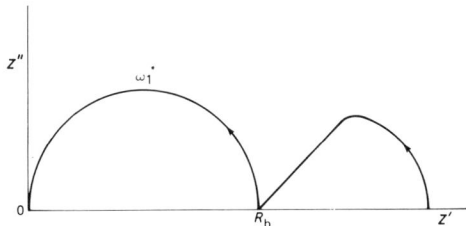

**Figure 8** Equivalent circuit and predicted complex-plane impedance spectrum for the situation of two mobile species, one being non-blocked, $(0, \infty)$. (The arrows indicate the direction of increasing frequency.)

*Blocking Electrodes with No Specific Adsorption (0,0)*. The first case discussed by Armstrong is that where the metal electrode is considered to be blocking to both anions and cations. The model proposed is shown in Figure 9. The metal surface is adjacent to a layer of cations and anions (layer 1) which is in turn adjacent to another layer of anions and cations (layer 2). The number of anions in the two layers is the same, equivalent to a charge of $q_-$ C cm$^{-2}$, and equal to that which would be found in the bulk of the solid. In layer 2, electroneutrality holds, so that the cation charge ($q_{2+}$) is equal to $q_-$. In layer 1, however, there is a deviation from electroneutrality, *i.e.* $q_{1+} + q_- = q_s$, so that there is a net charge $q_s$ on the electrolyte side of the interphase. This charge $q_s$ is balanced by a charge $q_m = q_s$ on the metal surface. In layer 1 it is assumed that the cations are present in normal lattice sites and that energetically these sites are only slightly disturbed by the presence of the metal surface. Armstrong justifies his assumption that the space charge extends only one atomic distance by the fact that in these solids the Debye length is always less than the size of an atom. There is also the requirement that $|q_m| < 20\,\mu$C cm$^{-2}$, so that the deviation from stoicheiometry in the first layer is never sufficient to change sufficiently the chemical potential of the cation, *i.e.* $|(q_m/q_-)|$ must be less than 0.1, where $q_-$ might be 200 $\mu$C cm$^{-2}$.

The model of Figure 9 leads directly to the equivalent circuit and the complex-plane impedance spectrum of Figure 10, where $C_g$ is a geometric capacitance which arises from the finite dielectric constant, $\varepsilon_r$, of the solid electrolyte and $R_b$ is its bulk resistance. (The bulk dielectric constant and conductivity can be assumed to be independent of frequency over a limited frequency range. It should be noted, however, that $\varepsilon_r$ reflects the behaviour of the ions during acceleration as well as the fixed parts of the crystal lattice.) The impedance is supposed to be measured between the metal electrode and a metal wire reference electrode placed at a distance *l* from the metal electrode (*l* is typically > 1 mm).

For this case,

$$C_g = \frac{\varepsilon_0 \varepsilon_r A}{(l - r_1)} \approx \frac{\varepsilon_0 \varepsilon_r A}{l} \qquad (18) = (12)$$

$$R_b = l/A\sigma \qquad (13)$$

$$C_{dl} = \varepsilon_0 \varepsilon_r' A / r_1 \qquad (19)$$

$$\omega_1^* = (R_b C_g)^{-1} \qquad (14)$$

Here, $\sigma$ is again the electrolyte conductivity, $\varepsilon_r'$ is the effective dielectric constant of the electrode–first layer region, and $r_1$ is the distance between the centres of charge of the cations in the first layer and the electrode surface. The capacitance $C_g$ is not placed across the total cell since the displacement current between the metal electrode and layer 1 would then be counted twice over. Thus, $C_g$ can be thought of as arising from the summation of the double-layer-type capacitances between layer 1 and layer 2, layer 2 and layer 3, *etc.* At low frequencies, when $C_g$ has no influence, the equivalent circuit would be simply $C_{dl}$ in series with $R_b$, and this has been demonstrated experimentally. By this model, the value of $C_{dl}$ predicted can never exceed $\approx 50\,\mu$F cm$^{-2}$, and it should be independent of the electrode/electrolyte

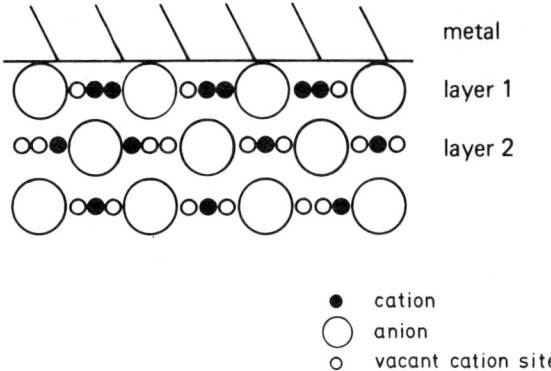

**Figure 9** *Model for the metal/solid electrolyte interphase, in the blocking situation, with no specific adsorption.*

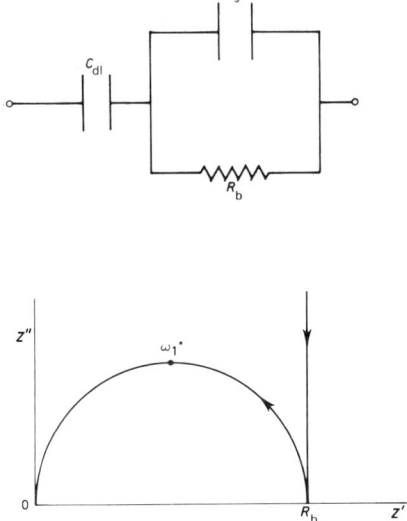

**Figure 10** *Equivalent circuit and resulting complex-plane impedance spectrum for the model of Figure 9. (The arrows indicate the direction of increasing frequency.)*

potential difference. It may, however, be more realistic to expect that $C_d$ will slowly vary with applied potential, since it is obvious that the equilibrium positions of metal cations in the first layer will be, to some extent, a function of $q_m$.

*Blocking Electrodes with Specific Adsorption (0,0).* The second possibility considered by Armstrong is that, in layer 1 of Figure 9, because of the proximity of the metal surface, some abnormal positions become possible for the electrolyte cations, as in Figure 11. They may be positions which are simply closer to the electrode than the normal ones, but in addition a fraction, $\lambda$, of the cation charge, $Z_+e$, may be neutralized by partial electron transfer from the metal. Since there is no equilibrium between metal cations in the bulk of the metal and the different metal cations in the bulk of the electrolyte, this is still the blocking case. If $\Gamma_A$ is the number of cations in abnormal sites (in mol cm$^{-2}$) then the charge transferred from the metal to the cations will be $-\lambda\Gamma_A Z_+ F$ C cm$^{-2}$. It is convenient to define a quantity $Q$ such that $Q = q_m - \lambda Z_+ F$, where $q_m$ is the charge remaining on the metal surface. Figure 12 shows the equivalent circuit and impedance spectrum in the complex plane for this situation. At limiting low frequencies, a double-layer capacitance will be found as

$$(C_{dl})_{\omega\to\infty} = \frac{dQ}{dE} = \left(\frac{\partial Q}{\partial \Gamma_A}\right)\left(\frac{d\Gamma_A}{dE}\right) + \left(\frac{\partial Q}{\partial E}\right)_{\Gamma_A} \tag{20}$$

{It is likely that $(\partial Q/\partial E)_{\Gamma_A} = C_\infty$ will still be independent of potential and $\approx 50\,\mu\text{F cm}^{-2}$. However, the term $[(\partial Q/\partial \Gamma_A)(d\Gamma_A/dE)] = C_{ad}$ is not limited by simple parallel-plate capacitor considerations and could quite easily lead to values of $C_{dl}$ as high as $1000\,\mu\text{F cm}^{-2}$. All that can be said with certainty is that there is a limit on the magnitude of $\int (E/E_0) C_{dl} dE$, where $E_0$ is the potential of zero charge, since only one monolayer of specifically adsorbed cations is possible.}

In the equivalent circuit of Figure 12 there are two capacitances constituting $C_{dl}$, corresponding to the two parts of equation (20). This is because there is a possibility that specific adsorption may be an activated process requiring a characteristic time $\tau = C_{ad} R_{ad}$ for completion. The terms $C_g$ and $R_b$ are given by equations (18) and (13) respectively, and $\omega_1^*$ is given by equation (14). Also

$$\omega_2^* = (C_{dl} R_A)^{-1} \tag{21}$$

*Non-blocking Electrodes* ($\infty$). Armstrong next considers the situation where the possibility of metal–cation exchange, between the electrode and the electrolyte, exists. The structure of the double layer which would be expected depends upon the size of the exchange current for cation transfer between the metal and the cations in

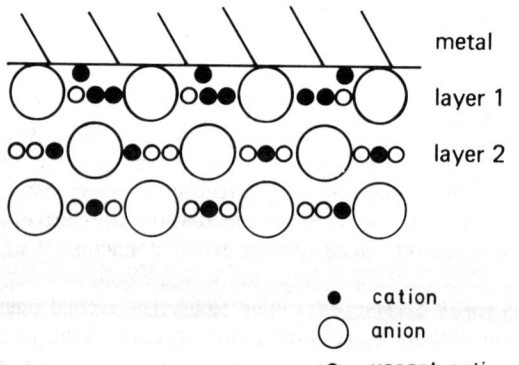

**Figure 11** *Model for the metal/solid electrolyte interphase, in the blocking situation, with specific adsorption.*

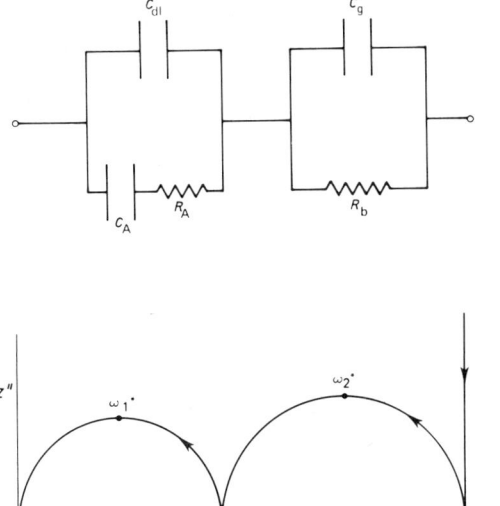

**Figure 12** *Equivalent circuit and resulting complex-plane impedance spectrum for the model of Figure 11. (The arrows indicate the direction of increasing frequency.)*

the first layer adjacent to the surface. For a substance which has a conductivity of $10^{-1}\Omega^{-1}\text{cm}^{-1}$, the charge-transfer resistance for cation transfer between two adjacent cation layers is $\sim 10^{-6}\Omega\text{cm}^2$, since the sum of these charge-transfer resistances must lead to the macroscopic conductivity. Thus, if the charge-transfer resistance for cation transfer between the metal and layer 1 is very much greater than this value, which is very likely, then the equilibrium structure of the double layer described earlier, and shown in Figure 9, will still hold, with the transfer of charge between metal and electrolyte simply introducing a slight perturbation of the system. Since the electrode/electrolyte potential difference can only be changed by changing the concentration of metal cations in the first layer, the normal Warburg impedance cannot exist, so that the equivalent circuit for this situation is that of Figure 13. This Figure also shows the resulting impedance spectrum in the complex plane for this case. Here, $R_{ct}$ is a charge-transfer resistance between the electrode and the electrolyte, assuming the reaction to consist of one step. It should be noted, however, that the assumption that the transfer of, say, a silver ion between silver metal and an electrolyte is a single-step reaction is likely, in most cases, to be a gross simplification. In fact, it could well be a two- or three-step reaction with intermediate adatoms. In such a case, the interfacial impedance will be much more complex than that corresponding to a single resistance in parallel with the double-layer capacitance. A further possibility is that two- or three-dimensional nucleation is involved.

Once again, equations (13), (14), and (18) apply, and $\omega_2^*$ is given by

$$\omega_2^* = (C_{dl}R_{ct})^{-1} \tag{22}$$

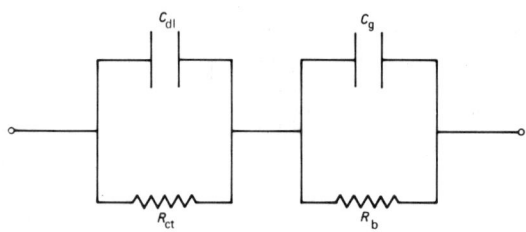

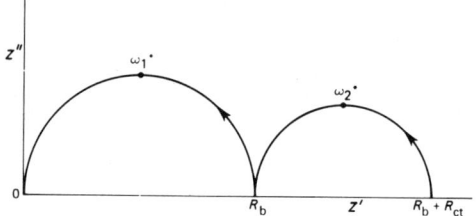

**Figure 13** *Equivalent circuit and resulting complex-plane impedance spectrum for the non-blocking case. (The arrows indicate the direction of increasing frequency.)*

For the case when the charge-transfer resistance for cation transfer between the electrode and electrolyte is less than that between adjacent layers in the bulk of the electrolyte, the interfacial impedance becomes negligible and the equivalent circuit reduces to $R_b$ in parallel with $C_g$ (Figure 6).

It is interesting to note that for charge transfer across the electrode/solid electrolyte interface, where the charge being transferred is the mobile cation of the electrolyte, one would *not* expect a Tafel relationship between the current and the overpotential. This is because the concentration of cations in layer 1 must be varied if the potential is to be varied. Instead, the relationship would be, for metal deposition, and using the model of Figure 9

$$i = \Gamma_{c+} \exp\left(\frac{\alpha n F \eta}{RT}\right) \qquad (23)$$

where $\Gamma_{c+}$ is the concentration of cations in layer 1. This can be rearranged to give

$$i = i_0\left[1 + \left(\frac{C_{dl}\eta}{|q_-| - \Delta E C_{dl}}\right)\exp\left(\frac{\alpha n F \eta}{RT}\right)\right] \qquad (24)$$

where $\Delta E$ is the potential difference between the point of zero charge and the reversible $M \rightarrow M^+$ potential. ($C_{dl}$ has been taken to be independent of potential in this case.) Under many circumstances the term in square brackets in equation (24) will approximate to unity, and a normal Tafel relation will be found. Equation (24) can be regarded as the analogue of the Frumkin $\phi_2$ correction in aqueous electrochemistry.

# The Application of A.C. Impedance Methods to Solid Electrolytes

*Situations where a Normal Warburg Impedance can Arise.* In the cases of blocking electrodes with non-specific and with specific adsorption and of non-blocking electrodes, a conventional Warburg impedance cannot arise. However, there are certain situations where the equivalent circuit and the complex-plane impedance spectrum of Figure 14 (except for the effects of finite cell thickness) would be appropriate. These include:

(i) A solid electrolyte with immobile anions and two different sorts of cations, one of which is present at a low concentration, the two ions having similar mobilities. Such an electrolyte might be Na–$\beta$-alumina with a small fraction of the sodium ions substituted by silver ions. For the interphase Ag/Ag–Na–$\beta$-alumina the circuit of Figure 14 is to be expected, since Na$^+$ cations can act as a supporting electrolyte for Ag$^+$. (For the interphase Pt/Ag–Na–$\beta$-alumina at blocking potentials the equivalent circuit would be that of Figure 12.) The terms $C_g$ and $R_b$ of Figure 14 are given by equations (18) and (13), and equation (14) again applies: $\omega_2^*$ is once more given by equation (22).

(ii) A solid electrolyte where the products of the electrode reaction can diffuse without being influenced by electrical fields. This includes the Pt/Ag$_4$RbI$_5$ system between 640 and 680 mV anodic to Ag|Ag$^+$, when the diffusion of electrochemically generated iodine gives rise to a Warburg impedance.

*Complex Reaction Mechanisms.* Theoretical models have also been proposed for more complex reaction mechanisms, including two-dimensional nucleation and growth reactions on solid electrolytes.[17, 18]

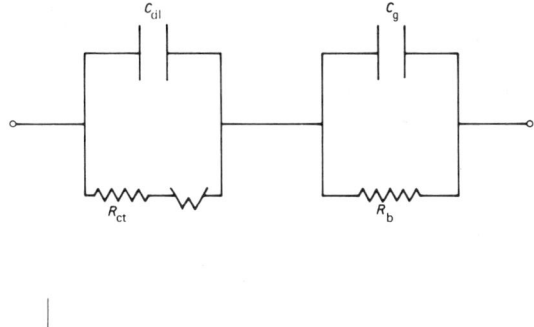

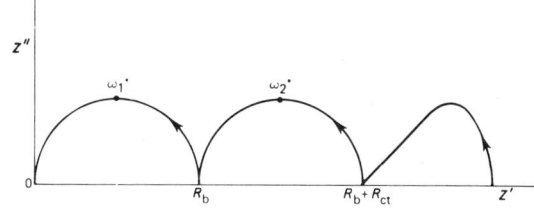

**Figure 14** *Equivalent circuit and resulting complex-plane impedance spectrum for a situation where a normal Warburg impedance can arise. (The arrows indicate the direction of increasing frequency.)*

[17] R. D. Armstrong and A. S. Metcalfe, *J. Electroanal. Chem. Interfacial Electrochem.*, 1976, **71**, 5.
[18] R. D. Armstrong and A. A. Metcalfe, *J. Electroanal. Chem. Interfacial Electrochem.*, 1977, **84**, 209.

**A Comparison of the Two Theoretical Models.**—From the preceding discussion, it can be seen that the conclusions reached by Macdonald and Armstrong are very similar. The major difference is that Macdonald assumes the mobile ions to be point charges and able to approach the electrodes within any distance, no matter how small, while Armstrong takes into account the finite size of the ion and the distance of closest approach of the ion to the electrode. This leads to an important difference in the interpretation of the double-layer capacitance between the two models. Another difference which arises is in the drawing of the equivalent circuits. Macdonald represents $C_g$, the geometric capacitance, as being across the whole cell, whereas Armstrong places it between the two double-layer capacitances of the cell, arguing that to place it across the whole cell involves inclusion of the $C_{dl}$ terms twice. A further point worth stressing is that Macdonald's theories, whilst exact, only apply to solid electrolytes which have very low concentrations of mobile charge carriers ($< M/1000$), and cannot be expected to apply to silver rubidium iodide or to $\beta$-alumina, which have much greater concentrations of mobile charge carriers.*

## 4 The Effect of Sample State upon the Observed Impedance

The theoretical models of Macdonald and Armstrong apply only in the case of a solid electrolyte in the form of a single crystal, with perfectly flat electrodes. It is, therefore, necessary to consider other types of sample. We shall first of all consider the ideal situation of a single crystal with perfectly flat electrodes, followed by other possibilities which may occur.

**Single Crystal.**—*Perfectly Flat Electrodes.* The impedance of single crystals of materials with high ionic conductivity, using perfectly flat electrodes, is expected to show simple behaviour. Figure 15(a) shows a physical representation of a single crystal with two perfectly flat, blocking electrodes parallel to each other. It is assumed that if conduction takes place in one plane only, as in the case of the $\beta$-aluminas, then this plane is orthogonal to the plane of the electrodes. The anticipated equivalent circuit and resulting complex-plane impedance spectrum for this situation are shown in Figures 15(b) and 15(c) respectively.

*Rough Electrodes.* Surface roughness at the electrode is likely to affect the measurement of the conductivity of solid electrolytes. The a.c. response to a rough electrode in contact with aqueous electrolytes has been discussed in detail by de Levie.[19, 20]

Since the use of solid electrolytes often involves higher resistivities, it is likely that surface roughness effects will be more pronounced. The anticipated complex-plane impedance spectrum for a perfectly flat blocking electrode, omitting the semi-circle due to $C_g$ and $R_b$, is shown as line (a) in Figure 16. When surface roughness becomes significant, the line will show deviation from the vertical. For a particular electro-

---

\* *Note added in proof:* More recently, Macdonald and Franceschetti have explicitly taken account of compact double layer effects, giving a generalized model which has the point-charge and finite-ion-size models as limits (*J. Electroanal. Chem. Interfacial Electrochem.*, 1979, **99**, 283).

[19] R. de Levie, *Electrochim. Acta*, 1965, **10**, 113.
[20] R. de Levie, *Adv. Electrochem. Electrochem. Eng.*, 1967, **6**, 329.

# The Application of A.C. Impedance Methods to Solid Electrolytes

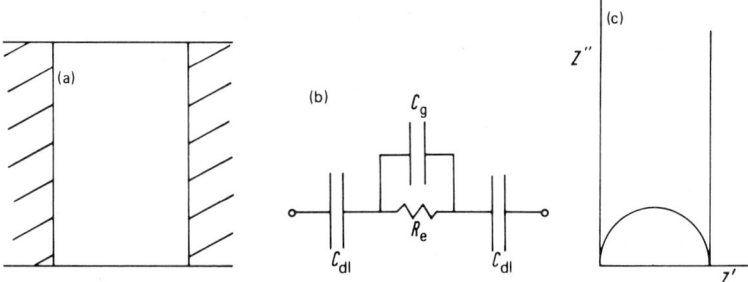

**Figure 15** Impedance of a single crystal with perfectly flat electrodes.
(a) Physical representation.
(b) Anticipated equivalent circuit.
(c) Behaviour in the complex impedance plane.

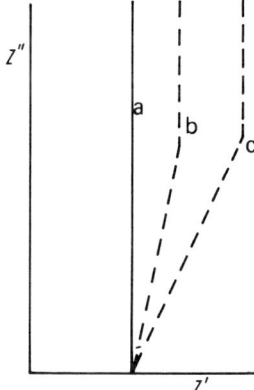

**Figure 16** The effect of surface roughness on the impedance of a single-crystal electrolyte.

lyte, and using the same frequency range, it is expected that the deviation will increase with increasing roughness. This is shown by (b) and (c) in Figure 16.

**Compressed Powder or Sinter.**—*Voidage.* In a compressed powder or sinter, the voidage is often as much as 5%. This results in the density being correspondingly lower than that of the single crystal. The conductivity will also be affected, by purely geometrical considerations. Provided that no other complications arise, the lowering of the conductivity is likely to be simply related to the amount of voidage.

*Particle Orientation.* When an anisotropic solid is in the form of either a sinter or a compressed powder, it is likely that various orientations will be adopted by the individual particles. As a result, the conductance of each particle in a direction perpendicular to the electrodes will be different. The differing orientations of the particles will lead to a 'spread' of conductances. Figure 17(a) shows the situation diagrammatically for five particles only. Figure 17(b) is the equivalent circuit of a

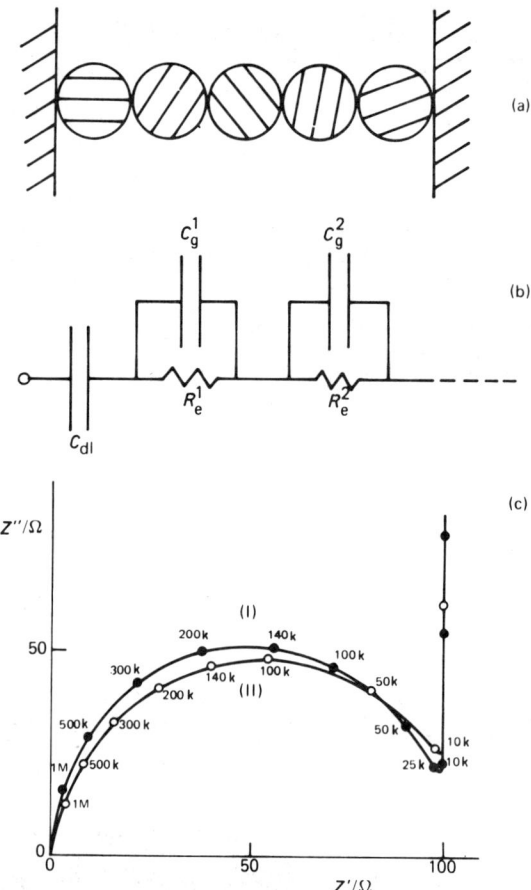

**Figure 17** *The impedance of a sinter with mis-orientated particles.*
*(a) Physical representation.*
*(b) Anticipated equivalent circuit.*
*(c) Behaviour in the complex plane.*

section across the sample. The inter-particle impedance is assumed to be zero. This is, in fact, a simplified arrangement since, in a real system, many conduction paths are possible. Figure 17(c) shows a complex-plane impedance plot for ten particles. Curve I was obtained when $R_b^1 = R_b^2 = R_b^3 \ldots = 10\,\Omega, C_g^1 = C_g^2 = C_g^3 \ldots = 100\,\text{nF}$, and $C_{dl} = 2\,\mu\text{F}$. This therefore represents a system in which the orientation of all the particles is the same. Curve II represents a non-uniform system and was obtained using the following values: $R_b^{1-4} = 10\,\Omega$, $R_b^{5-7} = 1\,\Omega$, $R_b^{8-10} = 19\,\Omega$, and $C_g^{1-10} = 100\,\text{nF}$. It is unlikely that $C_g$ will vary to any significant extent; $C_{dl}$ is again equal to $2\,\mu\text{F}$. Only a slight distortion of the semi-circular part of the curve was observed when various particle orientations were considered such that the conductance of the individual particles varied by a factor of 20 times. For considerable distortion to occur, it appears that a 'spread' of conductances greater than 20 times is required, and this type of distortion is likely to be most evident in sinters of materials such as the $\beta$-aluminas, where conduction takes place essentially only in one plane. In such a system, the conductivity of a sinter is expected to be approximately $\frac{2}{3}$ that of a single crystal that is orientated in the conducting mode, assuming a random distribution of orientations in the sinter.

*Inter-Particle Impedance.* In the previous section, any impedance between particles was ignored. The effect of inter-particle impedance can be examined by considering a section across the sample in which all the particles have the same orientation and the individual inter-particle impedances have the same value. This impedance may arise from a 'constriction resistance' of the type described by Bauerle,[21] or from a constriction resistance due to the small area of contact between the grains and an associated voidage, or from a true charge-transfer resistance between particles. Figure 18(a) shows a physical representation of the system and Figure 18(b) is the anticipated equivalent circuit.

Armstrong *et al.* have simulated this behaviour[22] to show the expected shape in the complex plane and its variation with electrode/electrolyte double-layer capacitance and inter-particle resistance. The results obtained for a system consisting of eleven particles in series are shown in Figures 18(c) and (d). In both figures, $C_g = 0$, $R_b = 0.91\,\Omega$, and $C_a = 10\,\mu\text{F}$. The effect of varying $C_{dl}$ is shown in Figure 18(c), where $R_a = 10\,\Omega$ and $C_{dl} = 2\,\mu\text{F}$ (curve *a*), $20\,\mu\text{F}$ (curve *b*), and $200\,\mu\text{F}$ (curve *c*). The consequences of varying $R_a$ are shown in Figure 18(d), where $C_{dl} = 20\,\mu\text{F}$ and $R_a = 0$ (curve *a*), $1\,\Omega$ (curve *b*), and $10\,\Omega$ (curve *c*).

In the simulations, $C_g$ was set equal to zero. In a real system, $C_g$ will be finite, and if $C_g R_b$ becomes comparable with $C_a R_a$ there will be an overlap of time constants, and the two impedances will be indistinguishable in the complex plane. This situuation is likely to arise when the second type of constriction resistance is present. In the other cases considered, however, the time constant associated with inter-particle effects is likely to be larger than $C_g R_b$.

*Grain-boundary Conduction.* It is possible that conduction along the surfaces of the grains in a compressed powder may enhance its conductivity. This will give rise to an additional resistance and capacitance in parallel with the bulk resistance and the geometric capacitance of the particle. Figure 19(a) shows a physical representation of such a case, and the expected equivalent circuit for such a situation in which it is assumed that there is no inter-particle impedance is shown in Figure 19(b). Two simulations of the circuit are shown. In the first, (a), the contribution to the impedance made by grain-boundary conduction is zero, *i.e.* as in a single crystal. In the second, (b), the grain-boundary conduction has an impedance equal to that of the bulk. It can be seen that extrapolation of the straight-line portion of the impedance leads to an erroneously low value for the bulk resistance, compared with that of the single crystal. In addition, calculation of the dielectric constant of the solid ionic conductor on the basis of a simple parallel-plate condenser model leads to a value which is higher than expected.

[21] J. E. Bauerle, *J. Phys. Chem. Solids*, 1969, **30**, 2657.
[22] R. D. Armstrong, T. Dickinson, and P. M. Willis, *J. Electroanal. Chem. Interfacial Electrochem.*, 1974, **53**, 389.

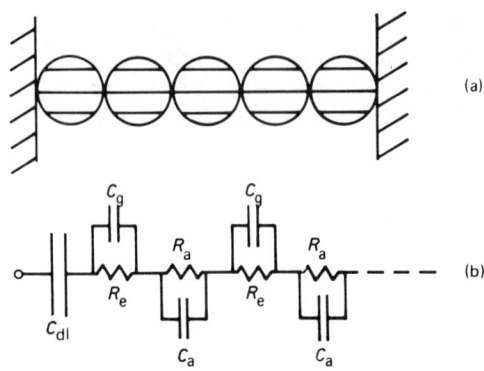

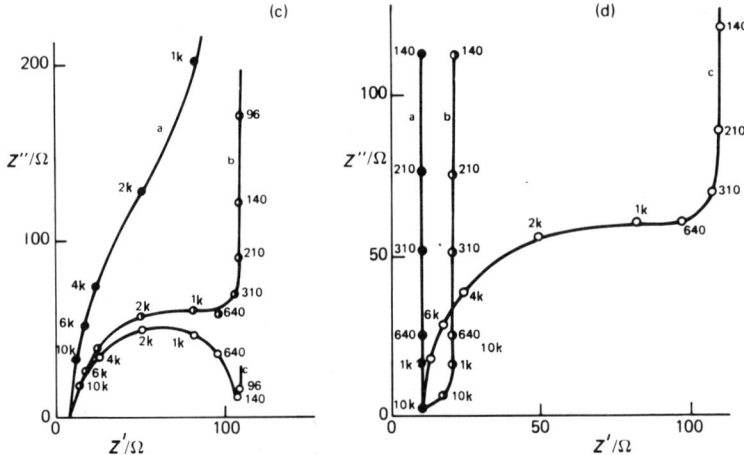

**Figure 18** *The impedance of a sinter with inter-particle impedances.*
  *(a) Physical representation.*
  *(b) Anticipated equivalent circuit.*
  *(c) Behaviour in the complex impedance plane, showing the effect of varying the electrode/electrolyte double-layer capacitance.*
  *(d) Behaviour in the complex impedance plane, showing the effect of varying the resistance between particles.*

## 5 Instrumentation

The simplest experimental arrangement for measuring the impedance of an electrochemical cell involves the use of an a.c. bridge. The cell, which is in one arm of the bridge, is balanced against decade capacitance and resistance boxes in another arm. If, at the balance point of the bridge, the components of the impedance are $R_s$ and $C_s$, then $Z' = R_s$ and $Z'' = 1/\omega C_s$. This type of experimental procedure is extremely laborious, since a balance point must be obtained at each frequency of study, and to

# The Application of A.C. Impedance Methods to Solid Electrolytes

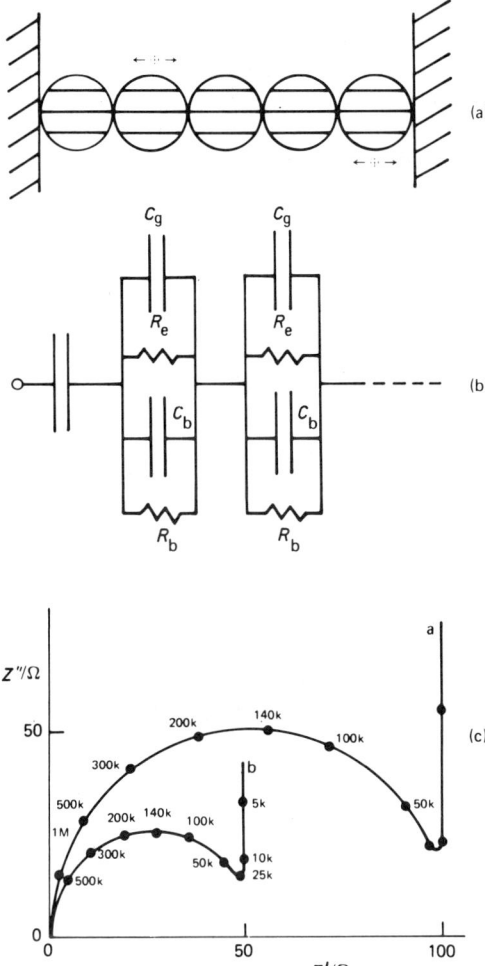

**Figure 19** *The impedance of a sinter with grain-boundary conduction.*
*(a) Physical representation*
*(b) Anticipated equivalent circuit*
*(c) Behaviour in the complex impedance plane*

obtain such a balance point takes several minutes, even for an experienced operator. This type of system will operate satisfactorily at frequencies up to 30 kHz, although frequencies below 100 Hz are difficult to deal with unless a very sophisticated detection system is used.

A considerable improvement can be made by measuring, directly, the in-phase and quadrature components of the potential and the current, using phase-sensitive detectors. A typical arrangement for such a device is shown in Figure 20. This gives a d.c. output that is related to both the amplitude and the phase of the sine-wave input. By suitable phase shifts, one can obtain the real and imaginary parts of the impedance. Thus, in this case, no balancing is involved, but to obtain an impedance spectrum with approximately five points for each decade in frequency down to 1 Hz may take two hours or more. Below this frequency, the experimental time increases rapidly, due to the time needed for accurate averaging by the phase-sensitive detector. Furthermore, the data obtained generally need to be processed numerically before the impedance can be obtained.

A more sophisticated experimental arrangement is based upon the Solartron 1170 Series frequency-response analysers. This system removes a great deal of the tedium involved in simple impedance measurements and allows a large frequency range to be covered more quickly, and usually more accurately. Armstrong et al. have covered the use of such a system in detail.[23] Essentially, the frequency-response analyser consists of a programmable generator which provides the perturbing sinusoidal signal, a correlator to analyse the response of the system, and a display to present the results. Figure 21 shows a typical experimental arrangement, using a Solartron 1174 frequency-response analyser.

As discussed in Section 2, the fundamental response of a system to a sinusoidal perturbation of the form $\Delta E \sin \omega t$ will be of the form $\Delta i \sin (\omega t + \theta)$.

The frequency-response analyser analyses the response of the system by a correlation process to determine $\Delta i$ and $\theta$. The correlation process has the advantages of rejecting all harmonics present in the system output and minimizing the effects of random noise. A single measurement at a particular frequency can be made by programming the generator with the required frequency and signal amplitude. More usually, however, the generator is programmed to sweep through a large frequency range by choosing suitable values of the maximum frequency (up to 1 MHz with the model 1174 and up to 10 kHz with the 1172), the minimum frequency (down to 0.1 mHz), and the number of points per decade at which measurements are to be taken. The instrument will then take measurements sequentially, in either direction, at equally spaced intervals, either on a logarithmic or a

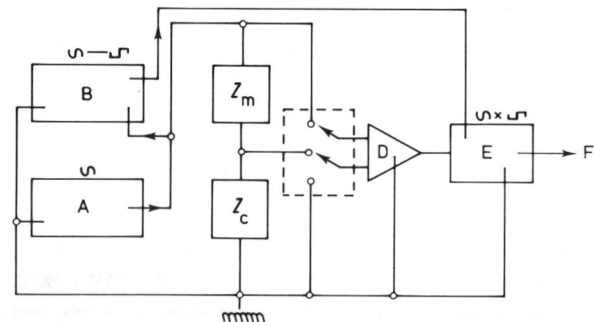

**Figure 20** *Block diagram of a typical phase-sensitive type of impedance analyser. A, sine-wave generator; B, sine-to-square converter; D, buffer amplifier; E, phase-sensitive detector; F, read-out.*

[23] R. D. Armstrong, M. F. Bell, and A. A. Metcalfe, *J. Electroanal. Chem. Interfacial Electrochem.*, 1977, **77**, 287.

# The Application of A.C. Impedance Methods to Solid Electrolytes

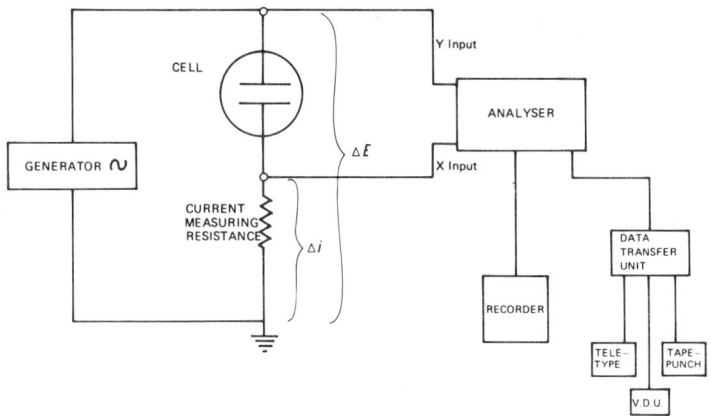

**Figure 21** *Schematic diagram of the experimental arrangement for automated impedance measurements using the Solartron 1174 frequency-response analyser.*

linear scale, over the required range. The response is given once a measurement has been completed and can be displayed in any of three possible notations: amplitude ($A$) and phase angle ($\theta$) relative to the output signal, log $A$ and $\theta$, or the real and imaginary parts of the impedance. The results, together with the measurement frequency, are then transferred to either a teletype printer or a tape punch, and they can also be plotted simultaneously on an $xy$ recorder to give the complex-plane impedance spectrum directly. The tape-punch facility is particularly useful as it allows the results to be fed directly into a computer for subsequent analysis. Alternatively, it is possible to link the system directly to a computer or a programmable calculator, which can then be used to control the system as well as for on-line processing of data.

Using the Solartron frequency-response analyser, measurements can be performed very quickly. For instance, in an ideal situation, the frequency range from 10 kHz to 10 Hz can be covered in 30 s, at ten points per decade. However, it must be emphasized that at frequencies below 10 Hz, the time for a single measurement is equal to the period of the signal ($10^3$ s at $10^{-3}$ Hz), and this becomes the major contribution to the total experimental time.

The methods of impedance determination which have been mentioned up to this point can all be considered as direct methods. The indirect determination of the transfer function was first performed by a numerical and graphical Laplace transform of the response of the current to a potential step[24] (or the potential response to a galvanostatic pulse).[25] Although this graphical treatment can be as time-consuming as the bridge measurements, an essential advantage is that a whole spectrum can be obtained in one measurement. Pilla and Doblhoffer[26, 27] used digital computing techniques to evaluate directly the Laplace and Fourier transforms, using a potentiostatic pulse of small amplitude. It should be noted that, in order to achieve

[24] M. D. Wijnen, *Recl. Trav. Chim. Pays-Bas*, 1960, **72**, 1203.
[25] E. Levart and E. Poirier d'Ange d'Orsay, *J. Electroanal. Chem. Interfacial Electrochem.*, 1966, **12**, 277; 1968, **19**, 335.
[26] A. A. Pilla, *J. Electrochem. Soc.*, 1970, **117**, 567; 1971, **118**, 1295.
[27] K. Doblhoffer and A. A. Pilla, *J. Electroanal. Chem. Interfacial Electrochem.*, 1972, **39**, 91.

accuracy comparable with that of a bridge, a large number of step responses must be averaged before the Fourier transform is found. Creason and Smith[28–30] have automated this approach, using on-line computing techniques involving the so-called 'fast Fourier transform algorithm'. They compare the various test signals possible and discuss the efficiency of the transform, the occurrence of extraneous noise, and the properties of the investigated system.[29] With Smith's equipment, measurement of an impedance spectrum over the frequency range 1 kHz to 10 Hz takes 2 s. The minimum theoretical time for such a spectrum is $\tau = 1/\omega_{min}$, where $\omega_{min}$ is the lowest radial frequency employed. Various authors have used this technique to determine the faradaic admittance, and the signal waveforms used include aperiodic transients[26, 27, 31, 32] and pseudo-random white noise.[28, 29, 33–35] Another indirect technique involves the analysis of the spontaneous fluctuations of a system around its steady state.[33, 36–38]

## 6 Some Experimental Factors which Affect the Analysis of Impedance Measurements

Several factors can affect the experimental results which are obtained in a.c. impedance studies, and a number have been discussed by Dickinson and Whitfield.[39] The most important points are summarized below:

(1) The first factor which they discuss involves the cell design and the choice of materials. In most measurements made on solid electrolytes, the cell is held together in some form of clamp, which is separated from the cell by some type of insulating material. These, together, set up an impedance which is in parallel with the actual impedance under study. The importance of this impedance depends upon the values of its components relative to the cell impedance. This additional parallel circuit is also an ideal loop for picking up radiated noise. Only careful cell design and selection of materials can reduce the effect of this parallel circuit to an acceptable level.

(2) The input impedance of the detection system, including input resistance, input capacitance, lead capacitance (and sometimes resistance), and their frequency dependence, can be of considerable importance; this may give rise to some additional structure in the complex impedance plane. The contribution from the leads can be minimized by keeping them as short as is practical.

[28] S. C. Creason and D. E. Smith, *J. Electroanal. Chem. Interfacial Electrochem.*, 1972, **36**, App. 1; 1972, **40**, App. 1.
[29] S. C. Creason and D. E. Smith, *J. Electroanal. Chem. Interfacial Electrochem.*, 1973, **47**, 9.
[30] S. C. Creason and D. E. Smith, *Anal. Chem.*, 1973, **45**, 2401.
[31] H. P. Van Leeuwen, D. J. Kooijnan, M. Sluyters-Rehbach, and J. H. Sluyters, *J. Electroanal. Chem. Interfacial Electrochem.*, 1969, **23**, 475.
[32] O. Dupre La Tour, J. Farcy-Bravacos, E. Levart, P. Malaterre, and D. Schumann, *J. Electroanal. Chem. Interfacial Electrochem.*, 1972, **39**, 241.
[33] P. Bindra, M. Fleischmann, J. W. Oldfield, and D. Singleton, 'Intermediates in Electrochemical Reactions', Oxford, 1973; *Faraday Discuss. Chem. Soc.*, 1974, **56**, 180.
[34] M. Ichise, Y. Nagayanagi, and T. Kojima, *J. Electroanal. Chem. Interfacial Electrochem.*, 1974, **49**, 187.
[35] G. Blanc, I. Epelboin, C. Gabrielli, and M. Keddam, *Electrochim. Acta*, 1975, **20**, 599.
[36] G. C. Barker, *J. Electroanal. Chem. Interfacial Electrochem.*, 1969, **21**, 127.
[37] B. Grafov, *Elektrokhimiya*, 1970, **6**, 188.
[38] G. Blanc, C. Gabrielli, and M. Keddam, *Electrochim. Acta*, 1975, **20**, 687.
[39] T. Dickinson and R. Whitfield, *Electrochim. Acta*, 1977, **22**, 385.

(3) The current-measuring resistance (see Figure 21) is often assumed to be a pure resistor. This may not, in fact, be the case, and the resistor may also have some capacitance or inductance. Careful selection of resistors can make the capacitance very close to zero. A useful method of checking this is to measure the current-measuring resistors on an a.c. bridge and to select resistors which have very low or zero capacitances. However, despite such precautions, stray capacitance effects can arise from solder joints and lengths of wire in the circuit containing the current-measuring resistor.

It can be beneficial to check a system by using it to measure resistors and capacitors of known value and thus to determine values of additonal impedances which may be present. This procedure will also allow improvements to be made on a system. It is also good practice to take a series of measurements using current-measuring resistances of different values.

The choice of electrode material can also have some fairly remarkable effects upon the impedance and lead to misinterpretation of the results. For example, if a metal electrode is applied in the form of a paste, and the organic constitutents are not completely removed, extra structure may appear in the impedance spectrum,[40] and one may be led to incorrect conclusions. Evaporation *in vacuo* or sputtering of electrodes onto the electrolyte removes the possibility of such erroneous results.

In measurements using blocking electrodes, the finish on the electrolyte surface is very important. Armstrong and Burnham have made a detailed study of the effect of surface roughness on the impedance of the Au/Na–$\beta$-alumina interface, using both single-crystal Na–$\beta$-alumina and polycrystalline Na–$\beta/\beta''$-alumina.[41] Impedance measurements were made with materials which had been polished to different degrees, using diamond pastes, and variations in the angle of the interfacial impedance line were studied. Figure 22 shows typical impedance spectra for (a) a single-crystal and (b) a sintered disc, both polished to 1 $\mu$m, and for (c) a sintered disc polished to 25 $\mu$m. The results showed that, in the case of the polycrystalline material, deviation of the slope from 90° is caused by the pulling out of crystallites from the electrolyte surface as it is polished or roughened. This pitting did not occur with the single crystal, and the slope of the impedance spectrum remained constant, to within 2°, over the range of surface finish of 1—90 $\mu$m.

The measured series capacitance, $C_s$, was shown to be markedly affected by polishing, both with the single crystal and with the polycrystalline material. When the slope of the impedance spectrum is close to 90°, $2C_s/A$ is equal to the double-layer capacitance per unit area. This was found to decrease with increasing roughness of the electrolyte surface, but where the slope deviated greatly from the vertical, the measured series capacitance is no longer a simple double-layer capacitance, and it increased again. In the paper, Armstrong and Burnham suggest possible explanations for this change in measured capacitance.

Another factor which can have an adverse effect on the results obtained is the failure to maintain uniform pressure on the cell. Figure 23 illustrates such an effect. Curve (a) shows a normal complex-plane impedance spectrum for the cell Au|Na–$\beta/\beta''$-alumina (polycrystalline)|Au in which uniform pressure has been

[40] R. P. Buck, D. E. Mathis, and R. K. Rhodes, *J. Electroanal. Chem. Interfacial Electrochem.*, 1977, **80**, 245.
[41] R. D. Armstrong and R. A. Burnham, *J. Electroanal. Chem. Interfacial Electrochem.*, 1976, **72**, 257.

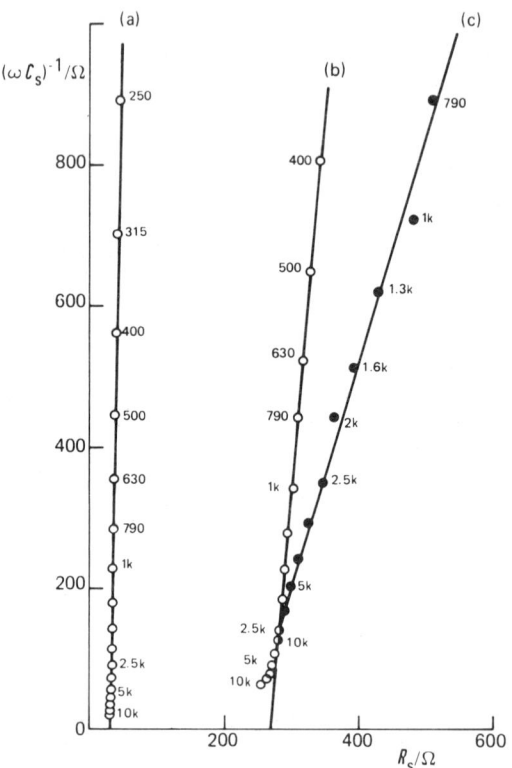

**Figure 22** *Comparison of complex-plane impedance spectra obtained with (a) single-crystal sodium–β-alumina, with gold electrodes, and (b) polycrystalline sodium–β/β''-alumina, with gold electrodes, both polished to 1 μm, and with (c) polycrystalline material polished to 25 μm. Frequencies/Hz.*

exerted to bring the brass contact plates up to the gold electrodes. Curve (b) shows the effect of lowering the contact pressure on one side of each electrode. It can be seen that the spectrum is shifted to a higher resistance and some distortion is introduced into it, particularly in the interfacial region. The distortion may be wrongly taken as being due to surface roughness effects, so care must be taken to exert sufficient, evenly distributed pressure on the cell.

## 7 Review of Some Experimental Results

**Stabilized Zirconia.**—The solid-state dissolution of oxides such as CaO or $Y_2O_3$ in $ZrO_2$ introduces a concentration of oxide ion vacancies that is determined by the 'oxygen deficit' of the solute compared with $ZrO_2$. The resulting 'doped' or stabilized zirconia is an extrinsic oxygen-ion-conducting solid electrolyte. Thus, CaO dissolves as $CaOV_O^{2-}$ and $Y_2O_3$ as $Y_2O_3V_O^{2-}$ (where $V_O^{2-}$ represents an oxygen-ion vacancy), introducing vacancies which carry the oxygen-ion current so that, at elevated temperatures, the conductivity of stabilized zirconia is almost exclusively

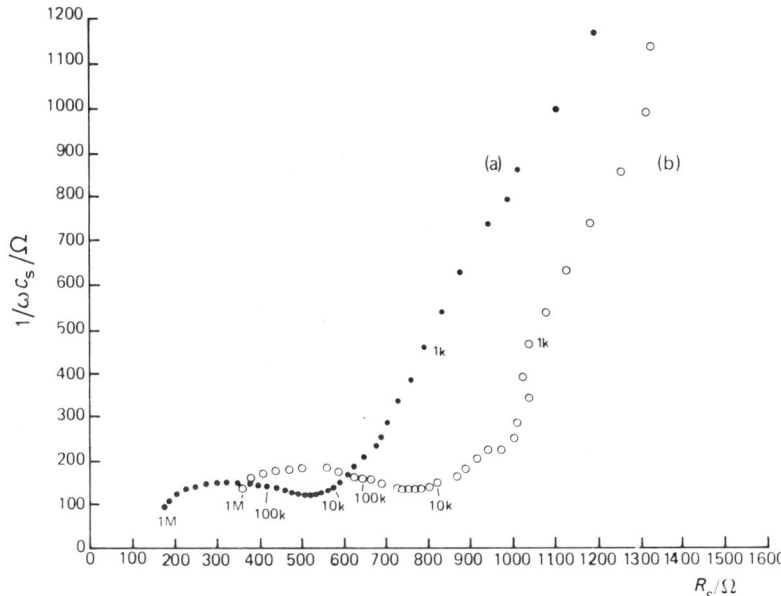

**Figure 23** *Complex-plane impedance of the cell* Au|Na–$\beta/\beta''$-alumina|Au, *showing the effects of exerting uneven pressure upon the cell: (a) normal spectrum with evenly distributed pressure; and (b) spectrum obtained after reducing the pressure on one side of each electrode.*

due to the motion of oxygen vacancies. The concentration of oxygen vacancies is typically $\sim 4\,\text{mol}\,l^{-1}$. Since the concentration of ionic defects is fixed by the contents of dopant, the effect of temperature is merely to increase the ionic mobility. Impedance studies have been made on both yttria- and calcia-stabilized zirconia.

Bauerle[21] made pioneering studies of the cell $O_2$, $Pt|(ZrO_2)_{0.9}(Y_2O_3)_{0.1}|Pt$, $O_2$, using a complex admittance method. A typical example of the complex-plane admittance spectrum obtained with the above cell, for a sample of moderate purity, is shown in Figure 24(a). The equivalent circuit for the cell, as proposed by Bauerle, is given in Figure 24(b) (an extra capacitance term for $C_g$ should be included across $R_3$). In the complex impedance plane this equivalent circuit would give rise to three semi-circles. Using samples with varying ratio of electrode area to length, and by measuring the d.c. conductivity of specimens (thus excluding electrode effects), Bauerle was able to ascertain which parts of the equivalent circuit of Figure 24(b) correspond to the electrode region and which parts correspond to the bulk properties of the electrolyte. The results showed that $R_1$ and $C_1$ correspond to electrode-polarization processes while $R_2$, $C_2$, and $R_3$ correspond to processes occurring in the bulk of the specimen.

Bauerle proposed that $C_1$ represents the double-layer capacitance of the electrode and $R_1$ represents an effective resistance for the overall electrode reaction $\tfrac{1}{2}O_2$ (gas)$+2e^-$ (Pt)$\rightleftharpoons O^{2-}$ (electrolyte). The capacitance $C_1$ was shown to be essentially independent of electrode preparation, temperature, and partial pressure of oxygen.

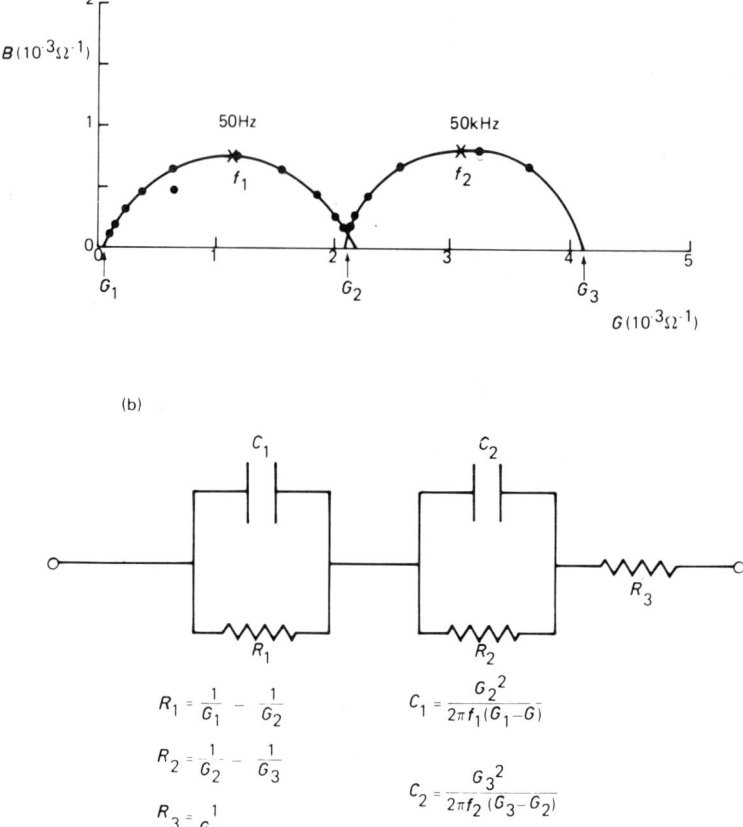

**Figure 24** (a) Complex-plane admittance spectrum for the cell $O_2,Pt|(ZrO_2)_{0.9}(Y_2O_3)_{0.1}|Pt,O_2$ at 600 °C.
(b) Proposed equivalent circuit

However, $R_1$ was found to be very sensitive to variations in such parameters. The results obtained suggested that the electrode resistance $R_1$ depends upon the partial pressure of oxygen *via* an adsorption–desorption equilibrium of $O_2$ molecules on the porous platinum surface, and it depends upon temperature *via* a thermally activated process which is most probably either the dissociation of oxygen molecules (on platinum) or an electron-transfer reaction at the electrode surface.

In his measurements, Bauerle used electrolyte specimens of moderate purity and a specimen of high purity. As expected, the electrolyte parameters $R_2$, $C_2$, and $R_3$ were independent of the partial pressure of oxygen. The factor $C_2$ was also shown to be independent of temperature. However, the quantities $R_2$ and $(R_2+R_3)$ showed very similar temperature dependence, thus suggesting that the resistances $R_2$ and $R_3$ arise from the same type of physical process. With the specimen of high purity, the high-frequency semi-circle, which Bauerle assigned to the electrolyte proper, was absent. Thus, the high-purity sample behaved as a simple resistance.

Bauerle suggests that the most likely explanation for these results is that the normal conduction of the $O^{2-}$ ion of the electrolyte is partially blocked at the grain boundaries by an impurity phase which is present there. On this basis, $R_3$ would correspond to the resistance within the grains, $R_2$ would correspond to a 'constriction' or inter-granular resistance, and $C_2$ would correspond to the capacitance across the impurity phase region. The fact that $R_2$ and $C_2$ arise from impurity regions is confirmed by the absence of the second semi-circle in the case of the high-purity sample. Microprobe studies carried out on the less pure samples also indicated the presence of a second phase in the grain boundaries.

Schouler and his co-workers[42–44] have measured the impedance of the cell $O_2$, $Ag|(ZrO_2)_{0.91}(Y_2O_3)_{0.09}|Ag,O_2$; they obtained only one semi-circle in the complex impedance plane. This semi-circle was assigned to interfacial phenomena and these results are, therefore, in good agreement with those of Bauerle.

Beekmans and Heyne have measured the impedance of various samples of calcia-stabilized zirconia.[45] Microprobe analysis of the samples was also carried out, and the authors attempted to correlate these results with the impedance behaviour obtained, and the known composition of the various samples. The results obtained indicate that a 'second phase' (as found by Bauerle), present between the grains from which the ceramic is composed, may greatly affect the electrical properties of the material. The amount and composition of this second phase determine whether or not a particular sample of zirconia is suitable for electrochemical applications. The second phase is formed, during the firing process of the ceramic, by the impurities present. These impurities are sometimes deliberately introduced by the manufacturer in order to improve some characteristic of the product or else to lower the sintering temperature. Beekmans and Heyne proposed the equivalent circuit of Figure 25(a) to represent the impedance of a zirconia sample. Here, $R_2$ is the inter-granular resistance arising from the second phase, $C_2$ is the inter-granular capacitance from the second phase, and $R_3$ is the resistance within the grains. The equivalent circuit of Figure 25(a) can be simplified to that of Figure 25(b). Networks of this type [Figure 25(b)] were shown, by the authors, to give impedance behaviour which was very similar to that of the zirconia samples, with the values of $R_2$ and $C_2$ depending upon the composition of the second phase.

From comparison of microprobe analysis results with impedance behaviour, Beekmans and Heyne were able to show direct dependence of $R_2$ and $C_2$ upon the presence and composition of the second phase. They were also able to make some general deductions about the composition of the second phase. For example, it was shown that a high silica content in the second phase strongly promotes intergrain polarization while a high alumina content in the second phase does not affect the impedance behaviour of zirconia samples, but it does lower the sintering temperature.

Since the stabilized zirconias have such a high concentration of mobile charge

---

[42] E. Schouler, M. Kleitz, and C. Déportes, *J. Chim. Phys. Phys.-Chim. Biol.*, 1973, **70**, 923.
[43] E. Schouler, G. Giroud, and M. Kleitz, *J. Chim. Phys. Phys.-Chim. Biol.*, 1973, **70**, 1309.
[44] E. Schouler and M. Kleitz, *J. Electroanal. Chem. Interfacial Electrochem.*, 1975, **64**, 135.
[45] N. M. Beekmans and L. Heyne, *Electrochim. Acta*, 1976, **21**, 303.

(a)

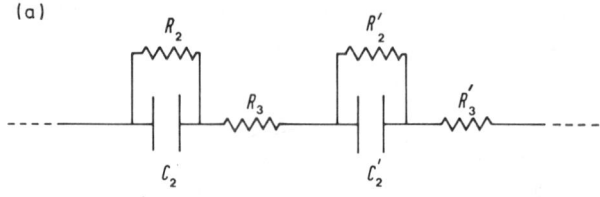

(b)

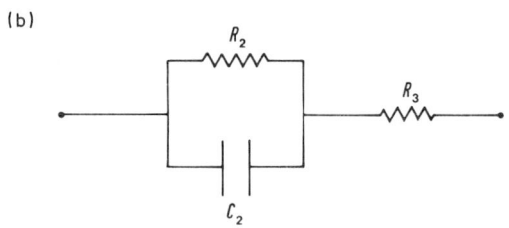

**Figure 25** *Equivalent circuits to represent the impedance of a zirconia sample.*

carriers, it is expected that the double-layer capacitance at a metal/zirconia interface should be determined by finite-ion-size effects. This would result in the double-layer capacitance being independent of both temperature and variations in the potential difference between the electrode and the electrolyte. At the present time, however, insufficient information is available to be able to say if this is in fact the case.

**The β-Aluminas.**—A great deal of attention has been given to the β-alumina type of solid electrolytes, in view of their potential as conductors of sodium ion in the sodium–sulphur cell.[46,47] In practice, the electrolyte employed in such cells is in the form of a polycrystalline ceramic mixture of sodium–β-alumina and sodium–β″-alumina. Sodium–β-alumina has the ideal composition $Na_2O \cdot 11Al_2O_3$, while Na–β″-alumina has the ideal composition $Na_2O \cdot 5Al_2O_3$. However, both materials deviate from ideality, with the β-form containing excess sodium (typically 25%) and the β″-form being sodium-deficient.

The general features of the β-alumina structure have been determined by Bragg *et al.*[48] and Beevers *et al.*,[49,50] a refinement of the structure being made by Peters *et al.*[51] In both forms, all of the sodium ions lie in a plane of relatively low atom density, and these planes are separated from one another by blocks of close-packed oxygen atoms ('spinel blocks') that are four oxygen atoms thick. These 'spinel blocks' extend infinitely in a direction perpendicular to the *c* axis of the crystals.

The major difference between β- and β″-alumina is that β-alumina contains two

---

[46] R. Bauer, W. Haar, H. Kleinschmager, G. Weddigen, and W. Fischer, *J. Power Sources*, 1976/77, **1**, 109.
[47] G. May and I. W. Jones, *Metall. Mater. Technol.*, August 1976, p. 427.
[48] W. L. Bragg, C. Gottfried, and J. West, *Z. Krist*, 1931, **77**, 255.
[49] C. A. Beevers and M. A. S. Ross, *Z. Krist.*, 1937, **97**, 59.
[50] C. A. Beevers and S. Brohult, *Z. Krist.*, 1936, **95**, 472.
[51] C. Peters, M. Bettman, J. Moore, and M. Glick, *Acta Crystallogr. Sect. B*, 1971, **27**, 1826.

'spinel blocks' that are related by a two-fold screw axis in a hexagonal unit cell, while $\beta''$-alumina contains three 'spinel blocks', related by a three-fold screw axis. Most of the polycrystalline materials which have been produced also contain small amounts of magnesium, lithium, or both, since these elements have been shown to stabilize the $\beta''$-form, which has a higher conductivity than the $\beta$-form.

Many different solid electrolytes with a structure of the $\beta$-alumina type have also been produced, either by substituting different mobile cations for $Na^+$ (e.g. $Tl^+$, $Ag^+$, $Li^+$, or $K^+$) or by substituting other elements for aluminium in the 'spinel blocks' (e.g. Ga or Fe).

Since these compounds have very high concentrations of mobile charge carriers (in Na–$\beta''$-alumina the concentration of $Na^+$ is approximately $10\,mol\,l^{-1}$, but in the conduction planes the local concentration of $Na^+$ is 2—3 times greater), it is expected that finite-ion-size effects will be very important in the a.c. impedance behaviour of such electrolytes.

Armstrong et al. first reported the complex-plane impedance spectrum of single-crystal Na–$\beta$-alumina[52] with gold (blocking) electrodes, obtaining a straight line which made an angle of 89° with the real axis, close to the theoretical value of 90°. The spectrum obtained was consistent with the equivalent circuit of Figure 10, with the straight line arising from the double-layer capacitance, and, from the results, values of $C_{dl}$ and $R_b$ were calculated. Figure 22(a) shows a typical spectrum.

In a paper in 1973,[53] Armstrong and co-workers investigated the impedance of the sodium/Na–$\beta$-alumina interphase at elevated temperatures, using polycrystalline Na–$\beta/\beta''$-alumina. This study demonstrated, for cases in which the electrolyte surface was completely wetted by the molten sodium electrodes, that the transfer of $Na^+$ ions across the electrode/electrolyte interface was very rapid (with an exchange current at 150 °C in excess of $500\,mA\,cm^{-2}$). This fact, together with the lack of any interfacial impedance in the impedance spectrum, shows the sodium electrodes to be non-blocking. Figure 26 shows a typical complex-plane impedance

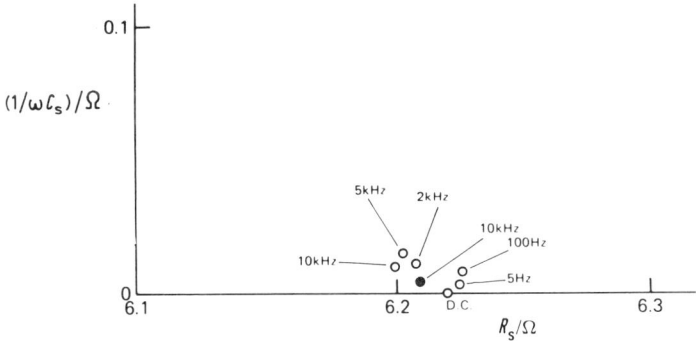

**Figure 26** *Complex-plane impedance spectrum, at 150° C, for the cell Na|polycrystalline Na–$\beta/\beta''$-alumina|Na, using a polished electrolyte disc.* ● *Wheatstone bridge:* ○ *p.s.d.*

[52] R. D. Armstrong, T. Dickinson, and P. M. Willis, *J. Electroanal. Chem. Interfacial Electrochem.*, 1976, **67**, 121.
[53] R. D. Armstrong, T. Dickinson, and J. Turner, *J. Electroanal. Chem. Interfacial Electrochem.*, 1973, **44**, 157.

spectrum for the cell Na|polycrystalline Na–$\beta/\beta''$-alumina|Na at 150 °C. Here, the a.c. impedance shows only scatter about the d.c. resistance value, corresponding to the total resistance of the sample.

Blocking electrode measurements have also been made on polycrystalline Na–$\beta/\beta''$-alumina samples, using gold electrodes.[54] The complex-plane impedance spectrum in this case again shows a straight line at low frequencies, arising from $C_{dl}$. However, in this case, the line deviates from the vertical due to surface roughness effects. At higher frequencies, additional structure is present in the impedance spectrum, due to bulk and inter-granular effects. Possible sources of these impedances have been discussed in Section 4 of this review. Many workers have successfully been able to separate the bulk and inter-granular contributions to the impedance of polycrystalline Na–$\beta/\beta''$-alumina.[55–62]

The variation of $C_s$, the series capacitance of the cell Au|polycrystalline Na–$\beta/\beta''$-alumina|Au, with applied potential has been investigated over a range of potentials from $-3$ to $+8$ V *versus* the $I_2$|NaI couple.[63] The variation of $C_s$ with potential was shown to be very small (about 15% between 0 and $+8$ V), as shown in Figure 27,

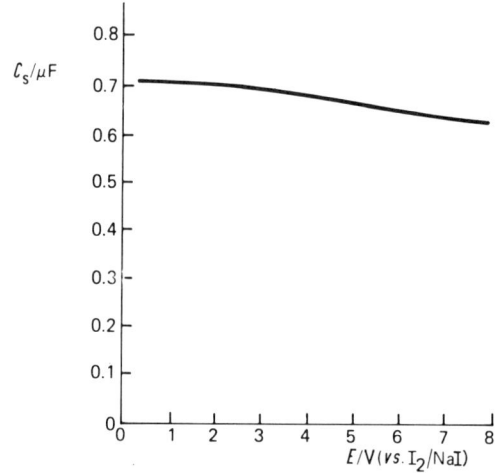

**Figure 27** *Dependence of $C_s$, the series capacitance of the cell Au|polycrystalline $\beta/\beta''$-alumina Au, on positive potential of the gold electrode (versus $I_2$|NaI).*

---

[54] R. D. Armstrong, T. Dickinson, and P. M. Willis, *J. Electroanal. Chem. Interfacial Electrochem.*, 1974, **53**, 389.
[55] G. C. Farrington, *J. Electrochem. Soc.*, 1974, **121**, 1314.
[56] G. C. Farrington, *J. Electrochem. Soc.*, 1976, **123**, 1213.
[57] R. W. Powers and S. P. Mitoff, *J. Electrochem. Soc.*, 1975, **122**, 226.
[58] I. M. Hodge, M. D. Ingram, and A. R. West, *J. Electroanal. Chem. Interfacial Electrochem.*, 1975, **58**, 429.
[59] R. J. Grant, M. D. Ingram, and A. R. West, *J. Electroanal. Chem. Interfacial Electrochem.*, 1976, **72**, 397.
[60] I. M. Hodge, M. D. Ingram, and A. R. West, *J. Electroanal. Chem. Interfacial Electrochem.*, 1976, **74**, 125.
[61] R. J. Grant, M. D. Ingram, and A. R. West, *Electrochim. Acta*, 1977, **22**, 729.
[62] A. Hooper, *J. Phys. D*, 1977, **10**, 1487.
[63] R. D. Armstrong, R. A. Burnham, and P. M. Willis, *J. Electroanal. Chem. Interfacial Electrochem.*, 1976, **67**, 111.

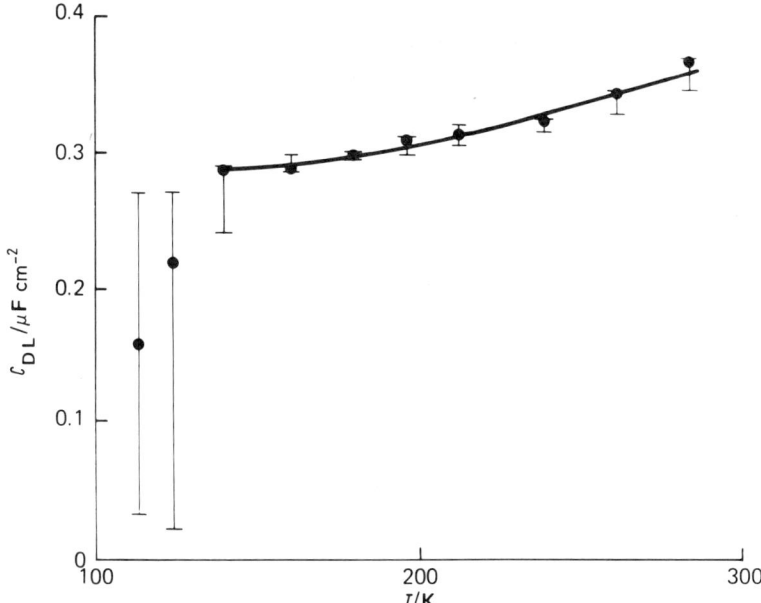

**Figure 28** *Dependence of $C_{dl}$, the double-layer capacitance of the cell Au|Na-$\beta$-alumina (single crystal)| Au, upon temperature. The points on the curve are the values of $C_{dl}$ calculated at 10 Hz, while the error bars indicate the approximate range of values of double-layer capacitance in the frequency range studied.*

thus supporting Armstrong's view that $C_{dl}$ arises from finite-ion-size effects rather than from a diffuse space charge within the electrolyte, as proposed by Macdonald.

Armstrong and Archer have studied the variation with temperature of the double-layer capacitance of the Au/Na-$\beta$-alumina interface.[64] The results obtained showed the double-layer capacitance to be essentially invariant with temperature (Figure 28), once again indicating that $C_{dl}$ arises from finite-ion-size effects, since if it arose from a diffuse space charge within the electrolyte one would expect $C_{dl}$ to decrease as the temperature increased.

Grant, Hodge, Ingram, and West have made a number of studies on both single-crystal and polycrystalline Na-$\beta$-alumina, using impedance and complex modulus techniques.[58—61] Their results indicate that the bulk effects of both single-crystal and polycrystalline materials, at low temperatures, cannot arise from a simple parallel combination of $C_g$ and $R_b$ but rather from a distribution of parameters, giving arise to a 'flattened' semi-circle with its centre below the real axis, in the complex impedance plane, and broadening of the peak in the electric modulus spectrum.

**Silver Rubidium Iodide (S.R.I.) and Related Compounds.**—In recent years, silver rubidium iodide, $Ag_4RbI_5$, has been the subject of a large number of electrochemical studies, including many a.c. impedance investigations. These impedance

[64] R. D. Armstrong and W. I. Archer, *J. Electroanal. Chem. Interfacial Electrochem.*, 1978, **87**, 221.

investigations have involved both two-electrode and three-electrode measurements.

The structure of $\alpha$-Ag$_4$RbI$_5$ has been investigated by Bradley and Greene[1] and by Geller.[2] Both studies showed S.R.I. to have a cubic stucture, and the more detailed work of Geller showed that in a unit cell, composed of four formula units, the sixteen Ag$^+$ ions are randomly distributed over 56 tetrahedral sites. The Ag$^+$ ion transport number in S.R.I. has been reported to be $1.00 \pm 0.01$,[65—67] which means that the conductivity of S.R.I. may be ascribed to the mobility of Ag$^+$ ions, with negligible contributions from Rb$^+$, I$^-$, or electrons.

The S.R.I./electrode interphase is of great interest to electrochemists because of the variety of reactions which are possible, depending upon the potential of the interphase and the type of electrode used. Armstrong *et al.* have investigated the anodic dissolution of silver into S.R.I.,[68—70] the deposition of silver from S.R.I. onto various electrodes,[71] and the discharge of iodide ions onto platinum and vitreous carbon electrodes.[72]

In a paper in 1974,[69] Armstrong and co-workers investigated the behaviour of the Ag/Ag$_4$RbI$_5$ interphase in the potential range 0 to +30 mV relative to the Ag|Ag$^+$ couple at a pressure of 200 kg cm$^{-2}$. A silver wire was used as a reference electrode. The results obtained were interpreted on the basis of a mechanism involving rate-determining nucleation and growth of holes in the metal surface at low overpotentials (<30 mV), changing to charge-transfer control at high overpotentials. The amount of information available from the early results was limited for two reasons. First, the impedance spectra could not be measured accurately down to low frequencies using the available instrumentation; and secondly, the theory of the electrocrystallization process which takes place was not available.

More recently, however, these measurements on the Ag/Ag$_4$RbI$_5$ interphase have been repeated, using a Solartron 1172 frequency-response analyser.[70] Also, Armstrong and Metcalfe have theoretically evaluated the impedance/frequency response for two-dimensional nucleation and growth under potentiostatic conditions.[17, 18] These two factors together have allowed a more useful discussion of the results obtained. The complex-plane impedance spectra obtained at 0, 15, and 25 mV *versus* Ag|Ag$^+$ are shown in Figure 29. At 35 and 50 mV the impedance spectra were not reproducible, probably because the erosion rate of the metal was too high for good contact to be maintained between the metal and the electrolyte.

It is possible to analyse the impedance spectra that are obtained in terms of the nucleation and growth mechanism. However, it is not possible to explain certain phenomena, such as the additional low-frequency capacitative behaviour at 25 mV. Steady-state results and impedance spectra gave rise to values for $n_c$, the number of atoms in the critical nucleus, which were in reasonable agreement with one another

[65] L. F. Topol and B. B. Owens, *J. Phys. Chem.*, 1968, **72**, 2106.
[66] H. Wedersich and W. V. Johnston, *J. Phys Chem. Solids*, 1969, **30**, 475.
[67] B. Scrosati, G. Germano, and G. Pistoia, *J. Electrochem. Soc.*, 1971, **118**, 86.
[68] R. D. Armstrong, T. Dickinson, H. R. Thirsk, and R. Whitfield, *J. Electroanal. Chem. Interfacial. Electrochem.*, 1971, **29**, 301.
[69] R. D. Armstrong, T. Dickinson, and P. M. Willis, *J. Electroanal. Chem. Interfacial Electrochem.*, 1974, **57**, 231.
[70] R. D. Armstrong and A. A. Metcalfe, *J. Electroanal. Chem. Interfacial Electrochem.*, 1978, **88**, 187.
[71] R. D. Armstrong, T. Dickinson, and P. M. Willis, *J. Electroanal. Chem. Interfacial Electrochem.*, 1975, **59**, 281.
[72] R. D. Armstrong, T. Dickinson, and P. M. Willis, *J. Electroanal. Chem. Interfacial Electrochem.*, 1973, **48**, 47.

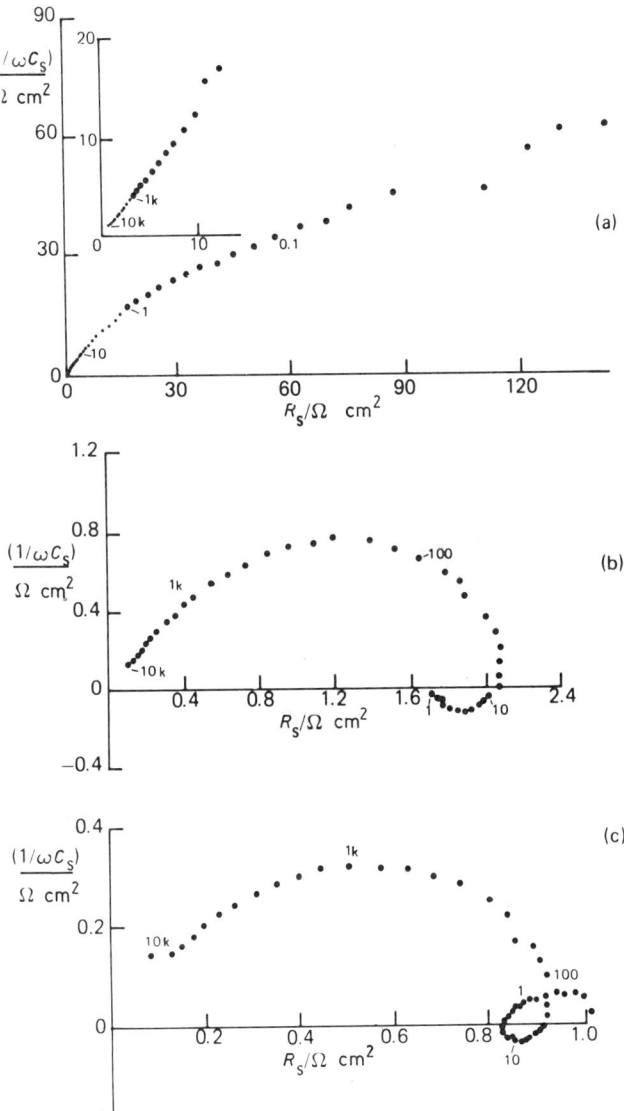

**Figure 29** *Complex-plane impedance spectra for the Ag/Ag$_4$RbI$_5$ interphase at anodic overpotentials of (a) 0, (b) 15, and (c) 25 mV versus Ag|Ag$^+$ at a pressure of 200 kg cm$^{-2}$. Figures represent frequencies/Hz (ten points per decade of frequency).*

and also consistent with the mechanism of dissolution being rate-determining nucleation and growth of two-dimensional holes in the electrode surface at low overpotentials, changing to charge-transfer control at an overpotential of about 35 mV, when $n_c$ becomes less than unity. The frequency of the maximum of the low-frequency shape was found to increase with potential, and this is also in agreement with the theory for two-dimensional nucleation and growth.

Armstrong and co-workers have also studied the behaviour of the interphase between copper electrodes and two different stabilized forms of ionically conducting CuBr.[73] A copper wire was employed as the reference electrode in three-electrode measurements. Impedance results, together with those from voltage sweeps, showed that a Tafel relationship exists between $i$ and $\eta$ for the Cu|Cu$^+$ reaction, provided that $\eta > 30$ mV. This was attributed to a slow charge-transfer reaction. The results showed a discrepancy between $i_0$ obtained by extrapolation and $i_0$ from the impedance measurements, and this was thought to be due to crystallization effects which became important at low values of $\eta$. This behaviour, therefore, parallels that found in the S.R.I. case.

The impedances of the Pt/Ag$_4$RbI$_5$, C/Ag$_4$RbI$_5$, and C+RbI$_3$/Ag$_4$RbI$_5$ interphases have also been studied at anodic potentials.[72]

Figure 30 shows typical complex-plane impedance spectra for the Pt/Ag$_4$RbI$_5$ interphase at varying anodic potentials versus Ag|Ag$^+$. Figure 30(a), at $+300$ mV, indicates an almost purely capacitative impedance, and the extrapolated resistance at high frequency corresponds to the resistance of the electrolyte between the working and reference electrodes. The slight deviation from purely capacitative behaviour may be due to surface roughness or else to the occurrence of a slight faradiac process. Figure 30(b) shows the spectrum obtained at $+650$ mV, and this is typical of the range $+640$ to $+670$ mV. This spectrum is composed of a high-frequency semi-circle which is coupled to a Warburg, or diffusion, impedance at low frequency (see Figures 8 and 14). This behaviour is consistent with a single-step charge-transfer process at its equilibrium potential. From the spectra in the range $+640$ to $+670$ mV, Armstrong *et al.* were able to obtain values for $i_0$ and hence show a pseudo-Tafel relationship between $i_0$ and $E$. At $+680$ mV, a dramatic change was observed in the impedance spectrum, as shown in Figure 30(c). This spectrum is consistent with the formation of a highly resistive layer on the electrode surface.

The C/Ag$_4$RbI$_5$ interphase showed very similar impedance behaviour to that of the Pt/Ag$_4$RbI$_5$ interphase at a potential of $+300$ mV versus Ag|Ag$^+$. Figure 31 shows the complex-plane impedance spectrum for this interphase at $+660$ mV versus Ag|Ag$^+$. In this case, the high-frequency semi-circle is very distorted and much larger than in the Pt/Ag$_4$RbI$_5$ case. Once more, a Warburg impedance was observed at low frequencies.

Figure 32 shows a typical complex-plane impedance spectrum for the C+RbI$_3$/Ag$_4$RbI$_5$ interphase at its open-circuit d.c. potential ($+666$ mV versus Ag|Ag$^+$). Again, a charge-transfer semi-circle is observed at high frequencies, changing to a Warburg impedance at low frequencies.

A pseudo-Tafel slope obtained in the Pt/Ag$_4$RbI$_5$ case is consistent with the discharge mechanism being

$$I^- \rightleftharpoons I\cdot + e^-$$

$$I\cdot + I^- \rightleftharpoons I_2 + e^-$$

[73] R. D. Armstrong, T. Dickinson, and K. Taylor, *J. Electroanal. Chem. Interfacial Electrochem.*, 1974, **57**, 157.

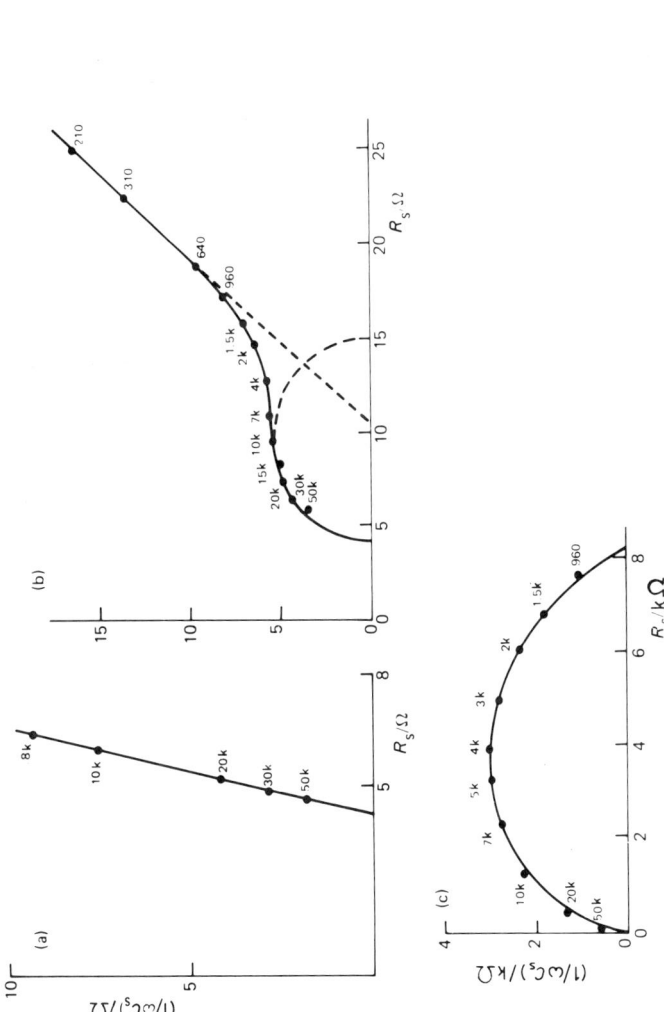

**Figure 30** Complex-plane impedance spectra for the $Pt/Ag_4RbI_5$ interphase at potentials versus $Ag|Ag^+$ of (a) 300 (b) 650, and (c) 685 mV. Figures represent frequencies/Hz.

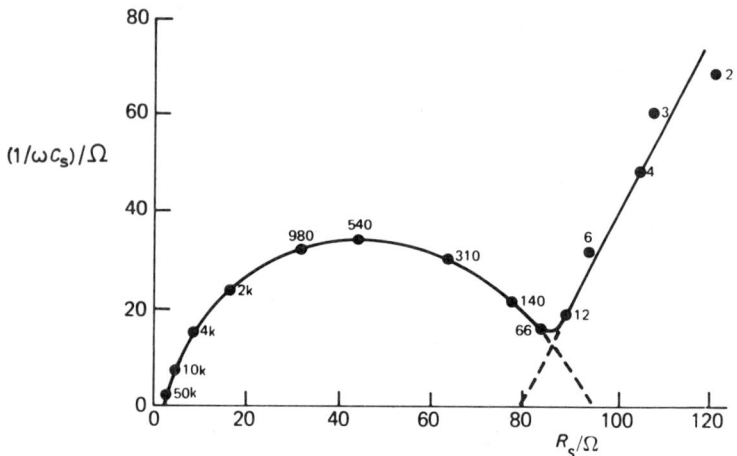

**Figure 31** *Complex-plane impedance spectrum for the C/Ag$_4$RbI$_5$ interphase at a d.c. bias potential of 660 mV versus Ag|Ag$^+$. Figures represent frequencies/Hz.*

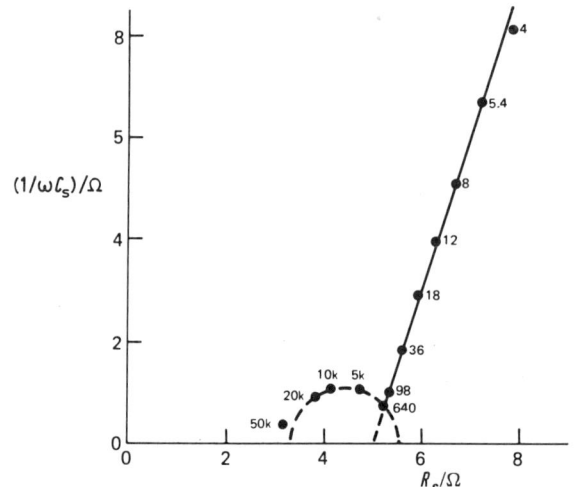

**Figure 32** *Complex-plane impedance spectrum for the C+RbI$_3$/Ag$_4$RbI$_5$ interphase at its open-circuit d.c. potential of 666 mV versus Ag|Ag$^+$. Figures represent frequencies/Hz.*

In the C/Ag$_4$RbI$_5$ case, the impedance behaviour indicated a much slower interfacial reaction, while in the C+RbI$_3$/Ag$_4$RbI$_5$ case a much smaller charge-transfer resistance, which is comparable to the Pt case, was observed. This was thought to be due to the catalytic effect of the graphite, in the electrode, upon the electrochemical oxidation of I$^-$ to I$_2$.

Armstrong *et al.* have also studied the anodic decomposition of Cu$_7$C$_7$H$_{15}$N$_4$Br$_8$

at a Pt electrode, using a copper wire reference electrode.[74] In this case, oxidation of $Cu^+$ to $Cu^{2+}$ was shown to occur, as opposed to oxidation of the halide, as in the case of S.R.I., and the results obtained were consistent with the mechanism being a first step involving oxidation at the electrode followed by diffusion of the newly formed $Cu^{2+}$ ions away from the electrode to a more stable position where they are less mobile.

The impedance of S.R.I. in a blocking situation has also been studied, using the cell $Pt|Ag_4RbI_5|Pt$.[54] The impedance was measured at zero pressure and at a pressure of $1000\,kg\,cm^{-2}$. At both pressures, a nearly vertical line was obtained in the complex plane, indicating a mainly capacitative impedance, with the high-frequency intercept on the real axis corresponding to the resistance of the electrolyte. Deviations from the vertical line were attributed to surface roughness effects. Double-layer capacitance values were also calculated and shown to increase with increasing pressure. This is consistent with a model in which there is greater electrode/electrolyte contact with increased pressure. Since the impedance at the two pressures was very similar, Armstrong *et al.* concluded that S.R.I. sinters at very low pressures, and particle–particle effects are impossible to detect.

With the solid electrolyte silver tungstato-iodide ($Ag_6WO_4I_4$), Armstrong, Dickinson, and Taylor were able to measure the impedance both in the presence of inter-granular effects and in their absence.[75] Two different forms of the electrolyte were studied; a polycrystalline sample and a vitreous modification. With both types of sample material, blocked and non-blocked investigations were made, using stainless steel and silver electrodes respectively. Figure 18 shows the anticipated effect of inter-particle impedances.

Using 18/8 stainless steel blocking electrodes, the pressure dependence of the impedance was determined for both materials. The behaviour of the polycrystalline modification (Figure 33) was very similar to S.R.I., as expected, and became independent of pressure at loads greater than $1 \times 10^3\,kg\,cm^{-2}$. The behaviour of the vitreous modification (Figure 34) was, however, more complicated. The impedance became independent only when the pressure exceeded $4 \times 10^3\,kg\,cm^{-2}$, and the shape of the plot at lower pressure was similar to that obtained with powdered Na–$\beta$-alumina.[54] These spectra are as expected from the equivalent circuit of Figure 18(b) when there is an overlap in the time constants. The inter-granular resistance would be expected to decrease with increasing pressure as the grains become more firmly contacted, and this is in fact seen in Figure 34.

Using the cell $Ag|Ag_6WO_4I_4|Ag$ and a silver wire reference electrode, Armstrong *et al.* measured steady-state current–voltage curves as a function of pressure. These measurements gave rise to Tafel plots which are typical for $Ag|Ag^+$ reactions found with other silver ion conductors. The pressure dependence was also found to be identical to that found for the impedance with blocking electrodes. Complex-plane impedance spectra were also measured at various overpotentials and applied pressures, using this cell. Figure 35 shows the impedance spectra obtained with the polycrystalline material. The spectra show a charge-transfer semi-circle, which

---

[74] R. D. Armstrong, T. Dickinson, and K. Taylor, *J. Electroanal. Chem. Interfacial Electrochem.*, 1975, **64**, 155.
[75] R. D. Armstrong, T. Dickinson, and K. Taylor, *J. Electroanal. Chem. Interfacial Electrochem.*, 1977, **78**, 45.

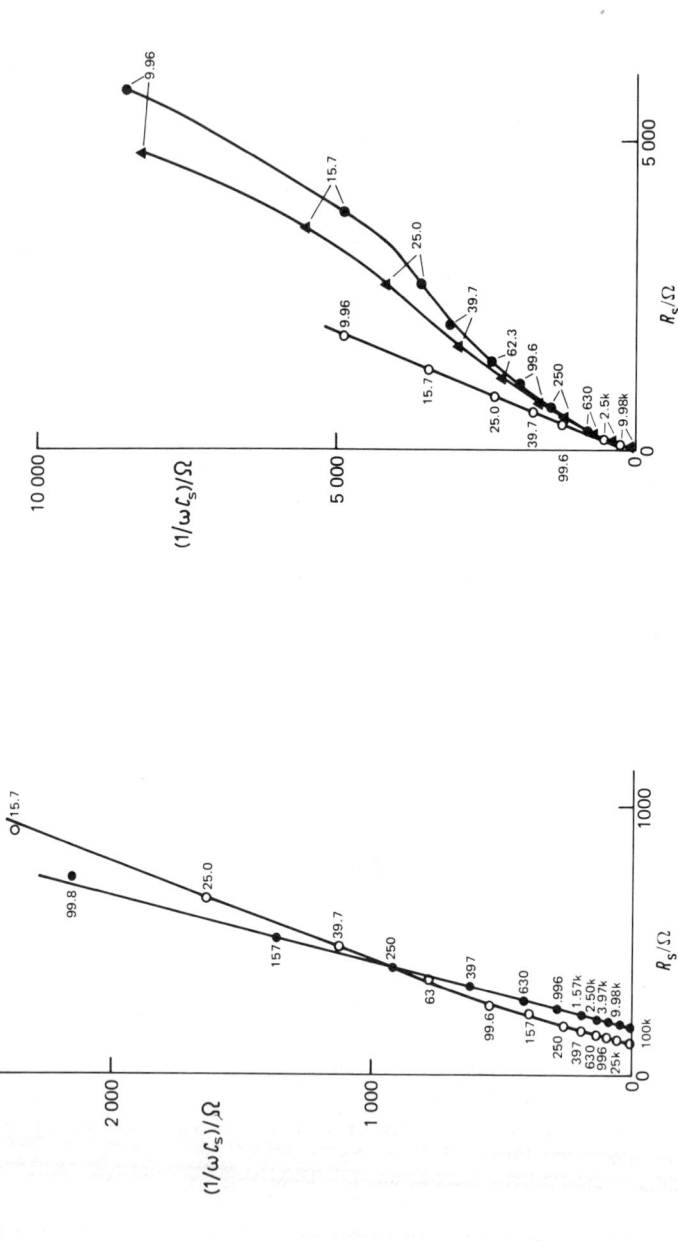

**Figure 33** *Impedance–pressure plots with stainless-steel electrodes on polycrystalline* $Ag_6WO_4I_4$. (●) 500 kg cm$^{-2}$; (○) 1000 and 1500 kg cm$^{-2}$. *Figures represent frequencies/Hz.*

**Figure 34** *Impedance–pressure plots with stainless-steel electrodes on vitreous* $Ag_6WO_4I_4$. (●) 500 kg cm$^{-2}$; (▲) 1000 kg cm$^{-2}$; (○) 4000 and 4500 kg cm$^{-2}$. *Figures represent frequencies/Hz.*

# The Application of A.C. Impedance Methods to Solid Electrolytes

becomes smaller as the overpotential increases. The other feature, an inductive semi-circle, was thought to be due to electrocrystallization effects, which are expected to predominate at low overpotentials.[17, 18]

The more complicated impedance spectra obtained with the vitreous material, at various pressures, are shown in Figure 36. At the lowest pressure [Figure 36(a)] two overlapping semi-circles occur. The second semi-circle is due to inter-granular effects while the first is the charge-transfer semi-circle due to the exchange reaction. As the pressure upon the electrolyte is increased, the inter-granular semi-circle becomes smaller, and at a pressure of $3 \times 10^3 \text{kg cm}^{-2}$ [Figure 36(c)] an inductive loop appears once again. Finally, at a pressure of $4 \times 10^3 \text{kg cm}^{-2}$ [Figure 36(d)] the spectrum contains only the charge-transfer semi-circle and an inductive loop. This shows that the pressure is high enough to remove inter-granular effects. At this pressure, the vitreous modification was shown to have impedance behaviour similar to that of the polycrystalline material.

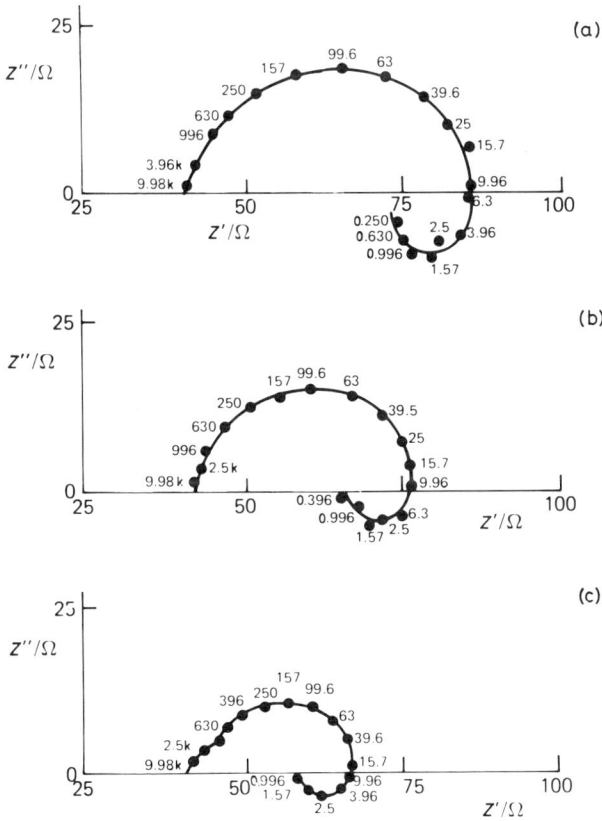

**Figure 35** Complex-plane impedance spectra for $Ag_6WO_4I_4$ (polycrystalline), using silver electrodes of area 0.2 cm$^2$. Pressure is 1000 kg cm$^{-2}$, and the figures represent frequency/Hz. (a) Overpotential is $+30$ mV; (b) overpotential is $+40$ mV; (c) overpotential is $+50$ mV.

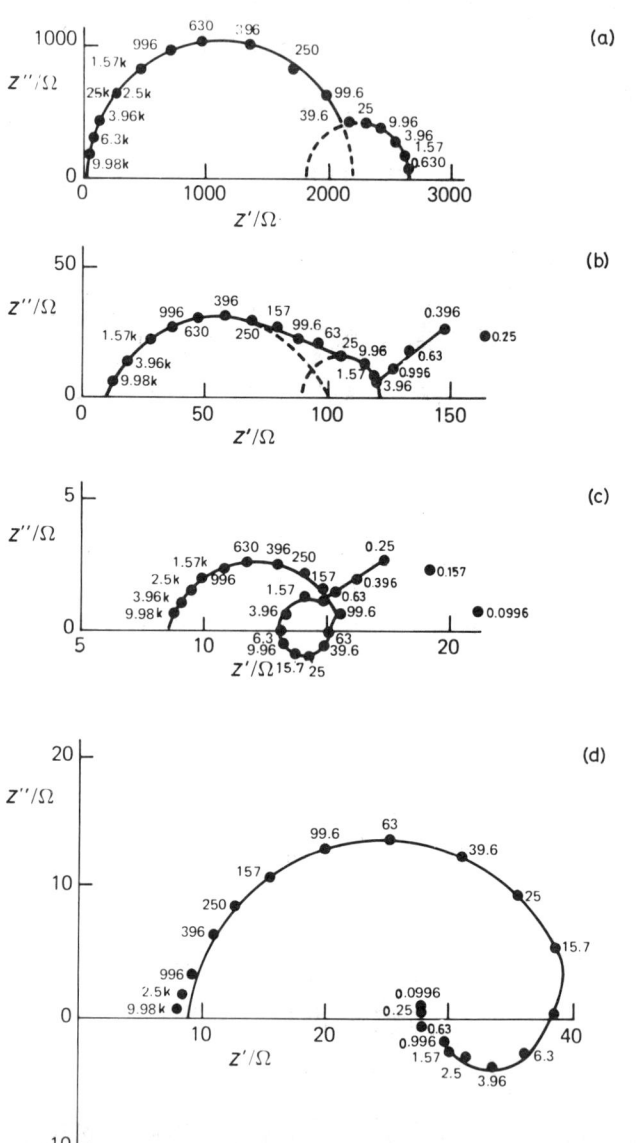

**Figure 36** *Complex-plane impedance spectra for* $Ag_6WO_4I_4$ *(vitreous), using silver electrodes. (a) The pressure is* $1 \times 10^3 \, kg\,cm^{-2}$, *and the overpotential* $+400\,mV$; *(b) pressure* $2 \times 10^3 \, kg\,cm^{-2}$, *overpotential* $+50\,mV$; *(c) pressure* $3 \times 10^3 \, kg\,cm^{-2}$, *overpotential* $+40\,mV$; *and (d) pressure* $4 \times 10^3 \, kg\,cm^{-2}$, *overpotential* $+40\,mV$.

**Silver Halides.**—The 'pure' silver halides are, at room temperature, extrinsic ionic conductors, *i.e.* the ionic conduction arises from defects in the material which, in turn, arise from inclusion of impurities in the crystal. Thus, conduction is by means of mobile defect species, primarily interstitial silver ions and vacancies. Funke has written a very comprehensive review of AgI-type solid electrolytes, including structural properties and ionic conductivity.[76] For silver halides, the concentration of mobile defect species is typically $4 \times 10^{-5}$ mol l$^{-1}$ at room temperature. Since these compounds have fairly low concentrations of mobile charge carriers, it is expected that the impedance of a cell with non-blocking electrodes on a silver halide electrolyte should agree very closely with that predicted by Macdonald.

Buck and co-workers have investigated the impedance of cells in which silver chloride is the electrolyte.[40] The electrodes used were graphite, silver, and electrolyte solution contacts, and measurements were made using single crystals of both purified silver halides and doped materials.

For the ideal non-blocking situation for an ionic conductor, without rate-limiting surface kinetics, and in which charge-carrier generation is sufficiently rapid to avoid concentration polarization, Macdonald predicts that the high-frequency complex-plane impedance spectrum should show a semi-circle centred on the real axis, with a radius of $R_b/2$, and that the equivalent circuit consists of $R_b$ and $C_g$ in parallel. Macdonald also predicts the absence of any further features in the complex plane.

All of the crystals used by Buck showed the expected semi-circle in the complex impedance plane at high frequency (above 100 Hz). However, in all cases, slight

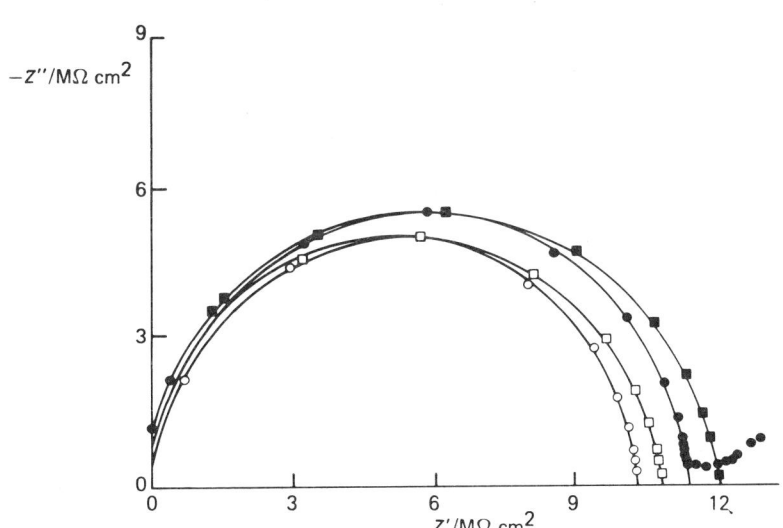

**Figure 37** *Complex-plane impedance spectra for single crystals of silver chloride.* (□) *First solution contacts;* (●) *graphite D.A.G. contacts;* (■) *vapour-deposited silver contacts;* (○) *second solution contacts.*

[76] K. Funke, *Progr. Solid State Chem.*, 1976, **11**, 345.

deviations from ideality occurred, causing depression of the semi-circle and deviation of measured points from the semi-circle. Polished crystals were shown to fit the theoretical predictions better than the rough crystals. These deviations from ideality may thus be due to surface roughness at the electrodes or else to inhomogeneity in the crystals under study. Only when using vapour-deposited silver electrodes did Buck find any evidence for an interfacial impedance (Figure 37). In all other cases, only one semi-circle was observed in the complex impedance plane, thus implying that the rate of generation of vacancies and interstitials is fast compared with the diffusion/migration of carriers.

Another prediction made by Macdonald is that $R_b$ should increase linearly, and $C_g$ should decrease inversely, with increasing crystal thickness. By employing crystals of varying thickness, Buck has shown that this, in fact, is the case.

Whilst Macdonald's detailed theories would be expected to apply in this case, unfortunately the high resistances of the samples did not permit Buck to investigate the range of frequencies where one would be expected to measure the interfacial impedance.

# 4
# The Electrical Double Layer

BY S. K. RANGARAJAN

*'Did you ever make real life into a drama?', said the Earl.*
*'Now just try. I've often amused myself that way'.*

The scope of this Report has been restricted to the principles of analysis and concepts underlying (I) the models for solvent structure at interfaces and (II) the theories of adsorption isotherms. Hence, literature published since 1974 pertaining to these two demarcated aspects alone is included here for discussion and comment.

The considerable number of publications that refer to related topics such as (a) improvements in the experimental study of charged interfaces in general, (b) classification and correlation of the experimental data for various electrodes and solvents, and (c) phenomenological coupling involving adsorption (*e.g.* effects on electrode kinetics, electrocrystallization, *etc.*) are omitted from the purview of this Report – not that they are uninteresting but because they are important enough to need more elaborate presentation.

## PART I: Models for Solvent Structure

### 1 What is the Information Sought?

Solvent structure at interfaces poses problems different from those met within the bulk phase.[1-4] This is made particularly interesting at *charged* interfaces, where the mediation by the electrode, aided by electrochemical variables such as charge ($\sigma$) or potential ($E$), is possible. Some of the questions connected with these investigations are:

(a) What is the extent of this region of homogeneity – a monolayer, a few, or many layers?
(b) How do we characterize this deviation from the bulk behaviour – by some order parameter reflecting the 'net' directional property of the polarization? By an index of 'broken hydrogen bonds',[5] or through the range of surface interactions?
(c) To what extent can one relate the observed, macro-information to the molecu-

---

[1] R. M. Reeves, in 'Modern Aspects of Electrochemistry', ed. J. O'M. Bockris and B. E. Conway, Plenum Press, New York, 1974, vol. 9, p. 239.
[2] B. E. Conway, *Croat. Chem. Acta,* 1976, **48**, 573.
[3] R. Parsons, *Croat.Chem. Acta,* 1976, **48**, 597.
[4] B. E. Conway, *Adv. Colloid Interface Sci.,* 1977, **8**, 91.
[5] J. O'M. Bockris and M. A. Habib, *J. Electroanal. Chem. Interfacial Electrochem.,* 1975, **65**, 473.

lar states of the solvent? In other words, how do we decouple the 'bulk' from the 'interface'?[6]

More specifically, in my view, information is to be sought at three levels:

I  *Average* properties, *e.g.* (1) the dipole potential and its dependence on thermodynamic variables[7–11] such as $\sigma$ and $T$, (2) electrostatic potential distribution,[12, 13] (3) the solvent excess entropy,[1, 14, 15] and (4) the entropy of formation of the interface.[1, 8, 15]

II *Structural* properties, *e.g.* (1) the nature of bonding at the interface, (2) the various molecular and electronic interactions operating therein, and (3) molecular distribution functions, especially the pairwise correlation functions.

III *Molecular* properties, *e.g.* (1) the chemisorption (bond) energy of the 'solvent–metal surface' system, (2) 'effective' polarizability and dipole moments, and (3) the characteristics of 'clusters', if they can be viewed as distinct chemical objects in the time and energy scales of our analysis.

Other aspects, by no means less interesting, such as (1) order–disorder transitions associated with solvents, (2) dielectric behaviour of the interface, and (3) correlations of interfacial properties with the metal on one hand[1, 16] and with the solvent on the other,[17, 18] are related to the above classification.

The literature indicates that the information becomes scarcer, on going down from level I to level III in the list given above. Sometimes, even the questions are not seen in a sharper focus. The present activity in the creation and study of models (see Sections 11 and 12) is a step towards retrieving (in part) structural information, and thus getting access to molecular 'parameters' (*e.g.* cluster size, metal–adsorbate energy, *etc.*). But there seems to be even less emphasis on level III, primarily because it is connected with the success in level II (there is a hierarchy in modelling). This is also due to the nature of experimental information available and the ability to pick out information pertaining to the above three levels from such studies.

Let us now look at the preliminary stages in describing and analysing the solvent structure. In view of recent strictures[6, 19–21] on the more conventional methodology, we shall first discuss the motivation and principles underlying the classical procedure.[22, 23]

[6] I. L. Cooper and J. A. Harrison, *J. Electroanal. Chem. Interfacial Electrochem.*, 1978, **86**, 425.
[7] S. Trasatti, *J. Electroanal. Chem. Interfacial Electrochem.*, 1977, **82**, 391.
[8] J. A. Harrison, J. E. B. Randles, and D. J. Schiffrin, *J. Electroanal. Chem. Interfacial Electrochem.*, 1973, **48**, 359.
[9] S. Trasatti, *J. Chem. Soc., Faraday Trans. 1*, 1974, **70**, 1752.
[10] S. Trasatti, *J. Electroanal. Chem. Interfacial Electrochem.*, 1974, **54**, 437.
[11] S. Trasatti, *J. Chim. Phys., Phys.-Chim. Biol.*, 1975, **72**, 561.
[12] M. A. V. Devanathan, *Trans. Faraday Soc.*, 1954, **50**, 373.
[13] J. R. Macdonald and C. A. Barlow, *Adv. Electrochem. Electrochem. Eng.*, 1976, **6**, 1.
[14] G. J. Hills and S. Hsieh, *J. Electroanal. Chem. Interfacial Electrochem.*, 1975, **58**, 289.
[15] M. A. Habib, in 'Modern Aspects of Electrochemistry', ed. J. O'M. Bockris and B. E. Conway, Plenum Press, New York, 1977, Vol. 12, p. 131.
[16] A. N. Frumkin, B. B. Damaskin, N. Grigorev, and I. A. Bagotskaya, *Electrochim. Acta*, 1974, **19**, 69.
[17] R. Payne, *Adv. Electrochem. Electrochem. Eng.*, 1970, **7**, 1.
[18] R. Parsons, *Electrochim. Acta*, 1976, **21**, 681.
[19] I. L. Cooper and J. A. Harrison, *Electrochim. Acta*, 1977, **22**, 519.
[20] I. L. Cooper and J. A. Harrison, *Electrochim. Acta*, 1977, **22**, 1361.
[21] I. L. Cooper and J. A. Harrison, *Electrochim. Acta*, 1977, **22**, 1365.
[22] P. Delahay, 'Double Layer and Electrode Kinetics', John Wiley–Interscience, New York, 1965.
[23] B. B. Damaskin, O. A. Petril, and V. V. Batrakov, 'Adsorption of Organic Compounds on Electrodes', Plenum Press, New York, 1971.

The Electrical Double Layer 205

## 2 The First Stages of Double-layer Modelling

Difficulties in studying such interfaces are many. Direct observations, as done in solid/vacuum interfaces, to provide relatively unambiguous electronic, molecular, and structural information are not easy to make in electrode/electrolyte interfaces. Various techniques,[24–43] e.g. ellipsometry,[25–29] reflection and transmission spectroscopy,[24, 30–33] Raman scattering and resonance,[34–38] photoemission,[39] and electrode immersion–emersion,[40–43] have, no doubt, been used to study adsorption. But even then, the specific information at the level of the interfacial solvent states is, if any, hidden.

The first stage in modelling is an appreciation of the extent to which the interacting elements of the system may be decoupled, with no serious loss of the faithfulness of representation. This depends on the length and energy scales of the model. Some of the difficulties in analysis lie in:
(A) the spatial extension of the interface (interacting layers);
(B) the various modes of interactions (molecular and electronic);
(C) the several ways in which the configurational states of the solvent can differ from the bulk; and
(D) deciding what information in the overall experimental observations is relevant to solvent models.

Historically, one recognizes strong short-range forces operating near the surface and the weaker long-range non-specific Coulombic interactions (hard-core repulsions apart) in the region further away from the surface.[44] Whereas the 'farther' region is populated by several chemical components present in the system, the 'inner' region is more discriminatory. Such a natural spatial division is reasonably served by postulating an outer Helmholtz plane (OHP), and the Gouy–Stern–Grahame model[22] carries this out. The regions resulting from such a separation can be

[24] A. Bewick and J. Robinson, *J. Electroanal. Chem. Interfacial Electrochem.*, 1975, **60**, 165.
[25] R. H. Muller, *Adv. Electrochem. Electrochem. Eng.*, 1973, **9**, 167.
[26] R. Parsons, *Sci. Prog. (London)*, 1977, **64**, 29.
[27] M. W. Humphreys and R. Parsons, *J. Electroanal. Chem. Interfacial Electrochem*, 1977, **82**, 369.
[28] K. Kunimatsu and R. Parsons, *J. Electroanal. Chem. Interfacial Electrochem*, 1979, **100**, 335.
[29] S. Gottesfeld and B. Reichman, *J. Electroanal. Chem. Interfacial Electrochem.*, 1976, **67**, 169.
[30] S. Gottesfeld and B. E. Conway, *J. Chem. Soc., Faraday Trans. 1.* 1974, **11**, 1793.
[31] J. D. E. McIntyre, *Adv. Electrochem. Electrochem. Eng.*, 1973, **9**, 61.
[32] W. R. Heineman and J. F. Goelz, *J. Electroanal. Chem. Interfacial Electrochem.*, 1978, **89**, 437.
[33] R. M. Lazorenko-Manevich, E. B. Brick, and Ya. M. Kolotyrkin, *Electrochim. Acta*, 1977, **22**, 151.
[34] R. L. Paul, A. J. McQuillan, P. J. Hendra, and M. Fleischmann, *J. Electroanal. Chem. Interfacial Electrochem.*, 1975, **66**, 248.
[35] D. L. Jeanmaire and R. P. van Duyne, *J. Electroanal. Chem. Interfacial Electrochem.*, 1977, **84**, 1.
[36] A. J. McQuillan, P. J. Hendra, and M. Fleischmann, *J. Electroanal. Chem. Interfacial Electrochem.*, 1975, **65**, 933.
[37] V. V. Marinyuk and R. M. Lazorenko-Manevich, *Sov. Electrochem. (Engl. Transl.)*, 1978, **14**, 381.
[38] G. Hagen, S. Glavaski, and E. Yeager, *J. Electroanal. Chem. Interfacial Electrochem.*, 1978, **88**, 269.
[39] Yu. A. Prishchepa, Z. A. Rotenberg, and Yu. V. Pleskov, *J. Electroanal. Chem. Interfacial Electrochem.*, 1975, **66**, 3.
[40] W. N. Hansen, C. L. Wang, and T. W. Humphreys, *J. Electroanal. Chem. Interfacial Electrochem.*, 1978, **93**, 87.
[41] W. N. Hansen, C. L. Wang, and T. W. Humphreys, *J. Electroanal. Chem. Interfacial Electrochem.*, 1978, **93**, 137.
[42] D. M. Kolb and W. N. Hansen, *Surf. Sci.*, 1979, **79**, 205.
[43] W. N. Hansen and D. M. Kolb, *J. Electroanal. Chem. Interfacial Electrochem.*, 1979, **100**, 493.
[44] B. E. Conway, *Sov. Electrochem. (Engl. Transl.)*, 1977, **13**, 695.

handled better, separately, from a theoretical angle,[45] since the interaction pictures are different in the two regions. This is equivalent to visualizing 'the whole' by either considering 'the inner' as the dominant and 'the OHP and beyond' as the perturbation, or *vice versa,* depending upon the strength of substrate–solvent interactions. Also, the space-charge layer theory (Gouy–Chapman) is better formulated (even if not solved correctly) than that of the inner region of solvent and specifically adsorbed states, where even the interaction picture is blurred. So much for factor (A).

The statistical-mechanical problems in describing 'the OHP and beyond' are different from those pertaining to the layer adjacent to the surface. Even though some interactions such as the hard-core effects and the Coulombic seem common, differences exist even in handling these! For example, the size effects are modelled as hard spheres in the 'bulk' region whereas the same notion for the so-called specific adsorbed molecule is treated as a 'lattice problem' (note that the equations of state are not the same for the 'localized' and 'mobile' cases). The Coulombic effects in the Gouy–Chapman case are analysed by employing the more familiar bulk solvent as the 'background continuum'. But the use of 'a continuum' or 'a background' in 'the inner region' is questionable, in view of the strong dependence of the polarization states of the interfacial solvent on field strength. Added to this, there exist novel interactions that are induced by the substrate or other adsorbates in the inner layer, and these many-body interactions will have to be modelled[56] while analysing the inner layer. The nature of mathematical approximations can also be different while considering the two regions. This is what was meant by saying that the interface can be subjected to closer scrutiny better once it has been separated than when considered as a single, complex unit.

To allow modelling the 'inner solvent region', it becomes essential to decouple other adsorbate effects, not by a *spatial* separation, but through an appropriate picture of interactions and their effects (factor B). Then, a look at the solvent configuration states through a more detailed model analysis becomes possible (factor C); (see Sections 11 and 12). Finally, the above process of reduction in models also provides an experimental counterpart, *viz.* to reduce the *observed (experimental)* data to a form that is relevant to solvent analysis (Sections 4 and 10).

Very briefly, we have sketched the reduction methodology in scientific analysis as applied to the double layer. It is not unlike other idealisations in different contexts; *e.g.,* replacing a 'continuum' by 'discreteness' and *vice versa,* or a 'frequency spread' by a central 'frequency' plus the 'broadening' mechanisms. There is, of course, an element of arbitrariness in all such simplifications.[6,44] The *ad hoc* feature here is the creation of artificially 'new interacting phases' in the process of handling the physics of an interphase![19–21] It is justified only if the new 'objects' are less coupled than those we started with. This may sometimes be overdone by introducing 'inner regions' within 'inner layers' and several 'outer layers (OHP)'. The reduction method then becomes counter-productive.

In this context, more immediate questions are: 'are we ready to incorporate properly the interactions between the inner-layer molecules and those outside?', 'can the solvent–adsorbate effects in the inner region be decoupled?', *etc.*

[45] S. K. Rangarajan, *J. Electroanal. Chem. Interfacial Electrochem.*, 1977, **82**, 93.

## 3 Primary Model Reduction

The theoretical procedure sketched above is represented schematically in Figure 1.

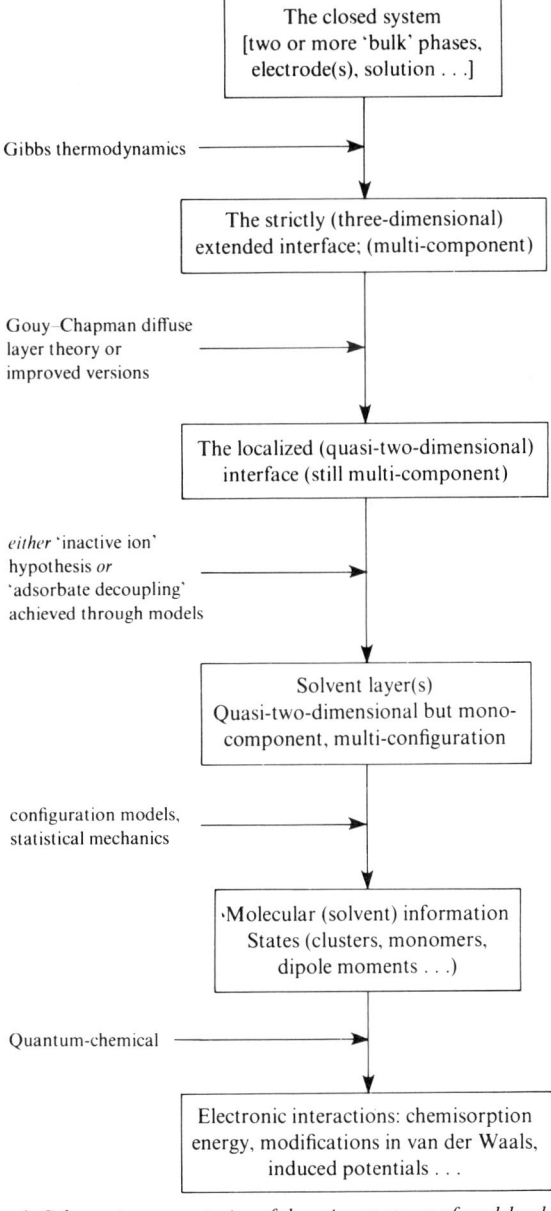

**Figure 1** *Schematic representation of the primary stages of model reduction.*

## 4 Primary Data Reduction

Corresponding to the scheme given in Figure 1, there is a scheme of data reduction too (see Figure 2).

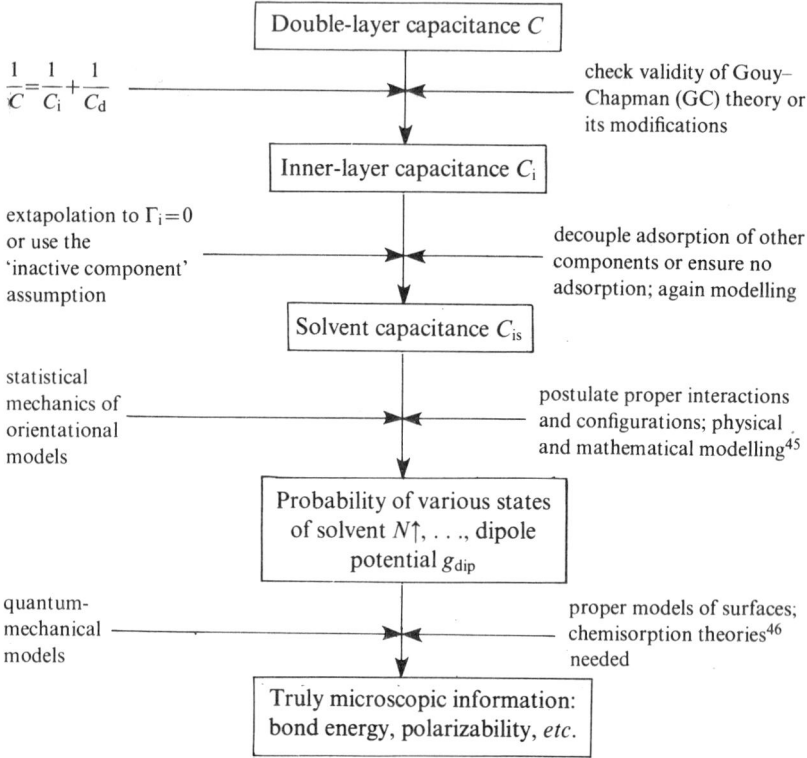

Figure 2 *Schematic description of data reduction*

## 5 Gouy Theory – Present Status

A cursory glance indicates the extent to which the models rely on diffuse-layer theory (GC); we do not discuss in this Report the interesting results on Gouy theory obtained from kinetic experiments used as double-layer probes.[47–49]

GC theory[22] has been accepted as a reluctant partner for long, beginning from the spill-over from the criticism of Debye–Hückel theory. Because of the molecular dimensions associated with the macro-calculations and the *ad hoc* nature of division

---

[46] R. Gomer, *Acc. Chem. Res.*, 1975, **8**, 420.
[47] F. C. Anson and B. A. Parkinson, *J. Electroanal. Chem. Interfacial Electrochem.*, 1977, **85**, 317.
[48] R. de Levie and M. Nemes, *J. Electroanal. Chem. Interfacial Electrochem.*, 1975, **58**, 123.
[49] M. J. Weaver and F. C. Anson, *J. Electroanal. Chem. Interfacial Electrochem.*, 1975, **65**, 711, 737, 759; 1977, **84**, 47.

of the interface,[50] its use has been severely criticized recently.[6, 16—18] Reservations expressed about Gouy theory fall under three classes: (a) on statistical-mechanical and self-consistency grounds (see this Section); (b) on what appropriate values to employ for parameters, e.g. dielectric constant, appearing in the classical theory (see Section 6), and (c) how the Gouy layer is coupled to the inner layer (e.g. boundary conditions of the Poisson–Boltzmann equation, co-adsorption of ions, etc.) (see Section 7).

The GC formulation has been the subject of several recent studies employing more proper correlation procedures, e.g. the mean-spherical[50—53] (MSA) and the hypernetted-chain approximations[54] (HNC). Particular mention must be made of reports by Blum,[50] Levine and Outhwaite,[51] and Henderson and Blum.[54] The primary task is to treat the ion-size effects consistently and, in view of the similarities with the Debye–Hückel theory, the advances in electrolyte theory have had an impact on the Gouy theory as well.

Levine and Outhwaite[51] consider the problem of the double layer from two angles, viz. the charging process (Guntelberg–Kirkwood–Loeb) and the MSA. Both of these lead to non-linear integro-difference-differential equations which are later suitably approximated. The formalism of Blum[50] does not resort to a division of the electrical double layer into two regions – the inner and the Gouy. Consequently, problems[51] arising out of an abrupt change in dielectric permittivity and distances of closest approach for ions to the metal must be handled more carefully. The criticism by Levine et al.[51, 55] refers to the above, as also to the use of an incorrect self-energy term and the improper evaluation of mean electrostatic potential.

Outhwaite et al.[55] have evaluated the potential distribution on the basis of their improved procedure.[51] The various stages in their analysis involve approximations for (1) the direct correlation function, $C_{ij}$, in the 'bulk' and (2) a consistently evaluated wall–ion direct correlation function, $C_{0i}$, Besides these, the relationships between (3) the radial distribution function, $g_{0i}(x)$, of an ion at a distance $x$ from the wall and the functions $\{C_{0i}, C_{ij}\}$ and between (4) the potential $\psi$ and $g_{0i}$ are also required.

A generalization of the Ornstein–Zernike equation has been used, besides the Poisson equation. The calculations assume the MSA (cf. $C_{0i}, C_{ij}$), ignore the details of the compact layer, employ a single scale factor for the size for all ions, and approximate the image terms (cf. the image effects[56] at distances of $\sim 3$—5 Å). The curves in Figure 3 compare the linearized GC theory with the modified version of Levine et al.[55] The oscillatory trend with $x$ for $\psi(x)$ may be seen in the MSA calculations.

Henderson and Blum[54] have shown how Gouy–Chapman theory can be deduced as an elegant application of HNC, by neglecting short-range terms.

50 L. Blum, *J. Phys. Chem.*, 1977, **81**, 136.
51 S. Levine and C. W. Outhwaite, *J. Chem. Soc., Faraday Trans. 2*, 1978, **74**, 1670.
52 C. W. Outhwaite, in 'Statistical Mechanics', ed. K. Singer (Specialist Periodical Reports), The Chemical Society, London, 1975, Vol. 2, p. 188.
53 D. Henderson and L. Blum, *J. Chem. Phys.*, 1978, **69**, 5441.
54 D. Henderson and L. Blum, *J. Electroanal. Chem. Interfacial Electrochem.*, 1978, **93**, 151.
55 C. W. Outhwaite, L. B. Bhuiyan, and S. Levine, *Chem. Phys. Lett.*, 1978, **64**, 150.
56 T. Takaishi, *Prog. Surf. Sci.*, 1977, **6**, 45.

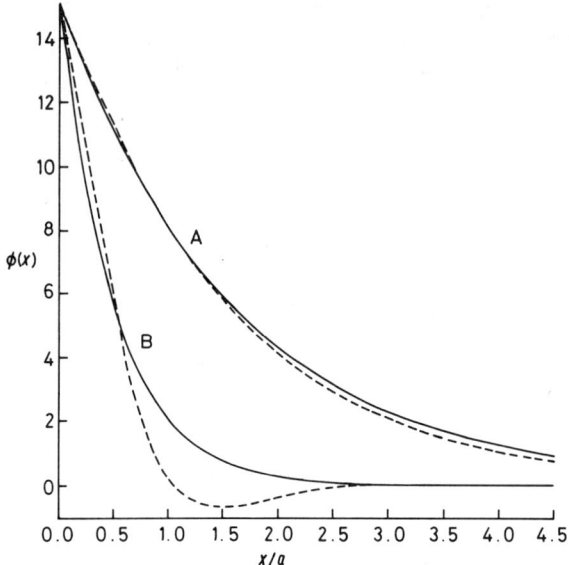

**Figure 3** Non-dimensional mean electrostatic potential $\phi(x) = e\psi(x)/kT$ for a 1:1 electrolyte. In the MSA(- - -) and GC(———) theories, the ionic concentrations are (A) 0.1976 mol $l^{-1}$ ($k_0 a = 0.621$) and (B) 1.968 mol $l^{-1}$ ($k_0 a = 1.96$); $k_0$ is the Debye parameter.[22] (Reproduced by permission from Chem. Phys. Lett., 1978, **64**, 150)

## 6 Gouy Theory – The Dielectric Constant

It is not that doubts about GC theory have been expressed on formal grounds alone: certain observations on the dielectric constant of the diffuse layer reveal anomalous features, not apparently reconciled. It has even been stated that 'the magnitude of the dielectric constant in the diffuse part of the double layer must be regarded as an unresolved problem'.[57-60] In the case of a mercury/ethylene glycol interface, an effective $\varepsilon$ which is 1.16 times the bulk value has been reported (cf. references 13, 61—67 for attempts to model the dielectric constant in the aqueous regions,

[57] V. A. Chagelishvili, D. I. Dzhaparidze, and B. B. Damaskin, Sov. Electrochem. (Engl. Transl.), 1977, **13**, 1110.
[58] Yu. M. Sokolov, G. A. Tedoradze, and R. A. Arakelyam, Sov. Electrochem. (Engl. Transl.), 1973, **9**, 537.
[59] R. A. Arakelyam, G. A. Tedoradze, V. I. Tatarinov, and A, P, Keleyan, Sov. Electrochem. (Engl. Transl.), 1974, **10**, 1424.
[60] D. I. Dzhaparidze and V. A. Chagelishvili, Proceedings of the 4th Symposium on Double Layer and Adsorption on Solid Electrodes, Tartu University, 1975, p. 89 (in Russian).
[61] J. O'M. Bockris and M. A. Habib, Z. Phys. Chem. (Frankfurt am Main), 1975, **98**, 43.
[62] S. Levine and W. R. Fawcett, J. Electroanal. Chem. Interfacial Electrochem., 1979, **99**, 265.
[63] S. Levine and K. Robinson, J. Electroanal. Chem. Interfacial Electrochem., 1973, **41**, 159.
[64] S. Levine, K. Robinson, and W. R. Fawcett, J. Electroanal. Chem. Interfacial Electrochem., 1974, **54**, 237.
[65] J. W. Perram and M. N. Barber, Mol. Phys., 1974, **28**, 131.
[66] F. P. Buff and N. S. Goel, J. Chem. Phys., 1972, **56**, 2405.
[67] J. R. Clay, N. S. Goel, and F. P. Buff, J. Chem Phys., 1972, **56**, 4245.

usually from a low value near the metal surface to its large bulk value, growing monotonically).[68]

A similar effect[69–71] has been observed for mercury/acetonitrile (AN) solutions, and for In–Ga eutectic/acetonitrile solutions, in the presence of thiourea. The value of $\varepsilon$ of the diffuse layer in this case[69] was almost double that of the bulk phase! The usual Parsons–Zobel test of plotting $1/C$ vs. $1/C_d$ gave linearity,[69–71] as expected, but the slopes were $< 1$, leading to the above values of $\varepsilon$. The other criteria, e.g. those based on independence of capacitance minimum on the salt concentration and the reconstruction (and comparison with experiments) of potential dependence, on the basis of GC theory (with higher $\varepsilon$) were satisfied.[69–71] Several tentative explanations, such as association and selective ionic solvation by solute, dipolar species[70] (e.g. thiourea in AN), have been invoked to explain the anomalies. In the latter case, a consequence is an enrichment of the diffuse double layer by the species that have a high dipole moment and that are 'carried' by diffuse-layer ions. These are not entirely convincing explanations, and need further analysis.

It must be mentioned that the anomalies reported above are exceptions rather than the rule. For example, recently a similar test[72] carried out for the mercury/dimethylformamide interface with TEAP (tetraethylammonium perchlorate) as the electrolyte gave excellent agreement between the dielectric constants in the bulk and the diffuse layer; the GC model was obeyed.

Testing GC theory with solid electrodes is not straightforward, considering factors like surface roughness, but even here the agreement has been reported as good (silver,[73] bismuth,[74] lead[74–77]). For a recent application of GC theory to a highly unsymmetrical electrolyte (1:12), refer to reference 78.

## 7 Gouy Theory – Further Phenomenological Modelling

Whereas it is imperative that the basis and the experimental fits with GC and its improved versions should be pursued, the contact with experiments will also be served well by more judicious phenomenological models. As an illustration, we point out the need to model better the coupling between the specific and the non-specific regions of the double layer. This includes the counter-ion effects too.[79–84] As another example, the OHP has been modelled[45,85] as an irregular

---

68 H. Behret, F. Schmithals, and J. Barthel, *Z. Phys. Chem. (Frankfurt am Main)*, 1975, **96**, 73.
69 A. M. Kalyuzhnaya, N. B. Grigorev, and I. A. Bagotskaya, *Sov. Electrochem. (Engl. Transl.)*, 1974, **10**, 1628.
70 N. B. Grigorev and Yu. M. Povarov, *Sov. Electrochem. (Engl. Transl.)*, 1976, **12**, 462.
71 N. B. Grigorev, A. M. Kalyuzhnaya, and I. A. Bagotskaya, *Sov. Electrochem. (Engl. Transl.)*, 1976, **12**, 411; 1975, **11**, 1469.
72 W. R. Fawcett, B. M. Ikeda, and J. B. Sellan, *Can. J. Chem.*, 1980, in the press.
73 T. Vitanov, A. Popov, and E. S. Sevastyanov, *Sov. Electrochem. (Engl. Transl.)*, 1976, **12**, 557.
74 Z. N. Ushakova and V. F. Ivanov, *Sov. Electrochem. (Engl. Transl.)*, 1976, **12**, 477.
75 G. Valette and A. Hamelin, *C.R. Hebd. Seances Acad. Sci., Ser. C*, 1974, **279**, 2953.
76 G. Valette and A. Hamelin, *J. Electroanal. Chem. Interfacial Electrochem.*, 1973, **45**, 301.
77 Yu. A. Prishchepa, Z. A. Rotenberg, and Yu. V. Pleskov, *Sov. Electrochem. (Engl. Transl.)*, 1975, **11**, 1430.
78 R. Parsons and R. Peat, *Trans. Soc. Adv. Electrochem. Sci. Technol.*, 1977, **12**, 187.
79 U. V. Palm, M. G. Väärtnõu, and M. A. Salve, *J. Electroanal. Chem. Interfacial Electrochem.*, 1978, **86**, 35.

surface whose topography is controlled by the 'states of adsorption'[1,3] ('up' and 'down'), which in turn are determined by molecular interactions. The distribution of potential in the diffuse layer is then evaluated by projection operator techniques,[86] the averaging having been carried out on the basis of the inner-layer Hamiltonian. This reveals a different type of coupling, and will be discussed elsewhere. An alternative model[85] considers the details of the configurations of the inner layer but considers the orientation of the solvent in the neighbourhood of ions in the diffuse layer as a perturbation of the homogeneity of the region. Such a form of coupling is also realistic.

In the context of adsorption of organic compounds, 'breakdown of equipotentiality of the OHP',[87,88] the steric factors associated with the double layer that are caused by large dipolar compounds,[89] and significant rearrangement of the diffuse layer[90] have been reported. More than one OHP and the equivalent circuits for such dual diffuse layers have been considered too.[91] In the light of several models of the solvent structure proposed recently (see Section 12) and the lack of parity of sites associated with the individual configurations, considerations similar to those given above become relevant.

## 8 Inactive Ions – Are there any?

Finally, we wish to emphasize that the next stage, *viz.* reducing the information on the inner layer to that of the solvent alone, is not easy if the electrolyte is 'surface-active'. Since the common anions, *e.g.* halides, pseudohalides, and sulphates, are adsorbed more readily than cations, the search for 'inactive ions' is usually confined to anions.

Fluoride ion is now a suspect in this regard,[1] but, either due to the 'weak' nature of adsorption or due to doubts[92] on the validity of the analysis leading to the above conclusion,[1] the ion $F^-$ is treated as non-adsorbed for all practical purposes. It is also believed that $F^-$ is 'inactive'[93] even at lead/methanol, bismuth/methanol, and silver (single crystal)/aqueous solution interfaces.[73]

[80] B. B. Damaskin, U. V. Palm, and M. G. Väärtnõu, *J. Electroanal. Chem. Interfacial Electrochem.*, 1976, **70**, 103.
[81] M. G. Väärtnõu and U. V. Palm, *Sov. Electrochem. (Engl. Transl.)*, 1977, **13**, 89.
[82] B. B. Damaskin, U. V. Palm, E. Petyärv, and M. A. Salve, *J. Electroanal. Chem. Interfacial Electrochem.*, 1974, **51**, 179.
[83] U. V. Palm, M. G. Väärtnõu, M. A. Salve, and E. K. Yuriado, *Sov. Electrochem. (Engl. Transl)*, 1977, **13**, 1248.
[84] M. G. Väärtnõu and U. V. Palm, *Sov. Electrochem. (Engl. Transl.)*, 1977, **13**, 1032.
[85] S. K. Rangarajan, in the press.
[86] S. K. Rangarajan and K. L. Sebastian, *Mol. Phys.*, 1978, **36**, 343.
[87] Yu. N. Kuryakov, B. B. Damaskin, and S. L. Dyatkina, *Sov. Electrochem. (Engl. Transl.)*, 1977, **13**, 1506.
[88] Yu. N. Kuryakov, B. B. Damaskin, and S. L. Dyatkina, *Sov. Electrochem. (Engl. Transl.)*, 1978, **14**, 386.
[89] A. N. Frumkin, B. B. Damaskin, and O. A. Petrii, *J. Electroanal. Chem. Interfacial Electrochem.*, 1974, **53**, 57.
[90] B. B. Damaskin and Yu. N. Kuryakov, *Sov. Electrochem. (Engl. Transl.)*, 1978, **14**, 802.
[91] A. B. Ershler and V. S. Krylov, *Sov. Electrochem. (Engl. Transl.)*, 1976, **12**, 1286.
[92] W. R. Fawcett, *Isr. J. Chem.*, 1979, **18**, 3.
[93] Z. N. Ushakova and V. F. Ivanov, *Sov. Electrochem. (Engl. Transl.)*, 1976, **12**, 477.

The Electrical Double Layer    213

Insignificant inner-layer activity of $ClO_4^-$ in In–Ga eutectic alloy/dilute aqueous solutions,[94] in acetonitrile, DMSO,[95] (also see references 17–92); $ClO_4^-$ in mercury/propylene carbonate and mercury/acetonitrile;[96] and $PF_6^-$ in mercury/ethylene carbonates[97] have also been reported. Many conclusions on the inactivity cited above must be taken as qualitative and, at best, semi-quantitative. Perchlorate ion, which was thought earlier[98] to be 'inactive' at the mercury/DMSO interface, later turned out not to be so, on more careful examination.[99]

## 9 Criticisms of the Model and Data Reductions (see Sections 3 & 4)

*'But what happens when you come to the beginning again?' Alice ventured to ask.*

Some doubts have been expressed about the validity[6, 44, 50] of even this preliminary stage. These doubts are related to some or all of the factors (A)–(D) mentioned in Section 2.

Krishtalik[100] had pointed out that dipolar spatial correlation-distance (range) and the thickness of the inner layer are comparable, and, on this basis, the effectiveness of orientational polarization models (see Sections 11 and 12) advocated for the solvent layer can be questioned.

On the other hand, Cooper and Harrison[6, 19—21] have questioned the very basis outlined in Figures 1 and 2. Some specific objections that are raised are to the use of (i) the concept of $\varepsilon$ as applied to a molecular-dimensional-region, inner layer having thus been 'elevated to the status of a dielectric phase'[19] and (ii) the applicability of the GC theory as well as the meaningfulness of its calculations.

Bockris et al.[5, 15] recognize the diffuse-layer-decoupling step but do not accept 'the reduced capacitance data' (in the case of mercury/aqueous NaF) as attributable to solvent alone.

This formalistic debate,[51, 101] it seems, will go on until a 'satisfactory' statistical-mechanical theory of charged interfaces, together with an adequate quantum-mechanical description of the electrode surface, is developed.

## 10 Experimental Observations for Verifying Solvent Models

The reduced data available for verifying solvent layer models are presented in various forms:[1, 15, 92] (1) inner-layer capacitance, $C_i$, studied as a function of potential, $E$, and temperature, $T$ (see Figure 4); (2) surface excess entropy[1, 8, 102]

$$\Gamma_s = -\left(\frac{\partial \gamma}{\partial T}\right)_{P,\mu,E_\pm}$$

[94] L. M. Dubova, V. G. Boitsov, and I. A. Bagotskaya, *Sov. Electrochem. (Engl. Transl.)*, 1978, **14**, 492.
[95] L. M. Dubova and I. A. Bagotskaya, *Sov. Electrochem. (Engl. Transl.)*, 1977, **13**, 49.
[96] W. R. Fawcett and R. O. Loutfy, *Can. J. Chem.*, 1973, **51**, 230.
[97] W. R. Fawcett and M. D. Mackey, *J. Chem. Soc., Faraday Trans. 1.*, 1973, **69**, 634.
[98] T. A. Severova, R. V. Ivanova, and B. B. Damaskin, *Sov. Electrochem. (Engl. Trans.)*, 1976, **9**, 838.
[99] B. B. Damaskin, T. A. Severova, and R. V. Ivanova, *Sov. Electrochem. (Engl. Transl.)*, 1976, **12**, 619.
[100] L. I. Krishtalik, *Sov. Electrochem. (Engl. Transl.)*, 1977, **13**, 1634.
[101] B. B. Damaskin, L. Kuznetsova, U. V. Palm, M. Väärtnõu, and M. Salve, *J. Electroanal. Chem. Interfacial Electrochem.*, 1979, **100**, 365.
[102] H. D. Hurwitz and C. V. D'Alkaine, *J. Electroanal. Chem. Interfacial Electrochem.*, 1973, **42**, 77.

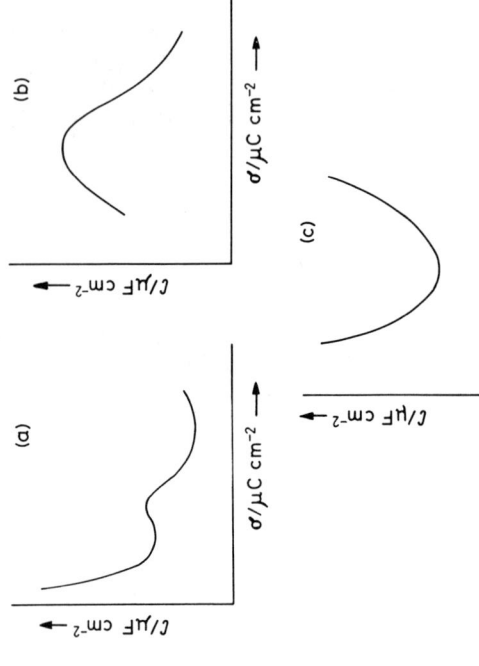

**Figure 5** Classification (schematic) of the capacitance behaviour of non-aqueous solvents at the mercury electrode, according to Parsons.[18]
(a) 'water-like' capacitance curves. Features are a hump and two minima, with more or less steeply rising sections at the extremities. Examples are water, formamide, N-monosubstituted formamides, related amide solvents, and DMSO.
(b) 'pure hump' curves. Examples are ethylene and propylene carbonates, sulpholane, γ-butyrolactone, and γ-valerolactone.
(c) 'water-unlike', inverted parabolas. Examples are alcohols, formic acid, ammonia, NN-dimethylacetamide, acetonitrile, acetone, ethylene glycol, diethylene glycol, and pyridine.

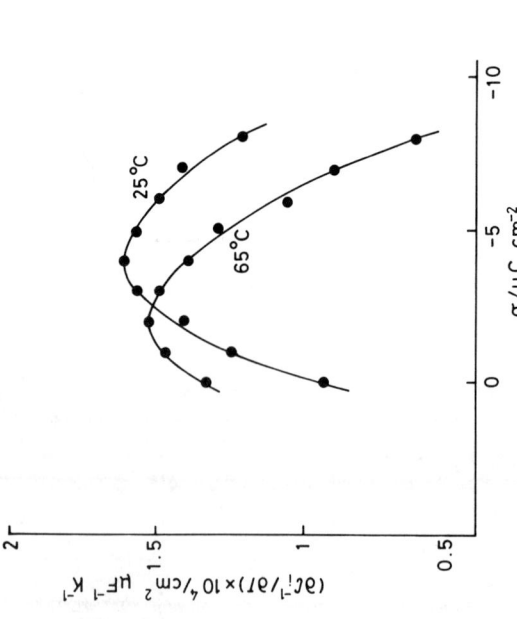

**Figure 4** Temperature coefficient of the reciprocal inner-layer capacitance, as a function of charge on the metal at two different temperatures
(Reproduced by permission from J. Electroanal. Chem. Interfacial Electrochem., 1977, **82**, 391)

*The Electrical Double Layer*

reduced to solvent excess entropy, $S^{*}$,[1,14] through

$$S^* = \Gamma_s - (\Gamma_+ \bar{S}_+ + \Gamma_- \bar{S}_-)$$

where $\Gamma_\pm$ are the ionic surface excesses and $\bar{S}_\pm$ the partial molar ionic entropies (accessible experimentally); (3) the entropy of formation of the interface,

$$\Delta S^\sigma = S^\sigma - m^\sigma{}_{Hg}\bar{S}_{Hg} - m^\sigma{}_{H_2O}\bar{S}_{H_2O} - m^\sigma{}_+ \bar{S}_+ - m^\sigma{}_- \bar{S}_-$$

where $S^\sigma$ is the total entropy per unit area, the $m$'s are the number of moles of various components in the interface, and $\bar{S}$'s are, as before, the partial molar entropies. The diffuse-layer contribution is negligible[8] near the potential of zero charge (PZC). (4) Surface excess volume due to the interface, $\Gamma_v$, may be reduced to that due to solvent, $V^*$, as

$$V^* = \Gamma_v - \sum \Gamma_i \bar{V}_i$$

the $\bar{V}$ being the partial molar volumes;

$$\Gamma_v = \left(\frac{\partial \gamma}{\partial P}\right)_{T,\mu,E_\pm}$$

The thermodynamic basis, interrelationship among the variables,[1,18,115] and evaluation of the various quantities have been discussed in detail (see refs. 1, 8, 14, 15, 102—105).

Discussions based on correlations[1,4,106-108] extensively make use of data in the form of the dipole potential at the PZC as a function of temperature or in the form of charge–potential curves (see Section 14).

While the capacitance has been measured for a large class of systems, the processing of $S^*$ or $\Delta S^\sigma$ has been restricted to a few only. The capacitance analysis seems to be preferred in view of its features (*e.g.* hump and/or cathodic minimum, anodic rise, *etc.*) in several cases (see Figure 5), the relative ease of measurement (*cf.* variables $P$ and $T$), and analysis (*cf.* corrections for diffuse layer, ionic components, *etc.*). It is also claimed that the shape of entropy-excess curves is less sensitive to models[3] (*e.g.* clusters *versus* monomers) and that they are rather featureless.

The varied characteristics that have been associated with the classical data of Grahame for mercury/aqueous sodium fluoride solutions are not universal, as evidenced from the classification of Parsons.[18] Figure 5 shows the dominant types obtained in the capacitance measurements in non-aqueous solvents.

It would be useful to obtain values for the surface excess entropy or volume for the above systems and compare them. One must accept the classification in Figure 5 with caution.[18,92] For, once a wider potential range (over which the electrode remains 'polarizable') becomes available for capacitance measurements, the cross-

---

[103] N. H. Cuong, C. V. D'Alkaine, A. Jenard, and H. D. Hurwitz, *J. Electroanal. Chem. Interfacial Electrochem.*, 1974, **51**, 377.
[104] D. Schumann, P. Vanel, and C. Bertrand, *J. Chim. Phys.*, 1977, **43**, 6.
[105] S. Trasatti, *J. Electroanal. Chem. Interfacial Electrochem.*, 1975, **64**, 128.
[106] S. Trasatti, in 'Modern Aspects of Electrochemistry', ed. J. O'M. Bockris and B. E. Conway, Plenum Press, New York, 1979, Vol. 13.
[107] S. Trasatti, *Adv. Electrochem. Electrochem. Eng.*, 1977, **10**, 213.
[108] A. N. Frumkin, B. B. Damaskin, I. A. Bagotskaya, and N. B. Grigoryev, *Electrochim. Acta*, 1974, **19**, 75.

over of some of the systems from one type to another becomes visible. It is also likely that new features[72] (*e.g.* two maxima at very positive and negative potentials, with a minimum in between) may manifest themselves. For example, it is noted[72] that such an accessibility of the potential range enabled a maximum to be seen at very positive potentials in the case of acetone[109] and dimethylformamide[72] on mercury electrodes. These systems were earlier[18] classified as type III (see Figure 5). Some observations on solid metals[92, 110] indicate how the absence of a sufficient potential range positive to the PZC has masked interesting features (see Figure 2 of ref. 110).

## 11 A Critique of Two-state Models

The notion of orientation, *per se*, of interfacial water has been invoked earlier in different contexts,[111, 112] and so has the notion of a cluster.[113, 114] However, the more recent models, be they two-state [Watts–Tobin, Levine–Bell–Smith (LBS), Bockris–Devanathan–Muller (BDM)], three-state [Damaskin–Frumkin (DF),[115] Bockris–Habib (BH),[15] Fawcett[92, 116]], or even four-state (Parsons[117]), go beyond this notion in their chemical identification and description of physical states. The basis for postulating a *finite set* of orientations to describe solvent at the interface rests on (1) the ability of the external field to influence the polarization states; (2) the specific nature of interactions with the substrate, even in the absence of the field (*cf.* preferred orientation at the PZC), which itself removes the 'degeneracy' or equivalence among the orientations; and (3) a 'discrete' approximation for all such possibilities.

*A priori*, this is a valid approximate approach, similar to replacing a continuum by an appropriate discrete medium or a distribution by its representative mean or median! But the crux is in 'identifying' the configurations and in the inclusion of proper interactions.

The two-state models considered hitherto referred to monomer states, and differ only in the choice of

(A) dipole moments in the two orientations (BDM *vs.* LBS), or

(B) the electrical variable $\sigma$ or $\varepsilon$ (Watts–Tobin *vs.* BDM; *cf.* the congruence problem in isotherms), or

(C) the model for the dielectric permittivity (LBS *vs.* others; *cf.* polarizability effects).

Moreover, the earlier models have been directed towards reproducing primarily the features of Grahame's data for NaF, and so the choice was not made strictly on an *ab initio* basis. The focal points of study were to understand the relative displacement of the extrema in the charge axis of (1) the capacitance, $C$, and (2) the

---

[109] Z. Borkowska, *J. Electroanal. Chem. Interfacial Electrochem.*, 1977, **79**, 206.
[110] A. N. Frumkin and B. B. Damaskin, *Electrochim. Acta*, 1974, **19**, 69.
[111] A. N. Frumkin, *J. Colloid. Sci.*, 1946, **1**, 290.
[112] J. O'M. Bockris and E. C. Potter, *J. Chem. Phys.*, 1952, **20**, 614.
[113] R. Parsons, *J. Electroanal. Chem. Interfacial Electrochem.*, 1964, **8**, 93.
[114] B. B. Damaskin, *Sov. Electrochem. (Engl. Transl.)*, 1965, **1**, 51.
[115] B. B. Damaskin, *J. Electroanal. Chem. Interfacial Electrochem.*, 1977, **75**, 359.
[116] W. R. Fawcett, *J. Phys. Chem.*, 1978, **82**, 1385.
[117] R. Parsons, *J. Electroanal. Chem. Interfacial Electrochem.*, 1975, **59**, 229.

derivative of capacitance with respect to temperature, or, more correctly, $\partial(1/C)/\partial T$, and (3) the entropy, as well as the zero-crossing of (4) dipole potential, $g_{dip}$, and (5) the order parameter, $R=(N\uparrow-N\downarrow)/(N\uparrow+N\downarrow)$, $N\uparrow$ and $N\downarrow$ being the numbers of dipoles oriented in the two directions $\uparrow$ and $\downarrow$. It is important to reconcile[118] the capacitance maximum occurring at positive charges with the entropy maximum observed[1, 103, 104] around $-4\ \mu\text{C cm}^{-2}$.

The above models lead to an expression for $R$ in the same form[92, 118]

$$R = -\tanh[(v_1\sigma' + v_2 R)/kT], \qquad (1)$$

where $v_1$, $v_2$ are constants involving dipole moments, polarizability, *etc.*, these being different in various models, and $\sigma'$ is the charge measured with respect to $\sigma_0$, the charge at which $R=0$.

In the LBS model, in contrast to the BDM and Watts–Tobin models, $g_{dip}$ is not strictly proportional to $R$, even though it is still a linear function of $R$. This modification alone[118] is sufficient to cause a displacement of the charge $\sigma_0$ (at which $R=0$) from $\sigma_g$ (at which $g_{dip}=0$).

The differences among the models are no doubt visible, but more interesting is the 'symmetry' underlying the Watts–Tobin, BDM, and LBS descriptions. The order parameter, $R$, is an odd function of $\sigma'$, *i.e.* $R(-\sigma')=-R(\sigma')$. Also, $(dR/d\sigma)$, and hence $(dg_{dip}/d\sigma')$, are even functions of $\sigma'$, as may be easily verfied from equation (1). This is not all. The (partial) derivatives of $R$ and $g_{dip}$ with respect to temperature are odd functions too, while their charge derivatives are even functions of $\sigma'$. Thus, it is rigorously proved, by-passing all the laborious algebra, that $C_i$ and $\partial(1/C_i)/\partial(1/T)$ will both have extrema at the same charge, *viz.* $\sigma'=0$; *i.e.*, $\sigma_m=\sigma_0$ in all the above models! It must also be noted that our arguments may be applied even when equation (1) is replaced by more general equations, satisfying some broad constraints on their behaviour with respect to $\sigma'$ and $T$. The above arguments are intended to emphasize that, by merely introducing more and more parameters, it is not always possible to simulate experimental features!

Two recent contributions[45, 119] on two-state models must be mentioned. Cooper and Harrison[119] discuss the mean-field results of an equivalent ferroelectric problem. This work emphasizes the inadequacies of the two-state descriptions; especially the reconciliation of experimental and model observations on the location of the extrema. That infinite or even negative capacitances are predicted by these models is noted.[119] This is due to the competition between the electrode charge and the solvent polarization in determining the inner-layer capacitance.

The other paper, by the present Reporter,[45] discusses an isomorphism existing at the level of Hamiltonians among apparently distinct physical situations such as ionic, dipolar, solvent, and mixed adsorption. Interactions and their mapping in the lattice picture are considered in detail. The Coulombic and short-range interactions are incorporated in this approach. It is possible to go beyond mean-field approximations systematically.

A lucid comparison of two-state models[118] by Parsons and a detailed review by Fawcett[92] are also available.

---

[118] R. Parsons, *Trans. Soc. Adv. Electrochem. Sci. Technol.*, 1978, **13**, 239.
[119] I. L. Cooper and J. A. Harrison, *J. Electroanal. Chem. Interfacial Electrochem.*, 1975, **66**, 85.

## 12 Three- and Four-state Models

*'I'm afraid I can't put it more clearly,'* Alice replied, very politely, *'for I can't understand it myself, to begin with: and being so many different sizes in a day is very confusing.'*

Since the experimental features remained unexplained by the available two-state models, further attempts were made towards the inclusion of more 'states' in the model. The Damaskin–Frumkin model[120] started this trend. Damaskin and Frumkin (DF) identified their states as 'up' and 'down' clusters, and a monomer, chemisorbed with the oxygen towards the metal. This is akin to a Watts–Tobin picture of ionic adsorption, with an ion replaced here (DF) by a monomeric solvent!

An improved version by Parsons[117] sought to introduce another orientation of the monomeric state. The four-state model[117] was able to predict better the capacitance *versus* temperature behaviour, besides the well-known capacitance *versus* charge curves of Grahame for sodium fluoride. The improvement was not only in the inclusion of one more state (the 'down' monomer) but also in accounting properly for the total number of water molecules.

A third important paper in this direction, by Damaskin,[115] considers the multi-site nature of the clusters and the dipole–dipole interactions,[121] omitted in the earlier considerations[117, 120] of the three- and four-state models. The agreement with the experiments was improved by a proper choice of parameters.

A recent version[122] of the four-state model, by Fawcett *et al.*, is an adaptation of an earlier contribution by Levine *et al.*[123] for mixed solvents. This takes into account polarizability effects and the multi-site requirements of clusters in a more systematic manner.

In this connection, two other contributions[116, 124] must be mentioned.

The first model, described by Bockris and Habib[15, 124] (BH), is also a three-state cluster model except that the cluster size is two molecules, in contrast to others[117, 120] where a cluster size of three of four molecules is considered in numerical calculations. The authors also assume that the dimer possesses no net dipole moment, unlike the clusters of the DF model. Whereas the Damaskin–Frumkin model permits two configurations for clusters and one for monomers, the BH model considers two ('up' and 'down') states for the monomer but only one for the dimer. There are some fundamental differences in analysis and interpretation too (see Section 13).

The other is the three-state model of Fawcett,[92, 116] which is not strictly a cluster model but similar to others, in analysis and the physical picture underlying it. A 'flat' or 'parallel' state is superimposed on the earlier 'up' and 'down' configurations for the solvent. Polarizability effects are considered but the site parity among all the three states is retained.

For the mathematical details of the several models sketched above, the reviews by Fawcett,[92] Habib,[15] and Conway[4] may be consulted.

[120] B. B. Damaskin and A. N. Frumkin, *Electrochim. Acta*, 1974, **19**, 173.
[121] B. B. Damaskin, U. V. Palm, and M. A. Salve, *Sov. Electrochem. (Engl. Transl.)*, 1976, **12**, 226.
[122] W. R. Fawcett, S. Levine, R. M. de Nobriga, and A. C. McDonald, *J. Electroanal. Chem. Interfacial Electrochem.*, 1980, in the press.
[123] S. Levine, K. Robinson, A. L. Smith, and A. C. Brett, *Discuss. Faraday Soc.*, 1975, **59**, 133 (see also pp. 172, 173).
[124] J. O'M. Bockris and M. A. Habib, *Electrochim. Acta*, 1977, **22**, 41.

## The Electrical Double Layer

The four-state model of Fawcett et al.[122] is the most complete version of the above, but the physical aspects, unfortunately, are masked by algebraic details.

The number of parameters required to fit these models has progressively increased too, detracting from the usefulness of the efforts to some extent. The basic parameters are (1) the energies of interaction with the substrate of the various states (not all of these need be different[117, 120]); (2) the dipole moments (especially of clusters; also note that the dipole moment of the monomer state is also considered to be an adjustable parameter[115] of $\sim 3.68$ debye); (3) dielectric parameters, in the case of models not including polarizability effects, and polarizability coefficients, in the case of Fawcett's three- and four-state models; and (4) the size ratios associated with the states. Besides these, there are also other parameters, e.g., the inner-layer capacitance in the absence of polarization effects (alternatively, the thickness of the layer[123]) and discreteness-effect coefficients[115] or effective co-ordination number factors.[116, 122]

However, the calculations of Bockris and Habib,[5, 15, 124] similar to the earlier BDM model, are different from the others. Attempts to obtain some of the parameters *ab initio* in their entropy calculations must be noted.

### 13 A Critique of Multi-state Models

With proper choice of parameters, satisfactory agreement of the theoretical models with the experimental data (see Section 10) has become possible. The data of Grahame for NaF with mercury electrodes, being the touchstone of these analyses, have been compared and found to be reasonably reproduced[117] at various temperatures and electrode charges. Also, the interfacial capacitance data with mercury electrodes and with formaldehyde and methanol as solvents also seem to be compatible with the four-state model of Parsons.[18] It is observed[18] that the Watts–Tobin model[1] is adequate in the case of ethylene carbonate. On the other hand, good agreement with the capacitance data of mercury, bismuth, and cadmium electrodes in aqueous solutions of 'inactive' electrolytes such as NaF, LiClO$_4$, and KF, respectively, has also been reported.[115, 120]

Fawcett has successfully applied his three-state model[92] to capacitance data for methanol,[116] ethylene carbonate,[116] and dimethylformamide.[72] The four-state model of Fawcett et al.[122] has also been shown to give a reasonable description of data for mercury/aqueous sodium fluoride, at various temperatures.

It seems, therefore, possible that we can understand (semi-quantitatively, at least) the various features associated with double-layer data on solvents (see Figure 5). The introduction of a cluster as a unit may be expected to reflect 'solvent association'[116] and to accommodate internal interactions of solvent molecules[3] or any residual hydrogen bonding.

But there are difficulties too. There are inconvenient predictions; for example, (1) a negative temperature coefficient[3, 81, 118] of the dipole potential at the PZC is predicted by all the models barring the four-state one of Fawcett et al.[122] and (2) a strong positive temperature coefficient in the cathodic region.[117] The choice of parameters can be embarrassing,[18] with unrealistic (relative to those reported in the 'bulk') dipole moments,[115] polarizabilities,[116] co-ordination numbers,[116] or energies of interaction with the substrate, and the cluster dipole moment varying with temperature.[122] It is tempting to explain away some of these or even to remove the anomalies by invoking features such as the variation of thickness of the inner layer, electrostriction,[13] changes in the density of water,[24] the influence of solvent molecules associated with the counter-ions,[44] etc., but in that process the models may become even more phenomenological and less molecular! (note also the $D_2O$–$H_2O$ results[18, 125]).

If the *clustering* tendency and the associated parameters (*e.g.* size, dipole moment, and polarizability) are uniquely decided by the *solvent*, they can (and must) be consistently employed for other electrodes; this test should be useful.

The Bockris–Habib dimer model has been criticized by Parsons[3, 118] and by Fawcett.[92, 116] Even though the agreement of the BH model with excess entropy of the solvent is excellent, this is considered insufficient to validate the model in view of the lack of sensitivity[3] of such data to the details of models. As noted by Fawcett,[116] the BH analysis for the relative frequencies of the dimer and monomer states and also the relative orientational polarization is inconsistent. However, it is worth examining to what extent these approximations of BH, though unwarranted, are serious. There is also, of course, the explanation (or the lack of it) of the BH model as pertaining to the hump that has been criticized.[15]

How does one evaluate these models, relative to each other? Though it is premature to make any conclusions, the criteria must be faithfulness to experiments and physical concepts, and simplicity and elegance in assumptions underlying them.

First, the size requirements of the cluster and the monomer should be considered as different. How does one account for these 'sizes'? It is wrong to suggest that any of the localized lattice adsorption models hitherto considered for the solvent treat them as hard spheres or discs. The solvent molecules have only been occupants in a Langmuir lattice[45] with which a notion of maximum density is associated. The question is then, what is it that stipulates the lattice[45] governing adsorption of solvent: the electrode (ionic) structure, with its symmetry? Or is it the size of the solvent molecules, presumably chemisorbed? This is even more important in this context than in the case of ionic or neutral organic adsorption, wherein the Langmuir lattice is tacitly taken as that of the solvent layer! The answer to this question is important, to carry out the statistical mechanics properly in the lattice. All the models studied hitherto consider a two-dimensional Langmuir basis[45] only.

## 14 Phenomenological Models and Correlations

An alternative approach to the above model analysis is a judicious combination of experimental data and correlations (see refs. 1, 4, 7, 105—108, 126—133) with physical properties, *e.g.* electronegativity and work function. This can also be informative, as shown by Trasatti.[106, 107, 126—129] Frumkin *et al.*[108, 130] have

---

[125] R. Parsons, R. M. Reeves, and P. N. Taylor, *J. Electroanal. Chem. Interfacial Electrochem.*, 1974, **50**, 149.
[126] S. Trasatti, *J. Electroanal. Chem. Interfacial Electrochem.*, 1978, **91**, 293.
[127] S. Trasatti, *J. Chim. Phys. Phys.-Chim. Biol.*, 1975, **72**, 561.
[128] S. Trasatti, *J. Chem. Soc., Faraday Trans 1*, 1974, **70**, 1752.
[129] S. Trasatti, *Gazz. Chim. Ital.*, 1976, **106**, 219.

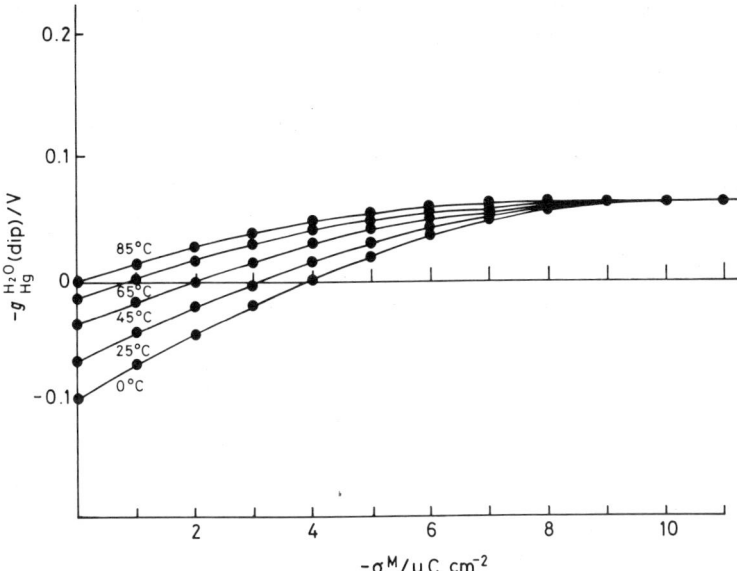

**Figure 6** *The contribution of the dipole of water to the potential of mercury at different temperatures, plotted as a function of charge.*
(Reproduced by permission from *J. Electroanal. Chem. Interfacial Electrochem.*, 1977, **82**, 391)

attempted, by similar methods, to discuss the hydrophilicity of metals, the work function, and their relationship to the structure of water in the inner layer. For a discussion of dipole potential, $g_{Hg}^{H_2O}$, as a function of temperature and related classification of metals as hydrophilic and hydrophobic, see references 7 and 105.

Trasatti[7] has been able to obtain the dipole potential, $g_{Hg}^{H_2O}$ (dip), for various values of the charge on the metal. The results, shown in Figure 6, are based on a procedure developed earlier.[105] The following values of $g_{Hg}^{H_2O}$ (dip) at the PZC have also been used for this purpose:[7]

| $T/°C$ | 0 | 25 | 45 | 65 | 85 |
|---|---|---|---|---|---|
| $g_{Hg}^{H_2O}$(dip)/V | 0.100 | 0.070 | 0.042 | 0.019 | 0.004 |

It is concluded from Figure 6 that the preferred orientation of water dipoles with the negative end towards the metal decreases as $T$ is increased at constant $\sigma$ or as $\sigma$ is made more negative at constant $T$: not surprising, of course.

The maximum contribution ($\sim 0.6$ V) is reached and is invariant with $T$. Also, the extrapolated intersection with the $\sigma$-axis, for each value of $T$, defines[105] the point of zero net orientation of the dipoles.

[130] A. N. Frumkin, B. B. Damaskin, N. B. Grigoryev, and I. A. Bagotskaya, *Electrochim. Acta*, 1974, **19**, 69.
[131] C. L. Gardner, *J. Electroanal. Chem. Interfacial Electrochem.*, 1975, **61**, 113.
[132] L. M. Dubova and I. A. Bagotskaya, *Sov. Electrochem. (Engl. Transl.)*, 1976, **12**, 770.
[133] I. A. Bagotskaya and A. M. Kalyuzhnaya, *Sov. Electrochem. (Engl. Transl.)*, 1976, **12**, 964.

It must be pointed out that the charge, $\sigma$, occurs in analysis not alone but as $(\sigma/kT)$. Thus the temperature effects shown in Figure 6 can be represented better with this modified abscissa.

Even after the reduction of the experimental capacitance to (inner layer) solvent capacitance, some further separation is possible. The potential difference in the inner layer is still composed of two components, due to 'free charge' on either of the 'substrates' and to dipoles; *i.e.* $1/C = (1/K_{ion}) - (1/C_{dip})$. It is the latter that is of immediate interest to solvent–metal interaction. Hence, correlations are sought[126] for $(g_{dip})_{\sigma=0}$ with the inverse of inner-layer capacitance $(1/C_i)$, as also with the enthalpy of formation, $\Delta H_f$, of the oxide MO.

Linear laws relating these variables are expected, and confirmed (see Figure 7). The physical significance of $(g_{dip})_{\sigma=0} = 0$ is taken as referring to the system with no specific metal–water interaction, and the corresponding $C_{i0}$ is 23.5 $\mu$F cm$^{-2}$. A similar extrapolation of $\Delta H_f^{\ominus} = 0$ gives $C_{i0} = 22.5$ $\mu$F cm$^{-2}$. The above plots are isothermal, and at $\sigma = 0$ the variation was achieved through altering the electrodes. Keeping the system the same (mercury–water), but varying the temperature, it is possible to correlate the resulting values of the capacitance $C_i$ with the corresponding $(g_{dip})_{\sigma=0}$. A linearity results again,[126] and the extrapolation to $g_{\sigma=0} = 0$ results in a value for $C_{i0}$ of 23.3 $\mu$F cm$^{-2}$, agreeing well with other extrapolations!

By a different extrapolation procedure, the capacitance, $K^i_{ion}$, due to 'free charges alone' is estimated as 17 $\mu$F cm$^{-2}$. The difference between the limiting capacitances due to dipole and 'free charges' is taken as indicating some reorientation of residual water. Some of the above results have been qualitatively related[126] to solvent cluster models.

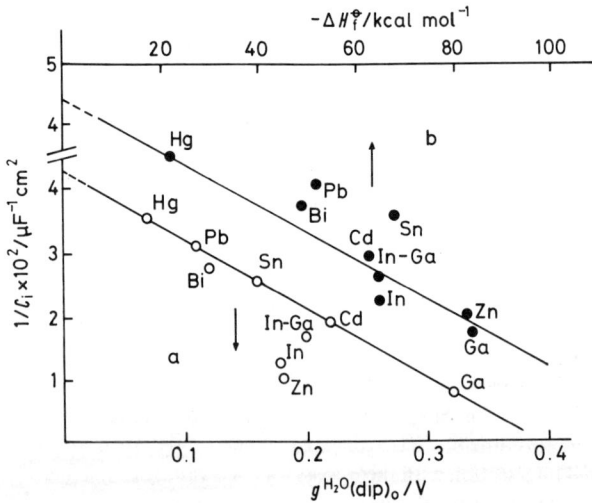

**Figure 7** *Inner-layer capacitance at the potential of zero charge, plotted as a function of (a) the surface potential associated with oriented water molecules at the electrode surface; (b) the enthalpy of formation of the oxide with formula MO.*
(Reproduced by permission from *J. Electroanal. Chem. Interfacial Electrochem.*, 1978, **91**, 293)

Certain difficulties in the above procedure for obtaining $K^i_{ion}$ and $C_{i0}$, especially for In–Ga, In, Zn, and Bi, have been noted.[126] The choice of $\Delta H_f^{\ominus}$ as an index of metal–oxygen affinity has also been criticized,[8] since $\Delta H_f^{\ominus}$ depends on the nature of bonds in oxides, these often being ionic. Besides, it is not clear why $g_{dip}$ and not its charge derivative is correlated with $1/C_i$. In most model systems[1,92] the fundamental relationship employed is $1/C_{dip} = \partial g_{dip}/\partial \sigma$, and this may not be strictly proportional to $g_{dip}$. A more quantitative link with model systems may be useful. For example, a comparison with two-state models would indicate that $(g_{dip})_{\sigma=0} = 0$ implies that $\Delta V = U\uparrow - U\downarrow = 0$, where $U$'s are solvent–metal interactions in the two orientations. Under this condition, we must remember that the $U$'s are not necessarily individually equal to zero. Another aspect that needs attention is the role of the 'inactive electrolytes'. The reduced capacitances could still depend upon the location of the OHP, as influenced by the invisible (!) electrolytes, and a correlation with the thickness parameter may be possible.

More details are available in recent reviews.[106,107]

An electrostatic description of the solvent layer has recently been proposed. Relaxing the hitherto imposed point-dipole-layer model, Oldham[134,135] replaces this by layers associated with the positive and negative charges of the symmetrical dipoles. The analysis is one-dimensional, the statistics are Boltzmanian, and the averaging is over all possible orientations, subject only to geometrical constraints. All the limitations of this model have been listed by the authors themselves,[135] and there is no need to repeat them here! The contact with the experiments is confined, at the moment, to the prediction of the hump.

Mohilner's discussion,[136–138] which is non-modelistic and thermodynamically inspired, centres around the standard electrochemical fugacity ratio, which is experimentally accessible as a function of charge. A maximum observed at $\sigma = -2.2\ \mu C\ cm^{-2}$ is believed to indicate a state when the inner-layer water is 'most bulk-like' than at any other value of $\sigma$ and is also supposed to form intermolecular hydrogen bonds. The above conclusion can only be considered tentative, and needs further justification.

The present activity in the (phenomenological) dielectric approach is little, with the exception of some reports on dielectric permittivity of the inner layer.[139,140]

## 15 Are Two-state Models Inadequate?

Since the study of multi-state models is motivated largely by the failure of earlier two-position models, it is natural to ask at this juncture just how inadequate the two-state models really are.

We first note that a high level of 'symmetry' is associated with the description of the two states in the early models. This is, in a sense, reflected in the merging of locations of zeros of $R$ and $g_{dip}$, the maximum of $C$ and its temperature coefficient,

[134] K. B. Oldham and R. Parsons, *Sov. Electrochem. (Engl. Transl.)*, 1977, **13**, 866.
[135] P. Dalrymple-Alford and K. B. Oldham, *Can. J. Chem.*, 1978, **56**, 861.
[136] D. M. Mohilner, *Bioelectrochem. Bioenerg.*, 1978, **5**, 185.
[137] N. Nakadomari, D. M. Mohilner, and P. R. Mohilner, *J. Phys. Chem.*, 1977, **81**, 244.
[138] N. Nakadomari, D. M. Mohilner, and P. R. Mohilner, *J. Phys. Chem.*, 1976, **80**, 1761.
[139] R. R. Salem, *Russ. J. Phys. Chem. (Engl. Transl.)*, 1974, **48**, 1526.
[140] R. R. Salem, *Russ. J. Phys. Chem. (Engl. Transl.)*, 1975, **49**, 304, 1199.

and the entropy extremum – all at the same point in the charge axis. A mere introduction of unequal dipole moments ($p\uparrow \neq p\downarrow$) was sufficient[118] to separate the charge locations of $R=0$ and $g_{dip}=0$. A different kind of effect is caused by the introduction of 'asymmetry' at the number of sites associated with each configuration. Obviously, the effect of this on the entropy is more than on the dipole potential, $g_{dip}$. Associating $x_1$ and $x_2$ as the site requirements of the two states and writing, say, $(N\uparrow x\uparrow + N\downarrow x\downarrow) = N_T$, one can introduce this asymmetry ($x\uparrow \neq x\downarrow$) heuristically and see the extent of removal of 'degeneracy'. Proceeding further, let us make the polarizability factors unequal for the two states ($\uparrow,\downarrow$). The role of the polarizability term – rather subdued in the LBS theory,[1] by the necessary requirement that $\alpha\uparrow = \alpha\downarrow$, arising from the chemical identity of the two states – becomes more transparent by permitting $\alpha\uparrow \neq \alpha\downarrow$. This may be realistic; for example, by identifying the two different states as a cluster and a monomer. Thus, a cluster–monomer model with one configuration only for each, but their dipole moments, polarizability, and site occupancy each being naturally different for the two states, predicts features that are not accessible to earlier two-state models and also obviates (partly at least) the need for higher-state models. Table 1 represents the above discussion schematically.

## 16 Perspectives

The discussions in Section 15 illustrate how, even within a restricted scope (*e.g.* a 'two-state-only' requirement in the above example), there is room for extensive modelling, both physical and mathematical.[45]

One may then ask, are our interactions realistic enough? The need to introduce clusters from '*outside*', as if *by a show of hand*, can itself be indicative of the unsatisfactory nature of our modelling. Clustering must preferably be seen to be an *effect*, and not introduced as a *cause*. As we remarked earlier, clusters are no strange objects, be it in theoretical models for water or in some magnetic systems: they are consequences of respective interaction pictures. This reveals that a search for proper interactions is worthwhile.

Suppose, for example, in a model on a Langmuir lattice, we arrange[141] three types of bonds randomly ($J$, $J'$, and $J''$, say), linking Ising spins[45] at the lattice points. In the limit of $J'' \to \infty$, clusters can be seen as a consequence, and interesting phase transitions can be predicted too.

Interactions, be they specific or directional, as in Lieb's vertex and $X$–$Y$ models, are realistic, especially in cases where there are components of the dipole moment that are parallel to the surface. It is not as if dipoles whose moments are parallel to the electrode do not respond to (or order) each other, even if they are apparently indifferent to the external field that is perpendicular to them! This important aspect has so far been ignored.

More important is to understand the role of the substrate in inducing (three-body) interactions that can order the adsorbates locally or over a long range. This possibility is worth looking into because the 'cluster' is not, *per se*, an object in solution but is invoked as a surface state, and is hence a product of surface interactions. This could explain the results that are specific to solvents and elec-

[141] S. Miyazima and K. Yonezawa, *Prog. Theor. Phys.*, 1974, **1**, 99.

**Table 1** *Progressive generalization of two-state models – levels of 'symmetry'*

| Model | Substrate interactions | Dipole moments* | Polarizability† | Site requirements | $g_{dip}$ | Remarks‡ |
|---|---|---|---|---|---|---|
| (a) | = | = | × | = | $pN_T R/\varepsilon$ | $\sigma_0 = \sigma_g = \sigma_c = \sigma_s = 0$ |
| (b) Watts–Tobin BDM (implicit) | ≠ | = | × | = | $pN_T R/\varepsilon$ | $\sigma_0 = \sigma_g = \sigma_e = \sigma_s \neq 0$ |
| (c) | ≠ | ≠ | × | = | $pN_T(R-\delta)/\varepsilon$ | $\sigma_0 \neq \sigma_g \neq 0$; $\sigma_0 = \sigma_c = \sigma_s \neq 0$ |
| (d) LBS | ≠ | ≠ | = | = | $[-a\sigma + pN_T(R-\delta)]/A_0$ | $\sigma_0 \neq \sigma_g \neq 0$; different from (c) only in the description of $\varepsilon$; still $\sigma_0 = \sigma_c = \sigma_s$ |
| (e) | ≠ | ≠ | ≠ | = | $[-\alpha\sigma + pN_T(R-\delta)]/A$ | $\sigma_0 \neq \sigma_g, \sigma_0 \neq \sigma_c, \sigma_0 = \sigma_s$ |
| (f) | ≠ | ≠ | ≠ | ≠ | more complicated; not reported here | $\sigma_0 \neq \sigma_g \neq \sigma_c \neq \sigma_s$ |

* In this column, = implies 'equal and opposite'; † an × means that polarizability effects are ignored; ‡ $A \neq B \neq C$ is to be taken as 'none of the quantities $A, B, C, \ldots$ is equal to each other, generally'.

In models (a)—(c), the dielectric permeability is to be treated as a phenomenological parameter. In model (d), $A_0 = 1 + (c_e/d^3)\alpha$, and $a = \alpha$; for model (e), $A = 1 + (c_e/d^3)[\frac{1}{2}(\alpha_1 + \alpha_2) + \frac{1}{2}(\alpha_1 - \alpha_2)R]$, and $\alpha = \frac{1}{2}(\alpha_1 + \alpha_2) + \frac{1}{2}(\alpha_1 - \alpha_2)R$

trodes. We reiterate that clustering is a characteristic not entirely of the solvent and is not independent of the two 'substrates' – the electrode and the quasi-two-dimensional OHP!

## PART II: Theories of Adsorption Isotherms

### 17 Interactions

> *'That is a great deal to make one word mean',*
> *Alice said in a thoughtful tone.*

The phenomenon of adsorption is governed by the various interactions among the constituents of the interface (see Figure 8), and the forms of adsorption isotherms hold the clue to the nature of these interactions. The path from the interactions to the isotherms is, usually, beset with approximations, both physical and mathematical;[45] Figure 9 illustrates the familiar route to isotherms from interactions.

Two physical approximations that are often resorted to are (1) idealizing the adsorption region as pseudo-two-dimensional and (2) reducing the description of adsorption as that of a pseudo-monocomponent system. Examples of familiar mathematical approximations are[45] (3) the so-called congruence hypothesis[22] and (4) the mean-field approximations resulting in the Frumkin isotherm[23] and also several treatments for 'discreteness effect' calculations.[13]

The extent of inaccuracies involved in approximations (1)—(4), though it has not been sufficiently realized, may be lessened by allowing the 'two-dimensional region' to interact suitably with the two substrates, *viz.* the electrode and the OHP, and by introducing certain 'background' parameters, such as the dielectric constant of the inner layer, registering the presence of other constituents, *e.g.* the solvent (also see Sections 22—24).

What distinguishes an electrochemical interface from other interfaces is the richness of the interaction picture (see Figure 10) and our ability to manipulate this. Figure 10 is a schematic representation.

There can be other schemes that emphasize more (or less) of some of the aspects shown in Figure 10.[44, 142—145] Conway,[44] for example, emphasizes the role of hydration co-spheres of ions in the inner layer interacting mutually and with the solvent in the inner layer, as is evident in Figure 11.

The pattern of adsorption sketched by Anson[142] focusses attention on the interfacial (rather than the solution) chemistry of the phenomenon. Five types are distinguished. These are qualitatively based on the behaviour of isotherms with charge and the nature of their interaction with adsorbed anions, if any, and also the substrate (see Figure 12); see also Table 2.

Dorfman[144] recommends an approach based on sigma ($\sigma$) and pi ($\pi$) interactions as the natural one for studying the capabilities and correlating the extent of adsorption.

[142] F. C. Anson, *Acc. Chem. Res.*, 1975, **8**, 400.
[143] E. A. Neves and F. C. Anson, *J. Electroanal. Chem. Interfacial Electrochem.*, 1976, **71**, 181.
[144] Ya. A. Dorfman, *Sov. Electrochem. (Engl. Transl.)*, 1977, **13**, 963.
[145] F. M. Kimmerle and H. Menard, *J. Electroanal. Chem. Interfacial Electrochem.*, 1974, **54**, 101.

# The Electrical Double Layer

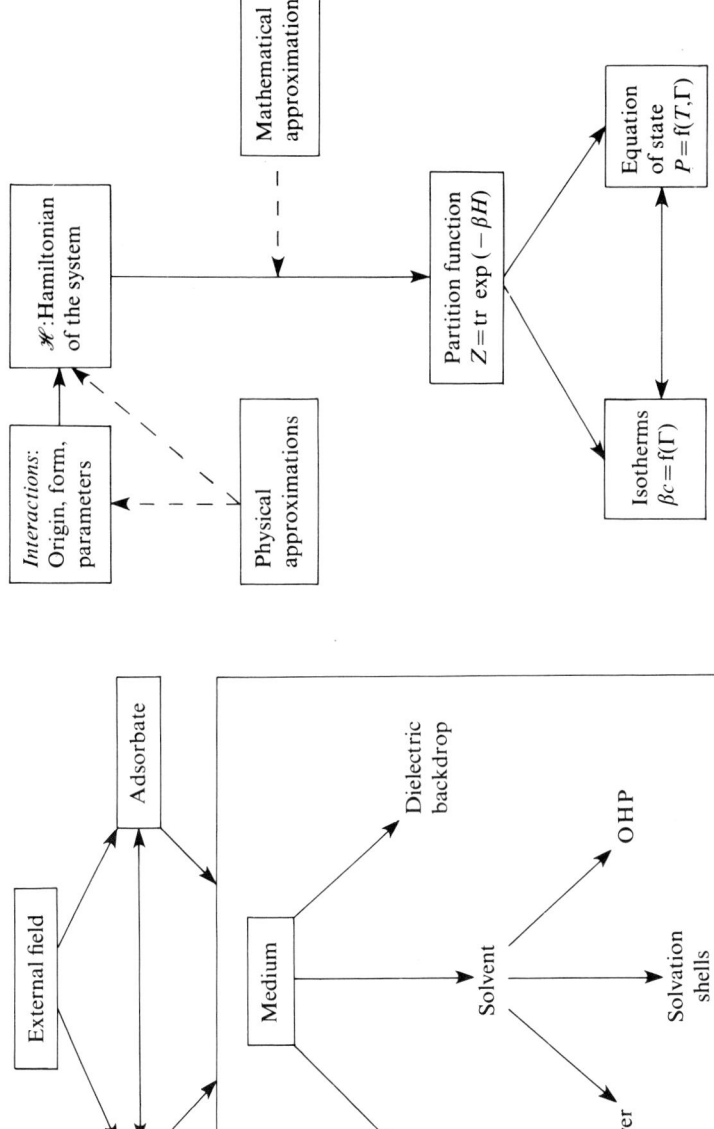

**Figure 8** *Various components of the interfacial system and their coupling (schematic)*

**Figure 9** *The nature of analysis; the route from interactions to isotherms. The arrows → indicate the operational sequence and ---→ denote stages during which approximations are introduced*

**(A)** *Particle–particle interactions*

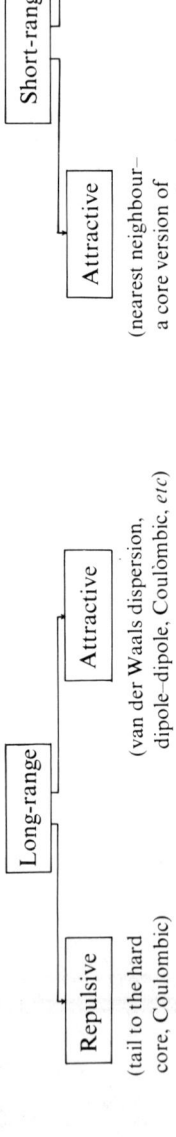

- **Long-range**
  - **Repulsive** (tail to the hard core, Coulombic)
  - **Attractive** (van der Waals dispersion, dipole–dipole, Coulombic, *etc*)
- **Short-range**
  - **Attractive** (nearest neighbour – a core version of van der Waals forces)
  - **Repulsive** (hard-core repulsion)

**(B)** *Particle–field interactions*

- **Direct effects** (*cf.* external field parameters, *e.g.* the electrode charge $\sigma_m$ or the inner-layer potential $\phi_{m2}$)
- **Indirect effects** (induced dipole moments, polarizability effects)

**(C)** *Particle–medium interactions*

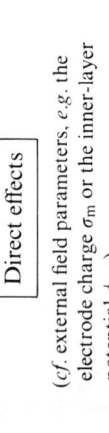
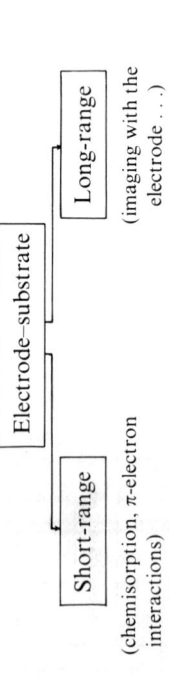
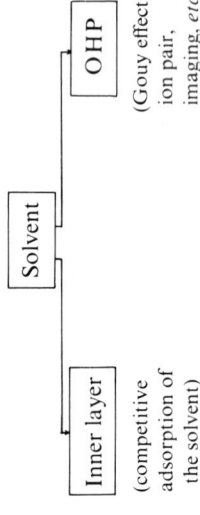

- **Electrode–substrate**
  - **Short-range** (chemisorption, $\pi$-electron interactions)
  - **Long-range** (imaging with the electrode ...)
- **Solvent**
  - **Inner layer** (competitive adsorption of the solvent)
  - **OHP** (Gouy effects, ion pair, imaging, *etc*)

**Figure 10** *The interaction picture*[45]

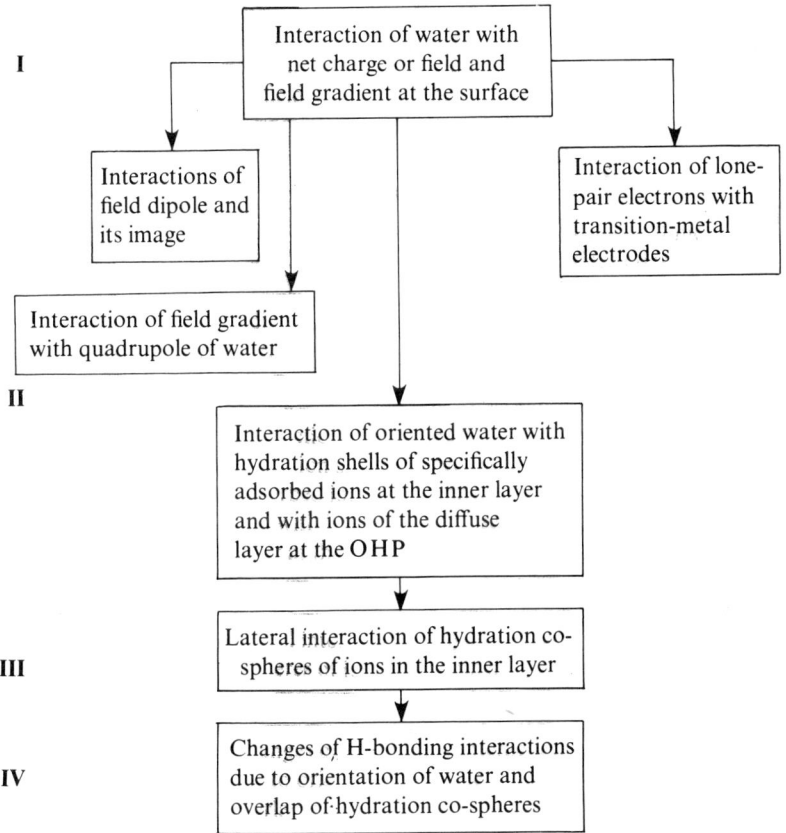

**Figure 11** *A scheme of interactions, according to Conway*[44]

IA Simple Coulombic or solvent structure-breaking factors: *e.g.*, $ClO_4^-$, $NO_3^-$, $H_2PO_4^-$, $PF_6^-$, $Cs^+$, $R_4N^+$, $R_3NH^+$, *etc.*

IB Covalent bonding with the electrode surface: *e.g.*, $I^-$, $Br^-$, $Cl^-$, $NCS^-$, $N_3^-$, $HS^-$, *etc.* (*cf.* guanidinium ions)

II Hydrophobic interactions and interactions of $\pi$-electrons and lone pairs of nitrogen with the surface: *e.g.*, organic ligands (bipyridyl, *o*-phenanthroline, *etc.*) which form complexes with metal ions

III Adsorption of cations of the triads which follow the transition metals, *e.g.* $Zn^{II}$, $Cd^{II}$, $Hg^{II}$, $In^{III}$, $Tl^{I}$, and $Pb^{II}$, as induced by anions of group IB in this classification

IV Adsorption of transition-metal cations, as induced by thiocyanate or azide anions

V Metal–metal bonding between the adsorbing complex and the metal surface

**Figure 12** *Classification of adsorption types by Anson*[142]

**Table 2** *Electrode–adsorbate interactions – some possibilities*

| Type | Examples | References | Remarks/related concepts |
|---|---|---|---|
| (a) Covalent | halides, pseudohalides | 1, 22 | Partial charge transfer/electrovalency (see Section 24) |
| (b) $\pi$-electrons and sharing of lone pairs of nitrogen | adenine, adenosine, ara-C, i-C, purine, pyrimidine derivatives, etc. | 23, 146, 147 | Orientational transitions; association (see Section 21); delocalization of electrons from the triple bond of the CN groups in the nitriles.[240] |
| (c) Enhancement of strength of ligand field on adsorption | fac-[Cr(OH$_2$)$_3$(NCS)$_3$], cis-[Cr(NH$_3$)$_4$(NCS)$_2$]$^+$, cis-[Cr(en)$_2$(NCS)$_2$]$^+$, trans-[Cr(NH$_3$)$_2$(NCS)$_4$]$^-$, [Cr(NCS)$_6$]$^{3-}$ | 142, 148 | Cationic complex behaving like an anion in its preference for a positively charged surface; electrode-induced adsorbate–adsorbate attraction. cf. 'solvophobic effect'.[149] |
| (d) Electrode-mediated surface complex states – bridging by adsorbed anions | Cations from the triads immediately following the transition metals, especially Zn$^{2+}$, Cd$^{2+}$, Hg$^{2+}$, In$^{3+}$, Tl$^+$, and Pb$^{2+}$, in the presence of anions listed under (a) | 142 | Competition between complexing tendencies and anion adsorption (cf. ref. 143) |
| (e) Surface crystallization | chloride, bromide, and iodide of both Tl$^+$ and Pb$^{2+}$ | 150—152 | Phase transitions similar to (d); surface solubility effect (see Section 21) |
| (f) Specific metal–metal bonding | [Rh(en)$_2$]$^+$ and some cyano-complexes of cobalt | 142 | Interesting similarity with their homogeneous reaction counterpart |

## 18 Electrode–Adsorbate Interactions

The role of substrate in adsorption phenomena is being increasingly realized, as reflected by the dictum[129] that 'the traditional approach to electrochemistry is thus reversed – no longer from the solution to the interface but from electrode to the interface'. Table 2[146–152] (which is only illustrative, and not exhaustive) lists a few details of this subset of interaction pictures as prevalent in the mercury/aqueous solution interface. Direct evidence for some of the interactions listed in Table 2 is absent, but can be inferred from isotherm analysis (cf. the Frumkin constant $a$) or comparable solution chemistry.[142]

## 19 Adsorbate–Adsorbate Interactions – Size Effects

The notions of saturation coverage, effective size, and hard-core repulsion are related. At moderate or high densities, it is imperative to incorporate the 'size effects' either in the continuum (hard disc) or the discrete (lattice) versions.[45] Invariably, the latter, viz. the lattice picture, is employed, as is evident from the popularity of the Frumkin or the Flory–Huggins[22,23] isotherms, with or without discreteness effects accounted for in them.

A recent report by Levine and Robinson[153] analyses the suitability of the isotherms based on the Flory–Huggins as well as the two-dimensional hard 'sphere'(?) (in the scaled-particle approximation) models, with particular reference to the Esin–Markov effect. The results may be termed inconclusive, on the basis of difficulties in obtaining agreement with the experimental data. The possibility of the scaled-particle theory leading to a negative term for the ion-size entropy, in the limit of adsorbate ions being much larger than the solvent molecules, is indicated.[60]

The 'size question' remains an open issue, as a perusal of recent literature would show. The ambivalent way in which the Frumkin or the Flory–Huggins model is employed for the various systems and the uncertainty in defining the state (and size) of those solvent molecules that are supposedly displaced by adsorbates are pointers to this state of affairs.

Let us recall that the adsorption itself is visualized as a solvent-displacement reaction, as shown in Scheme 1.

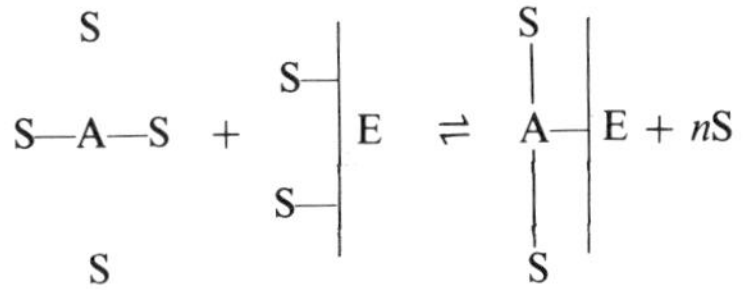

$$\text{S—A—S} + \begin{array}{c} \text{S} \\ \text{S—} \\ \text{S—} \\ \text{S} \end{array}\!\!\text{E} \quad \rightleftharpoons \quad \begin{array}{c} \text{S} \\ | \\ \text{A—E} \\ | \\ \text{S} \end{array} + n\text{S}$$

**Scheme 1**

[146] V. Brabec, M. H. Kim, S. D. Christian, and G. Dryhurst, *J. Electroanal. Chem. Interfacial Electrochem.*, 1979, **100**, 111.
[147] Y. M. Temerk, P. Valenta, and H. W. Nurnberg, *J. Electroanal. Chem. Interfacial Electrochem.*, 1979, **100**, 77.
[148] S. N. Frank and F. C. Anson, *J. Electroanal. Chem. Interfacial Electrochem.*, 1974, **54**, 55.
[149] M. Aihara and F. C. Anson, *J. Electroanal. Chem. Interfacial Electrochem.*, 1979, **99**, 55.
[150] H. B. Herman, R. L. McNeely, P. Surana, C.M. Elliott, and R. W. Murray, *Anal. Chem.*, 1974, **46**, 1258.
[151] C. M. Elliott and R. W. Murray, *J. Am. Chem. Soc.*, 1974, **96**, 3321.
[152] C. M. Elliott and R. W. Murray, *Anal. Chem.*, 1976, **48**, 259.
[153] S. Levine and K. Robinson, *J. Electroanal. Chem. Interfacial Electrochem.*, 1975, **58**, 19.

The question of adsorbate size is related to the size of the solvent S (or are there not many states of S to consider?) and the parameter $n$, an index of the *relative* range of the hard-core repulsion potential associated with A.

An approximate analysis of this problem in the continuum as a mixture of hard spheres (adsorbate and solvent) is possible, but little experimental support exists for such models. It is presumed, therefore, that improvements may be sought within the lattice framework. The success of lattice versions may be traced to the elegant way in which the idea of an area of mutual exclusion for the adsorbates is built into the picture by (1) defining a lattice with a maximum number of sites per unit area, (2) stipulating single occupancy, at most, of a site, and (3) permitting (depending upon the requirements) more sites per adsorbate molecule.

Thus, there are two scales associated with a lattice accommodating the adsorbate. The first, a primary one, is seemingly independent of 'the tenant' and the second is directly related to the size of the adsorbate. The modelling of the (primary) lattice itself is unclear:

(*a*) Is it governed by the symmetry of the substrate or its electron distribution?
(*b*) Is it the solvent monolayer structure that *is* the lattice?
(*c*) Is it merely the close-packing lattice of the adsorbates themselves?[45]

The answer is more likely to be (*b*), though somewhat modified by (*a*). But, as Sections 11–13 show, one is not so sure any more of what the size of the displaced solvent is! This dilemma, due to multi-state solvent, is acute near the PZC, where adsorption of neutral organic species is usually pronounced.

The task is made no easier by the published data analysis on this question. Some attempts to apply[149, 154—162] the Flory–Huggins version, based on appropriate $n$ values for ionic and neutral molecular adsorption, seem to be more an exception than the rule. It is claimed that the Frumkin isotherm fits well, even with some of the data which are supposed to obey the Flory–Huggins isotherm. It is reported that the Flory–Huggins model 'has performed excellently.........so long as the parameters ......... are all treated as freely variable and the user is not unduly concerned about physically curious values of the size ratio parameter $n$ resulting from optimisation'.[163] That seems to be a hard bargain, indeed!

Another aspect of the hard-core repulsion that is not completely covered by the hard-disc or Flory–Huggins picture is the aspect of symmetry or geometry. No analysis pertaining to geometries like 'rods' with various length to breadth ratio has been reported in the electrochemical context. The appropriateness of this model, especially to some systems exhibiting orientational transitions is obvious. For example, consider a linear adsorbate with three orientations:(1, 2) in the plane, and (3) normal, to the substrate. The 1 (or 2) positions need contiguous sites along 1(or

---

[154] B. E. Conway and H. P. Dhar, *Surf. Sci.*, 1974, **44**, 261.
[155] B. E. Conway and H. P. Dhar, *J. Colloid Interface Sci.*, 1974, **48**, 73.
[156] B. E. Conway and H. P. Dhar, *Electrochim. Acta*, 1974, **19**, 445.
[157] B. E. Conway, H. Angerstein-Kozlowska, and H. P. Dhar, *Electrochim. Acta*, 1974, **19**, 455.
[158] B. E. Conway, J.G. Mathieson, and H.P. Dhar, *J. Phys. Chem.*, 1974, **78**, 1226.
[159] J. O'M. Bockris and M. A. Habib, *J. Res. Inst. Catal., Hokkaido Univ.*, 1975, **23**, 47.
[160] M. J. Weaver and F. C. Anson, *J. Electroanal. Chem. Interfacial Electrochem.*, 1975, **60**, 19.
[161] A. A. Mousa, H. A. Ghaly, M. M. Abou-Romia, and F. El Taib Heakal, *Electrochim. Acta*, 1975, **20**, 489.
[162] E. Boukaram, R. Bennes, and D. Bellostas, *J. Electroanal. Chem. Interfacial Electrochem.*, 1977, **84**, 21.
[163] K. G. Baikerikar and R. S. Hansen, *Surf. Sci.*, 1975, **50**, 527.

# The Electrical Double Layer

2), while the normal requires one site only, say. The solvent occupies the 'vacant' sites, and can be treated either microscopically or as a background. The total number density in the three normal configurations is

$$\theta = \sum_{i=1}^{3} n_i N_T$$

where $N_T$ is the total number density and $n_i$ is that of state ($i$). The chemical potential, $\mu$, is related to $\theta$ in an involved manner.
If

$$n_1/N = n_2/N = s\theta$$

then[164]

$$\left(\frac{s}{1-2s}\right)\exp(\varepsilon_1 - \varepsilon_3) = \left(\frac{1-\theta[1+2(u-1)s]}{1-\theta(u-1)s}\right)^{u-1} \qquad (2)$$

and

$$\mu/kT = \ln\left\{\frac{(1-2s)\theta}{1-[1+2s(u-1)]\theta}\exp \varepsilon_3\right\} \qquad (3)$$

In the above, $\varepsilon_i$ ($i = 1, 2,$ or 3) is the energy of interaction (normalized with respect to $kT$) to the state $i$ with the substrates. Equation (2) relates $s$ and $\theta$ while (3) gives the isotherm, $viz.$ $\mu$ as a function of $\theta$. The general results for the case of ionic and neutral molecular adsorption have been obtained by the Reporter, and will be published elsewhere. The purpose in reproducing equations (2) and (3), which are valid for a simple, non-interacting model, is to emphasize the role of entropic terms in such asymmetrical, multi-site problems and also the inadequacy of simple Flory–Huggins or hard-sphere pictures.

## 20 Particle–Particle Interactions – Coulombic

Of the particle–particle interactions, an important component other than the hard core discussed in Section 19 is the Coulombic interactions among the ions and dipoles in the inner layer.

Two interesting questions associated with this are: (1) the familiar dichotomy, $i.e.$ micro- and macro-potential calculations, and (2) the imaging procedures.[13] Both these have been the subject of extensive analysis in the context of 'single-component' models.

With respect to the former question, the difficulty lies in the need to know correlation functions, for a given interaction picture, while the latter poses problems in deciding 'where' and 'how' the imaging is to be done. The solutions given are approximate and many, $e.g.$ (i) the 'cut-off' approach of Grahame and (ii) the hexagonal sub-lattice (where adsorbates are 'frozen' while the calculations are done!), are two examples.[13]

Recent experimental interest in joint adsorption and mixtures (see Sections 22 and 23), two-dimensional crystallization,[150–152] and potential distribution in lipid

---

[164] R. E. Boehm and D. E. Martire, *J. Chem. Phys.*, 1977, **67**, 1061.

bilayer membranes[165–168] provide motivation for the study of Coulombic interactions in a system of charges and dipoles confined to two different planes, under the influence of external fields.

The presence of two planes (I and II, say) introduces several possible combinations: anion in I and cation in II, or anions and cations in both the planes, or ions of one type in I and dipoles in the other, or dipoles in both I and II. It is easy to see that the theoretical formulations must be similar in all these cases, even if their predictions are not. This is true irrespective of fractional or monolayer coverage in the planes. At the conceptual level too, these are not very different from the case of the single plane. Details of calculations are, however, different, and more algebraic than the one-plane version.

Krylov and Myamlin extend the discreteness calculations to the two-component adsorption and analyse the cases of both ion(I)–ion(II)[169] and ion(I)–dipole(II)[170] systems. The relevance of such calculations to the formation of a 'surface crystalline layer' has been demonstrated by Krylov.[171] (See ref. 172 for similar calculations in kinetics.)

Both techniques, viz. the cut-off disc[169, 170] and the 'lattice' calculations,[171] are employed. The application of the 'cut-off' approach to this system is made by Krylov, assuming that the centres of the two discs lie in the same normal to the electrode and that the radii of the two regions are related (in the usual manner[13]) to the respective densities. The calculation of the electrostatic potential is then straightforward. The complicated expressions for the isotherm that follow camouflage the basic simplicity (and also the naïvety, in a way – see below) in the concepts underlying them. The resulting isotherms are indeed different from those of Langmuir or Frumkin, as they should be. But if our experience with the monocomponent discreteness calculation is any clue, there must be a reasonably valid approximation for this isotherm that resembles a two-component version of the Frumkin isotherm. In terms of inner-layer capacitances, $K_{ij}$, and the discreteness parameters, $\lambda_{ij}$, we may write this as

$$\beta_1 c_1 = \left(\frac{\theta_1}{1-\theta_1-\theta_2}\right)\exp(-2a_{11}\theta_1 - 2a_{12}\theta_2) \quad (4)$$

$$\beta_2 c_2 = \left(\frac{\theta_2}{1-\theta_1-\theta_2}\right)\exp(-2a_{21}\theta_1 - 2a_{22}\theta_2) \quad (5)$$

where

$$a_{ij} = \lambda_{ij} F/2RT K_{ij} \quad (i,j=1,2) \quad (6)$$

However, this has not yet been demonstrated.

The extension of the 'cut-off disc' method to two planes is not straightforward, unlike the case of the single plane. The centre of the disc is easily located in the *plane of the adsorption site* where the calculation is attempted. Where do we place the

[165] C. C. Wang and L. J. Bruner, *Biophys. J.*, 1978, **24**, 749.
[166] C. C. Wang and L. J. Bruner, *J. Membr. Biol.*, 1978, **38**, 311.
[167] R. Y. Tsien, *Biophys. J.*, 1978, **38**, 561.
[168] R. de Levie, *Adv. Chem. Phys.*, 1978, **37**, 99.
[169] V. S. Krylov and V. A. Myamlin, *Soc. Electrochem. (Engl. Transl.)*, 1977, **13**, 174.
[170] V. S. Krylov and V. A. Myamlin, *Soc. Electrochem. (Engl. Transl.)*, 1977, **13**, 1814.
[171] V. S. Krylov, *Sov. Electrochem. (Engl. Transl.)*, 1974, **10**, 1407.
[172] V. A. Myamlin and V. S. Krylov, *Sov. Electrochem. (Engl. Transl.)*, 1977, **13**, 1355; W. R. Fawcett and S. Levine, *J. Electroanal. Chem. Interfacial Electrochem.*, 1975, **65**, 505.

centre of the disc; in the counter plane? Is it on the normal to the planes, passing through the site in question? In other words, do the two centres become coincidental, on projecting onto the same plane? Will this be so, irrespective of the nature of the entities (cation, anion, or dipole) populating the planes?

The allocation of radii to the two discs is not obvious either. The internal distributions of the two planes are highly correlated with each other. This is especially so when the separation between the two planes is small ($\sim 1$ Å). The only constraint, consistent with the notion of planes of closest approach, is that the two types confine themselves to their respective planes. It is not clear, therefore, why the radii of the cut-off regions should be proportional to *the respective* densities prevailing in the two planes. The coupling is so strong with such small distances of inter-plane $\Delta$ (as compared to nearest-neighbour distances in either of the planes) that the spatial distributions in the individual planes are no longer 'private'. In the limiting case of $\Delta \to 0$, the cut-off regions should have the *same* radius $[\propto (\Gamma_1 + \Gamma_2)^{-\frac{1}{2}}]$, assuming the other parameters, *e.g.* charges or dipoles of the two types, to be the same, instead of an annulus (?). This limit is not at all unrealistic, and probably is realized in many situations of interest (the lipid membrane is an exception,[173] because the separation $\Delta$ is not small compared to other distances in the model). The other limit, $\Delta \to \infty$, also demonstrates how things can become different with the variation of parameters in the model. It is likely that Krylov's approach is more valid for large $\Delta$ than for small $\Delta$.

We revert now to the second issue, *viz.* imaging and the propriety of imaging in a medium whose dielectric permittivity is varying with distance.[59—62]

Bockris and Habib[61, 159] consider the diffuse layer as possessing a smoothly varying dielectric profile and have evaluated the interaction energy due to a single charge. The motivation was to model the diffuse layer properly, as well as to find out how far their original (BDM) thesis that the absence of a sharp boundary reduces imaging in solution is valid. To solve this problem, Bockris and Habib divide the dielectric zone into a large number of coupled layers of equal thickness. Then, passing over to the limit of a continuum from discreteness, the energy is evaluated. The diffuse layer is not explicitly taken into account in any other way.[60]

According to Bockris and Habib,[61, 159] their calculations reveal that the BDM model under-estimates by 1% only, in their 'single-imaging' model, whereas Levine *et al.* over-estimate (in their multiple imaging calculations[123]) by 40%.

Levine and Fawcett[60] have criticized the above approach[159] as being based 'on a questionable mathematical approximation'. They also demonstrate[60] how to interpret properly the significance of multiple imaging under conditions of dielectric variation[60, 63—65] and how the interaction between two ions adsorbed on the IHP is influenced by the diffuse layer.

In discreteness calculations, imaging in the metal – unlike imaging in the diffuse layer and the OHP – seems to be taken as well understood. But, strictly speaking, this is not true. To emphasize this point, we consider first imaging in a metal *surface*, rather than the usual *interface*.[174—176]

---

[173] R. de Levie and S. K. Rangarajan, unpublished observation.
[174] J. A. Appelbaum and D. R. Hamann, *Phys. Rev. B,* 1972, **6**, 1122.
[175] N. D. Lang and W. Kolm, *Phys. Rev. B,* 1973, **7**, 3541.
[176] S. Lundquist, 'Surface Science', International Atomic Energy Agency, Vienna, 1975, p. 331.

The image potential of charge $q$ at a distance $z_1$ from the surface is:[174]

$$U = -\left(\frac{q^2}{4(z_1-z_0)}\right) + \left[\left(\frac{\gamma}{8}\right)\frac{q^2}{(z_1-z_0)^3}\right] \qquad (7)$$

$\gamma$ being a constant [the square of the distance] that is of the order of atomic dimensions and $z_0$ the distance of the centre of mass of the induced charge from the surface:

$$z_0 = \int z\rho(z)dz / \int \rho(z)dz \qquad (8)$$

and $(-\rho)$ is the induced charge density.

An approximate analysis[174] shows that $z_0 \approx a_0$ for $z_1 \geqslant 5a_0$ and that its size depends on the density of the conduction electrons; $a_0 = 0.529$ Å. For $z_1 < 5a_0$, $z_0$ decreases sharply and can even become negative!

Two effects are now visible, *viz.* (i) the apparent 'shift' of the position of the surface (parameter $z_0$), (ii) the modification of the law for calculating the image energy; note the presence of the second term in equation (7).

When the distances considered, as in the inner-layer studies, are of molecular dimensions, it is necessary to consider the above effects more seriously. In the electrochemical context we have the metal covered with a monolayer solvent film, and thus the use of equation (7) is not obvious.

Recent Soviet work related to the above problem must be noted. The problem of the space-charge layer in the metal, originally analysed by Rice, has been reconsidered by Kuklin.[177] Other results pertaining to imaging in the metal and the interface between two media have also been reported.[178—184]

## 21 Phase Transitions

It is evident from the earlier discussion that adsorption at electrochemical interfaces is marked by (i) a versatile picture of interactions, (ii) accessibility of a wide range of external field (potential or charge), and (iii) realization of high densities (coverage $\theta \approx 1$). The phenomenon of phase transitions is a natural (if not a necessary) consequence of the above features. One may add to this the multi-component aspect, with the ever-present solvent influencing the scene quietly!

The many degrees of freedom present in this pseudo-two-dimensional system lead to many types of order parameters. Some of the familiar transitions are:

(*a*) gaseous (random) to condensed (ordered) phase (a monocomponent order–disorder);[88, 185—188]

---

[177] R. N. Kuklin, *Sov. Electrochem. (Engl. Transl.)*, 1978, **14**, 315.
[178] A. A. Kornyshev, A.I. Rubinshtein, and M. A. Vorotyntsev, *Phys. Status Solidi B*, 1977, **84**, 125.
[179] M. A. Vorotyntsev, A. A. Kornyshev, and A. I. Rubinshtein, *Sov. Electrochem. (Engl. Transl.)*, 1977, **13**, 1529.
[180] M. A. Vorotyntsev, *Sov. Electrochem. (Engl. Transl.)*, 1978, **14**, 781.
[181] M. A. Vorotyntsev, *Sov. Electrochem. (Engl. Transl.)*, 1978, **14**, 783.
[182] R. N. Kuklin, *Sov. Electrochem. (Engl. Transl.)*, 1977, **13**, 1550.
[183] N. S. Lidorenko, A. A. Izmestev, I. G. Medvedev, and G. F. Mechanik, *Dokl. Akad. Nauk. SSSR*, 1975, **223**, 639.
[184] R. N. Kuklin, *Sov. Electrochem. (Engl. Transl.)*, 1977, **13**, 1007.
[185] A. N. Frumkin, N. V. Fedorovich, B. B. Damaskin, E. V. Stenin, and V. S. Krylov, *J. Electroanal. Chem. Interfacial Electrochem.*, 1974, **50**, 103; A. R. Alumaa and U. V. Palm, *Sov. Electrochem. (Engl. Transl.)*, 1976, **12**, 291.

(b) one configuration of the adsorbate to another (orientational order–disorder);[145, 146, 169, 189—191]
(c) alloy-type order–disorder many-component system (Is condensation to be viewed as segregation?);
(d) the surface crystallization.[150—152]

Experimental observations pertaining to these are not all new,[23] as for, e.g., the case of two-dimensional condensation and orientational transitions: a few related results have been reported recently.

The following class of compounds that exhibit two-dimensional condensation can be arranged, in order of increasing activity,[88] , 2-oxa-adamantane < camphor < borneol < isoborneol < adamantanol < 3-hydroxy-homoadamantane.

An interesting observation on how the 'washing out' of the condensation (i.e., prevention of the phase change) can be brought about by the presence of ethyl alcohol has been made.[88, 192] This is obviously a case of competitive adsorption. The inference of 'washing-out', like that of condensation, is based on the magnitude of the Frumkin coefficient $a$ being $<2$ or $>2$. In the case of adamantanol, the attraction coefficient $a$ decreases from 5.5 (in the absence of alcohol) to 1.6 in the solution containing ethyl alcohol.[88, 192] It is inferred, as is to be expected, that more ethyl alcohol is needed to achieve this purpose, viz. the prevention of a condensed film forming, the higher the surface activity.

The nature of the phase (random or condensed) of the adsorbed layer, and *not* the coverage alone, seems to decide the model for the OHP and the double layer.[87, 91] A new equivalent-circuit representation has been suggested for this case. Independent diffuse layers pertaining to 'free' and 'covered' regions have been proposed. Employing a phenomenological analysis, Kuryakov et al.[87] have demonstrated a difference in behaviour even among those adsorbates that undergo condensation. It is shown that adamantanol, 3-hydroxy-homoadamantane, and borneol, which are the more 'strongly segregated' ones, seem to require that the modified version[91] of the diffuse layer effects be considered more than the less active members of this class of compounds, e.g. camphor and 2-oxa-adamantane.

Two-dimensional condensation behaviour has also been reported for phloroglucinol,[188] $d$-camphor 10-sulphonate,[187] and some other systems.

**Orientational Transitions.**—Several systems displaying interesting transitions have been studied recently by many workers, in particular by Dryhurst et al.[146] and by Nurnberg and his collaborators.[147]

Some of the systems studied are: adenine, deoxyadenosine, and deoxyadenosine

---

[186] J. Kuta, L. Pospisil, and I. Smoler, *J. Electroanal. Chem. Interfacial Electrochem.*, 1977, **75**, 407.
[187] A. C. Ramamurthy and S. Sathyanarayana, *J. Electroanal. Chem. Interfacial Electrochem.*, 1977, **73**, 253.
[188] S. Sarangapani, Ph.D. thesis, Madurai University, 1975; V. K. Venkatesan and S. Sarangapani, *Electrochim. Acta*, 1980, in the press.
[189] H. Nakadomari, D. M. Mohilner, and P. R. Mohilner, *J. Phys. Chem.*, 1976, **80**, 1971.
[190] U. Retter, H. Jehring, and V. Vetterl, *J. Electroanal. Chem. Interfacial Electrochem.*, 1974, **57**, 391.
[191] A. Jenard and H. W. Hurwitz, *J. Electroanal. Chem. Interfacial Electrochem.*, 1976, **70**, 27.
[192] E. V. Stenina, B. B. Damaskin, N. V. Fedorovich, and S. L. Dyatkina, *Dokl. Akad. Nauk. SSSR*, 1977, **236**, 400.

mononucleotides;[146, 193a] thymine;[193a] uracil;[194] methylated uracil derivatives;[195] and 1-β-D-arabinofuranosylcytosine (ara-C) and 1-β-D-ribofuranosylcytosine (r-C).[147] Detailed references to other contributions on the adsorption of adenine nucleotides, mono-oligonucleotides, and polynucleotides from the Julich school can be found in reference 147.

Two specific features observed in the above systems are presence of (a) 'dilute' and 'compact' layer regions, dictated by the concentration and the potential, and (b) capacitance 'pits', one or two in number.[146, 147, 196]

Interesting observations on the many types of structural changes of polynucleotides, depending on potential,[197] denaturing or unravelling of adsorbed DNA and other double-stranded polynucleotides,[198—201] adenine bound to the electrode through its two amino hydrogens, two types of 'perpendicular' layers, in the case of adenosine, and the transition from the *syn* to the *anti* conformation followed by a dipole-orientation transition in the *anti*-state[145] have been reported (see also references 202 and 203 on the last aspect).

Apart from the conventional Frumkin isotherm analysis and observing that $a > 2$ in the case of a compact layer or condensation,[185] no detailed theoretical analysis has, however, been reported.

The theoretical descriptions of transition phenomena have so far been based on a naive analysis of the Frumkin isotherm and finding out if the attraction parameter $a > 2$. If the saturation coverage shows anomalous variation with charge, or potential, or even with coverage, it is presumed that orientational transitions are present. Alternatively, the abruptness in plots of $\Gamma$ *versus* $\sigma$ (or $E$), capacitance 'dips' or 'pits', *etc.*, have also been used as diagnostic criteria for transitions. These phenomenological approaches have been extended recently in two directions: (1) to extend the analysis, at constant temperature, for two-component systems,[204, 205] (2) to investigate phase diagrams in a space spanned by temperature and other parameters for a *monocomponent* model.[206—209]

Some recent reports consider certain mathematical aspects of the theory of phase transitions.

Gurevich and Kharkats[206—209] analyse the effects of temperature variation on phase transitions, for monocomponent systems under the mean-field approxima-

---

[193] K. Kinoshita, S. D. Christian, and G. Dryhurst, *J. Electroanal. Chem. Interfacial Electrochem.*, 1977, (a) **83**, 151; (b) **85**, 377.
[194] V. Brabec, S. D. Christian, and G. Dryhurst, *J. Electroanal. Chem. Interfacial Electrochem.*, 1977, **85**, 389.
[195] V. Brabec, S. D. Christian, and G. Dryhurst, *Biophys. Chem.*, 1978, **7**, 253.
[196] Y. M. Temerk and P. Valenta, *J. Electroanal. Chem. Interfacial Electrochem.*, 1978, **93**, 57.
[197] P. Valenta and H. W. Nurnberg, *Biophys. Struct. Mech.*, 1974, **1**, 17.
[198] P. Valenta, H. W. Nurnberg, and P. Klahre, *Bioelectrochem. Bioenerg.*, 1974, **1**, 487; 1975, **2**, 245.
[199] V. Brabec and E. Palecek, *Stud. Biophys.*, 1976, **60**, 105.
[200] V. Brabec and E. Palecek, *Biophys. Chem.*, 1976, **4**, 79.
[201] E. Palecek, *Collect. Czech. Chem. Commun.*, 1974, **39**, 3449.
[202] D. Krznaric, P. Valenta, H. W. Nurnberg, and M. Branica, *J. Electroanal. Chem. Interfacial Electrochem.*, 1978, **93**, 41.
[203] V. Vetterl, *Bioelectrochem. Bioenerg.*, 1976, **3**, 338.
[204] E. M. Podgaetskii, *Sov. Electrochem. (Engl. Transl.)*, 1974, **10**, 643.
[205] E. M. Podgaetskii, *Sov. Electrochem. (Engl. Transl.)*, 1975, **11**, 1653.
[206] Yu. Ya. Gurevich and Yu. I. Kharkats, *Sov. Electrochem. (Engl. Transl.)*, 1978, **14**, 709.
[207] Yu. Ya. Gurevich and Yu. I. Kharkats, *Dokl. Akad. Nauk. SSSR*, 1976, **230**, 132.
[208] Yu. Ya. Gurevich and Yu. I. Kharkats, *Pis'ma Zh. Eksp. Teor. Fiz.*, 1976, **23**, 249.
[209] Yu. Ya. Gurevich and Yu. I. Kharkats, *J. Electroanal. Chem. Interfacial Electrochem.*, 1978, **86**, 245.

tion. The free energy of the system considered is

$$F(\theta) = W\theta - (\lambda\theta^2/2) - kT[\theta(1-m+\ln m\rho) - \theta\ln\theta - m(1-\theta)\ln(1-\theta)] \quad (9)$$

appropriate to the Flory–Huggins isotherm, when written in the form

$$\frac{kT}{\lambda} = \frac{\tilde{\theta}-\theta}{\ln[\rho m(1-\theta)^m/\theta]} \quad (10)$$

In equations (9) and (10),

$$\tilde{\theta} = W/\lambda \quad (-\infty < \theta < \infty) \quad (11)$$

and $\rho = \chi c$, where $m$ is the number of adsorption sites per particle, $c$ is the 'bulk' concentration, $\chi$ and $W$ are constants associated with the changes in entropy and free energy of the adsorbate during adsorption, respectively, and $\lambda$ is related to the interaction energy between two adsorbates. The richness of transitions in the parameter space $(\tilde{\theta}, \rho)$ is demonstrated. Note that $\lambda$ serves as a scale for the temperature, $T$, and has no other role. It is noted, in particular, that 'double transitions' may be displayed by asymmetric systems for which discontinuous adsorption and desorption at different temperatures may be found. This effect is more pronounced the larger the value of $m$, and vanishes in the limit $m=1$ (Frumkin).

Podgaetskii, on the other hand, has investigated[204, 205] the onset of transitions, under conditions of constant temperature but for two-component systems. This should be a natural generalization of the familiar prescription in the case of the Frumkin isotherm, *viz.* (attractive) $a > 2$. The analysis is algebraic, and involves several parameters. Extension of this analysis to the Flory–Huggins version for two components is straightforward.

Damaskin[210] considers a special case of two orientations of the same adsorbate with varying size requirements. The equations are

$$\beta_1 c_1 = \left(\frac{\theta_1}{1-\theta_1-\theta_2}\right)\exp(-2a_1\theta_1) \quad (12)$$

$$\beta_2 c_2 = \frac{\theta_2}{2(1-\theta_1-\theta_2)^2} \quad (13)$$

An analysis of these for the total surface excess

$$\Gamma = \Gamma_{1m}\theta_1 + \Gamma_{2m}\theta_2 \quad (14)$$

and the capacitance *versus* potential curves is possible in a relatively convenient form. The resemblance of these theoretical curves to those observed during the adsorption at a mercury electrode of heterocyclic compounds such as adenine, adenosine, deoxyadenosine,[145] and coumarin[161] has been noted.[210]

**Surface Crystallization.** A different kind of an interesting phase transformation is observed in the case of the adsorption of $Tl^+$ and $Pb^{2+}$ in aqueous halides. Different phenomena, what may be termed two-dimensional monolayer precipitation[150] and the formation of a crystalline bilayer,[151, 152] occur in the case of $Tl^+$ and $Pb^{2+}$ respectively. Large, abrupt changes in the surface excess $\Gamma_{Tl}$ and $\Gamma_{Pb}$, plotted against the concentrations of anions in solution, characterize these transi-

[210] B. B. Damaskin, *Sov. Electrochem. (Engl. Transl.)*, 1977, **13**, 690.

tions. 'Surface solubility products' have been evaluated for these conditions and observed to be less than their 'counterparts in the bulk'. An extensive analysis of the parameters characterizing the surface layers is carried out. That these phenomena cannot be considered as belonging to the 'anion-induced' category (see Table 2) is shown[152] through a careful study of the prediscontinuity regions.

## 22 Medium Effects

**Phenomena.**—Some of the phenomena falling into this category are:
(a) marked changes in the bulk activity of the organic compound resulting from the presence of the electrolyte; *e.g.*, 2-butanol *or* 2-methyl-2-propanol and $Na_2SO_4$,[189, 211] and urea–$Et_4NBr$ on mercury/aqueous solution[212] (see also refs. 213 and 214 for a general discussion of this aspect);
(b) effects arising out of weak, competitive adsorption of ions, subdivided into (i) a 'counter-ion effect' reflecting the specificity of the counter-ions (*e.g.*, adsorption of iodide[215] or nitrate[216] ions *vis à vis* $Na^+$ and $K^+$ as counter-ions) and (ii) formation of 'ion pair' at the interface (*e.g.*, adsorption of $Br^-$ in aqueous solution,[217] of $Cl^-$,[80] or $SCN^-$,[81, 84] in ethanol, and of $Cs^+$[82, 83] in methanol, on bismuth; see also ref. 79);
(c) co-adsorption of ions/dipoles that mutually enhances their adsorbabilities in relation to their adsorption when 'alone'; *e.g.*, adsorption of $SO_4^{2-}$ and of $Cd^{2+}$,[218] the mutual effect of $ClO_4^-$, $SO_4^{2-}$, $Cl^-$, and phosphate ions in their adsorption on ruthenium–titanium dioxide electrodes,[219, 220] competitive adsorption of thiourea and butyl alcohol or butylamine on mercury,[221, 222] and co-adsorption of salicylic acid and hydroquinone from methanol–water and from n-propanol–water mixtures;[223]
(d) inhibiting 'condensation'; *e.g.*, addition of ethanol *vis à vis* condensation of adamantanol (see Section 21);
(e) promotion of two-dimensional association; *e.g.*, adsorption of tribenzylamine in ethanol–water mixtures and the onset of a sudden transition, depending upon the percentage of ethanol;[186]
(f) specific 'electron couplings'; *e.g.*, adsorption of organic compounds on mercury electrodes in aqueous solutions containing $Tl^+$, the adsorbates being phenol[224]

[211] D. M. Mohilner and H. Nakadomari, *J. Electroanal. Chem. Interfacial Electrochem.* 1975, **65**, 843.
[212] F. M. Kimmerle and H. Menard, *Can. J. Chem.*, 1977, **55**, 3312.
[213] B. A. Abd-el-Nabey, A. de Battisti, and S. Trasatti, *J. Electroanal. Chem. Interfacial Electrochem.*, 1974, **56**, 101.
[214] A. de Battisti and S. Trasatti, *J. Electroanal. Chem. Interfacial Electrochem.*, 1974, **54**, 1.
[215] W. R. Fawcett and T. A. McCarrick, *J. Electrochem. Soc.*, 1976, **123**, 1325.
[216] W. R. Fawcett and J. B. Sellan, *Can. J. Chem.*, 1977, **55**, 3871.
[217] M. G. Väärtnõu, S. K. Inno, M. A. Salve, and U. V. Palm, Proceedings of the 4th Symposium on Double Layer and Adsorption on Solid Electrodes, Tartu (USSR), 1975, p. 59.
[218] S. Ya. Vasina, V.E. Kazarinov, and O. A. Petrii, *Sov. Electrochem. (Engl. Transl.)*, 1978, **14**, 63.
[219] V. E. Kazarinov and V. N. Andreev, *Sov. Electrochem. (Engl. Transl.)*, 1978, **14**, 577.
[220] V. E. Kazarinov and V. N. Andreev, *Sov. Electrochem. (Engl. Transl.)*, 1977, **13**, 588.
[221] S. L. Dyatkina, G. M. Rott, and B. B. Damaskin, *Sov. Electrochem. (Engl. Transl.)*, 1977, **13**, 807.
[222] S. L. Dyatkina and B. B. Damaskin, *Sov. Electrochem. (Engl. Transl.)*, 1976, **12**, 540.
[223] K. M. Joshi, S. R. Tendulkar, and M. R. Bagat, *J. Electroanal. Chem. Interfacial Electrochem.*, 1976, **68**, 303.
[224] A. N. Frumkin, N. S. Polyanovskaya, B. B. Damaskin, and O. Oskina, *J. Electroanal. Chem. Interfacial Electrochem.*, 1974, **49**, 7.

## The Electrical Double Layer

and aniline, n-butanol, and n-butylamine.[225] In the case of phenol, mutual reinforcement is observed due to interaction beween the $\pi$-electrons of the aromatic ring and the positive charge on the thallium atom, the effect vanishing as one proceeds from the aromatic series to butanol and butylamine (which do not possess $\pi$-electrons). The polarity of the Tl–Hg bond is very important;[226]

(g) non-monotonic nature of adsorption, caused by an interplay of a 'bulk effect', *viz.*, changes in chemical potential of a salt with the mole fraction of the organic component [see (a) above] and a 'surface effect', *viz.* preferential adsorption of the organic solvent at the electrode; *e.g.*, $SCN^-$ from NaSCN and NaF in water–acetone.[227]

The demarcation among (a)—(g) is not rigid [*cf.* (a) and (g); (d) and (e)] but is convenient.

It is a truism to state that the solvent is as much a part of the medium as the electrolyte, but the role of solvent (*i.e.*, water, in most experiments) has not been explicitly included in the above, even though its manifold ability to influence has been recognized (see reference 44 and Figure 11). A proper quantitative theory to account for the finer effects necessarily has to be a many-state description, linked appropriately with the models for solvent (see Part I of this chapter).

**Analysis of Mixtures.**—A related area of study is the analysis of mixtures of electrolytes. The original motivation for such studies through the Hurwitz–Parsons method was to maintain the diffuse layer the same while studying specific adsorption. This is useful if there is only one constituent in the mixture that is specifically adsorbed, the others having been assumed 'inactive' (see Section 8). Most of the recent studies that have employed mixtures are with respect to either the counterion effect (NaX + NaF *versus* KX + KF; $X = I^-$ or $NO_3^-$)[215,216] or ion association (KI + KF and LiA + LiClO$_4$ on Bi; $A = Cl^-$ or $I^-$):[79–83] see also references 228 (NaN$_3$ + NaF) and 229 (KCl + KF) for studies on polycrystalline silver.

With the inactivity of $F^-$ losing credibility[1] (and the adsorption at potentials away from the PZC certain for others), it is surprising that few attempts have been made to relax this assumption on $F^-$. An exception is an interesting report by Baugh and Parsons[230] on the simultaneous adsorption of guanidinium and chloride ions in the mixture containing NaCl. The importance of correcting for activity coefficients in mixtures[231,232] is realized by these authors. In the absence of the activity coefficients of guanidinium chloride, the values for tetramethylammonium chloride are used (but see ref. 233).

More detailed thermodynamic analysis of simultaneous adsorption should be useful. For example, Figures 13—15 show the capacitance–potential curves for the system (NaCl + NaNO$_3$), indicating (a) a potential region of almost constant

[225] N. S. Polyanovskaya and B. B. Damaskin, *Sov. Electrochem. (Engl. Transl.)*, 1978, **14**, 696.
[226] A. N. Frumkin, N. S. Polyanovskaya, and B. B. Damaskin, *J. Electroanal. Chem. Interfacial Electrochem.*, 1976, **73**, 273.
[227] W. Kemula, B. Behr, and J. Stroka, *J. Electroanal. Chem. Interfacial Electrochem.*, 1977, **75**, 3999; J. Stroka, Ph.D. Thesis, Warsaw University, 1975.
[228] E. R. Gonzalez, *J. Electroanal. Chem. Interfacial Electrochem.*, 1978, **90**, 431.
[229] A. V. Shelepakov and E. S. Sevastyanov, *Sov. Electrochem. (Engl. Transl.)*, 1978, **14**, 243.
[230] L. M. Baugh and R. Parsons, *J. Electroanal. Chem. Interfacial Electrochem.*, 1975, **58**, 229.
[231] S. Lakshmanan and S. K. Rangarajan, *J. Electroanal. Chem. Interfacial Electrochem.*, 1970, **27**, 127.
[232] S. Lakshmanan and S. K. Rangarajan, *J. Electroanal. Chem. Interfacial Electrochem.*, 1970, **27**, 170.
[233] G. Barone, V. Elia, U. Lepore, and D. Paparone, *Gazz. Chim. Ital.*, 1976, **106**, 567.

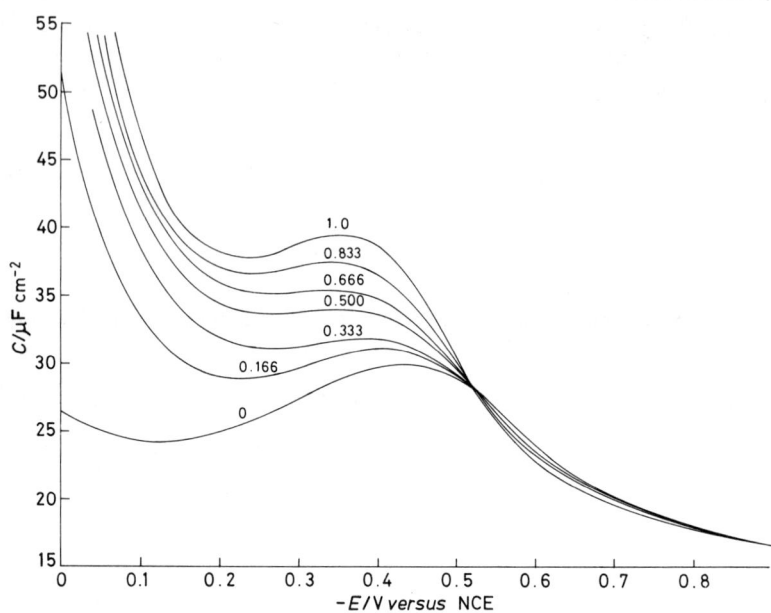

**Figure 13**

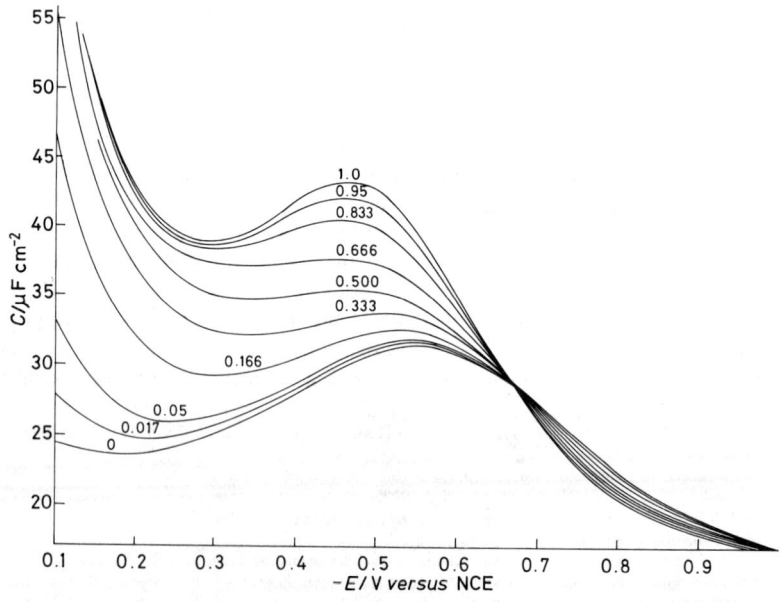

**Figure 14**

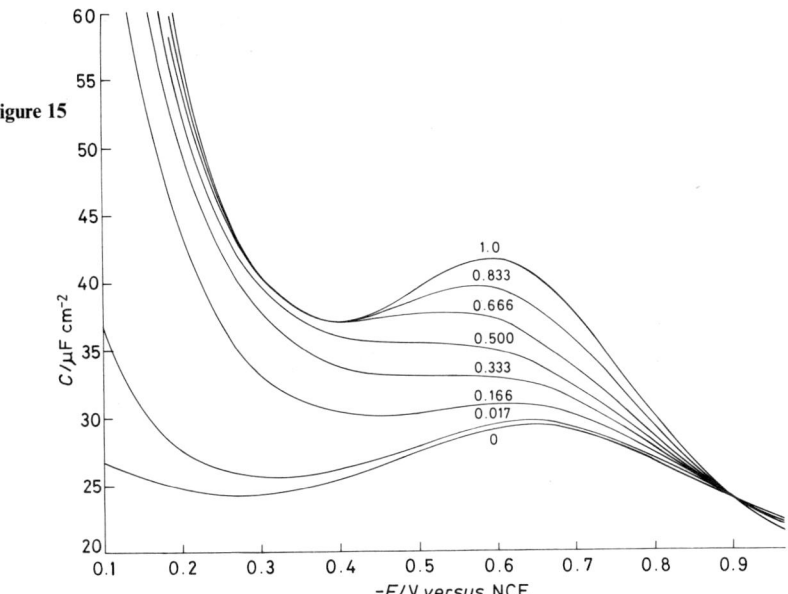

**Figure 15**

**Figures 13–15** *Differential capacitance of the mixture $mI$-NaCl+$(1-m)I$-NaNO$_3$; the values marked on the curves refer to mole fractions, m. Ionic strength, I, is 0.1 in Figure 13, 1.0 in Figure 14, and 5.0 mol l$^{-1}$ in Figure 15*

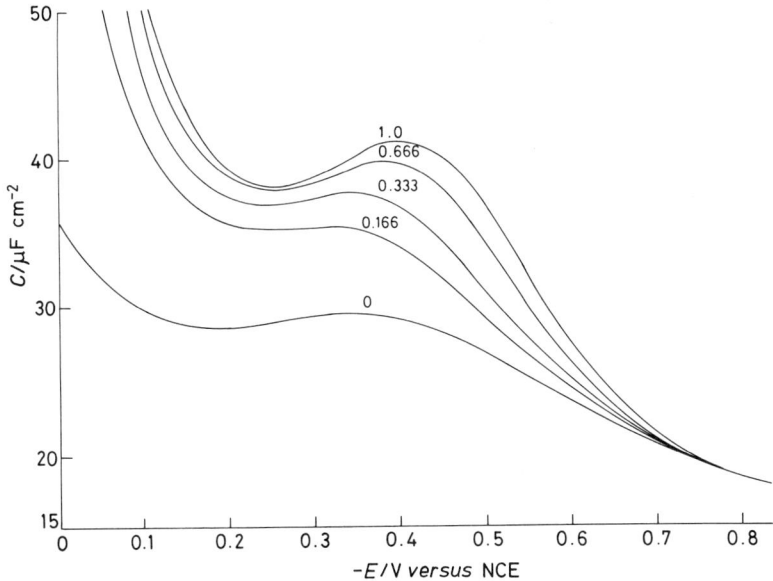

**Figure 16** *Differential capacitance of the mixture $mI$-NaCl+$(1-m)I$-NaOAc; the values marked refer to mole fractions, m. $I=0.3$ mol l$^{-1}$*

capacitance for certain molar ratios, that may be exploited to advantage in studies of non-linear relaxation, and (b) the 'crossing over' of the capacitance curves for varying molar ratios, at each ionic strength.[234]

The observation (b) indicates an inflection point for $\Gamma^{(1)} - [m_1/(1-m_1)]\Gamma^{(2)}$ with respect to the potential abscissa, $\Gamma^{(i)}$ ($i = 1, 2$) being the surface excesses of the two anions [note the generalization of the condition $(\partial C/\partial \mu)_E = (\partial^2 \Gamma/\partial E^2)_\mu$ for single salts[235] to mixtures[232]]. What makes this an attractive feature is that this is *not* a universal one (see, *e.g.*, Figure 16 for the mixture NaCl + NaOAc, where this is not observed[234]). A conjecture is that this behaviour is characteristic of the interplay of chemical ('structure-breaking' effects) and electrostatic (inner-layer capacitances associated with the ions?) covalent tendencies, and may be expected in the mixtures containing one anion of type IA and one of type IB (see Figure 12).

There is also a certain elegance in the parallel between all the routes of analysis for single salts and for mixtures. Figures 17 and 18 describe the analogues of analysis, and it is hoped that more results will be forthcoming on this aspect.

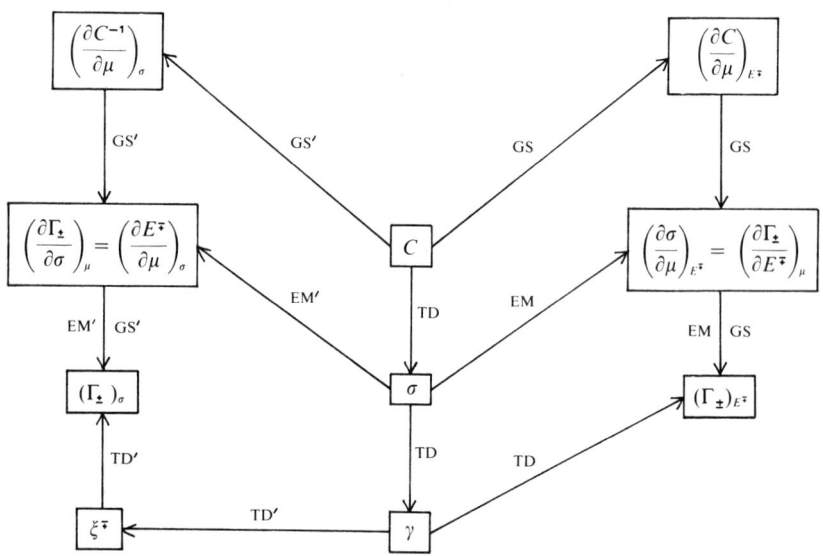

**Figure 17** *Various methods of analysis – single salts.*
    TD   *Thermodynamic method, classical.*
    TD'  *Thermodynamic method due to Parsons: function used is* $\xi^\mp = \gamma + \sigma E^\mp$
    EM  *Method via Esin–Markov coefficients, at constant* $E^\mp$
    EM'  *Analogue of EM, at constant* $\sigma$
    GS   *Grahame–Soderberg method*
    GS'  *Analogue of GS, at constant* $\sigma$
    *An arrow pointing upwards and connecting two variables indicates that a differentiation is involved in this process; similarly, one directed downwards suggests that an integration is involved. The procedure is to start anywhere and follow the arrow to reach* $\Gamma_\pm$

---

[234] S. Lakshmanan and S. K. Rangarajan, unpublished observations; S. Lakshmanan, Ph.D. thesis, Madras University, 1971.
[235] T. Kakiuchi and M. Senda, *J. Electroanal. Chem. Interfacial Electrochem.*, 1978, **88**, 219.

# The Electrical Double Layer

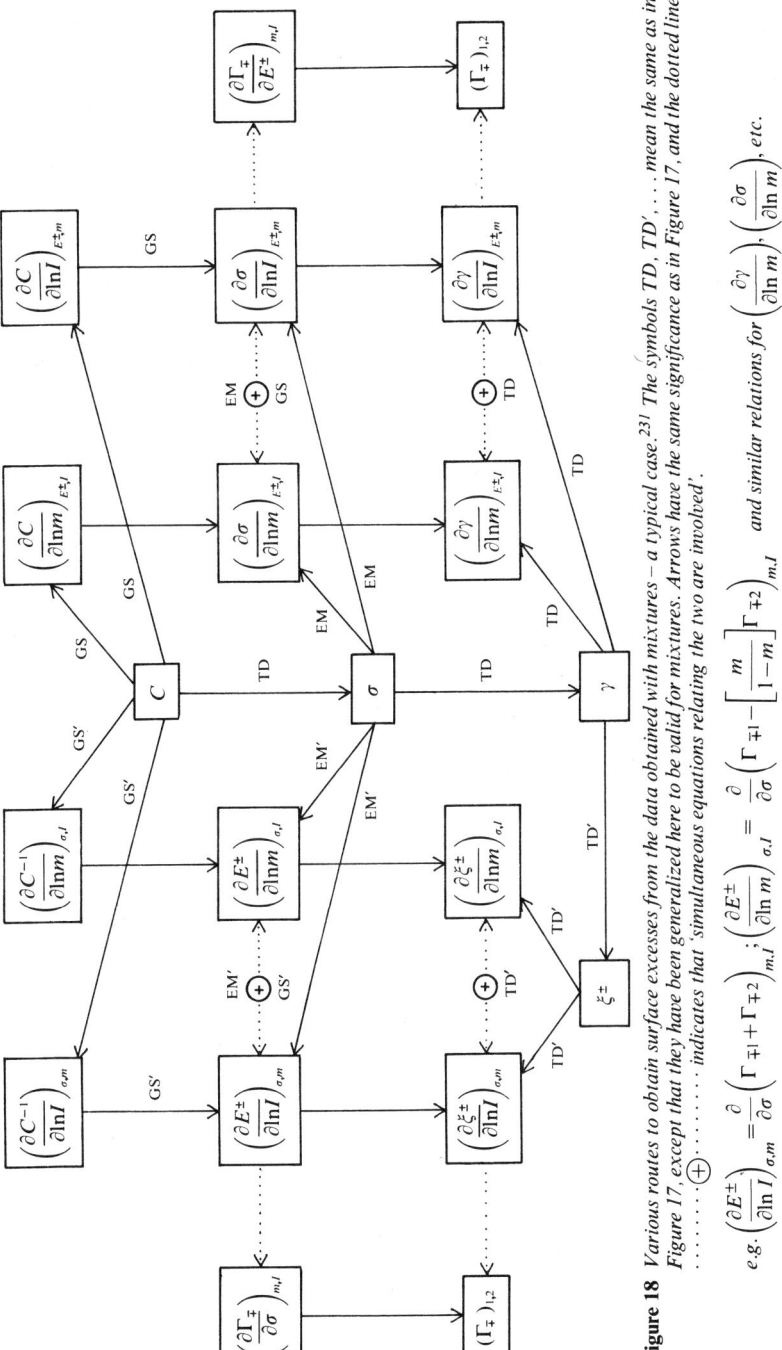

**Figure 18** *Various routes to obtain surface excesses from the data obtained with mixtures – a typical case.[231] The symbols TD, TD′, ... mean the same as in Figure 17, except that they have been generalized here to be valid for mixtures. Arrows have the same significance as in Figure 17, and the dotted line ......... indicates that 'simultaneous equations relating the two are involved'.*

*e.g.* $\left(\dfrac{\partial E^{\pm}}{\partial \ln I}\right)_{\sigma,m} = \dfrac{\partial}{\partial \sigma}\left(\Gamma_{\mp 1}+\Gamma_{\mp 2}\right)_{m,I}$ ; $\left(\dfrac{\partial E^{\pm}}{\partial \ln m}\right)_{\sigma,I} = \dfrac{\partial}{\partial \sigma}\left(\Gamma_{\mp 1}-\left[\dfrac{m}{1-m}\right]\Gamma_{\mp 2}\right)_{m,I}$ *and similar relations for* $\left(\dfrac{\partial \gamma}{\partial \ln m}\right)$, $\left(\dfrac{\partial \sigma}{\partial \ln m}\right)$, *etc.*

## 23 Multi-component Systems and the pseudo-Frumkin Isotherms

The possible phenomenological couplings [(a)—(g) of Section 22] caused by the medium, as also the solvent effects, having been listed, the question is *not* 'why do systems violate the Frumkin isotherm?', *but* 'how can these adsorbate systems be in agreement with the Frumkin isotherm?' Recalling that the Frumkin isotherm (FI) is a one-component model, one cannot explain away any discrepancies by pointing to experimental insensitivity alone! A plausible explanation – partly correct – can be found in the mean-field approximation that is implicit in the FI which blunts the finer details of the true interaction picture. A second possibility is that what is being obeyed is not the true FI but a pseudo-FI that carries in it certain parameters that characterize the medium; *e.g.*, the electrolyte concentration. The question then is, starting from a truly multi-component description, can one land on a pseudo-FI?

If the subscripts 1 and 2 denote the two adsorbates in equations (4) and (5), it is possible to eliminate $\theta$ and obtain

$$\beta_1 c_1 = F(\theta_1, \beta_2 c_2) \tag{15}$$

which, in general, will *not* be FI-like. It doesn't help much by rewriting it in the form of the FI[221,222,236]

$$\beta_1^* c_1 = \left(\frac{\theta_1}{1-\theta_1}\right)\exp(-2a_1^* \theta_1) \tag{16}$$

with $a_1^*$ now depending not only on $\beta_2$ but also on $\theta_1$. In such a case, equation (16) cannot even be called a pseudo-FI. However, in some limiting cases, for '2' at least, it is possible to demonstrate the *possibility of a pseudo-Frumkin isotherm*. Table 3 indicates a few such cases, covering joint adsorption, counter-ion, and medium effects. More complicated situations have been omitted.

It is interesting to note from Table 3 how not only the concentration $c_2$ but also the inner-layer potential or the charge enter the isotherm parameters $\beta_1^*$ and $a_1^*$, via $\beta_2 c_2$, thereby distorting the naive descriptions for $\Delta G_{ads}^{\ominus}$ (the free energy of adsorption), the saturation coverage, and the so-called interaction constant, $a$. So, if one is not aware of the coupling between adsorbates 1 and 2, surprises are in store while analysing! Peculiar potential dependencies in the apparent free energy of adsorption and the 'interaction constant', $a$ (which is actually an 'omnium gatherum'!) are to be expected.

In all cases (A)—(D) of Table 3, the component 1 obeys a pseudo-Frumkin isotherm (16) with the apparent parameters related to $\beta_2 c_2$, under the constraints imposed on the coverage $\theta_2$. Cases (A) and (D) are realizable in practice, depending upon the free energies of adsorption and the coverages, in the range of potentials studied. These pertain to competitive (joint) adsorption of ions/dipoles and explain, approximately, the origin of potential differences through $\beta_2$. Simple models for $\Delta G_{ads}^{\ominus}$ as linear or quadratic functions of charge may be violated, in view of this. Cases (B) and (C) in Table 3 are pertinent to a counter-ion, where one may model the cation '2' as being 'adsorbed' on anionic sites '1' (see reference 237, footnote 38). Such 'second-order adsorption' is likely in many more contexts, and must be carefully considered. Ion association, recently considered by Palm and collabora-

---

[236] B. B. Damaskin, A. N. Frumkin, and N. A. Borovaya, *Elektrokhimiya*, 1972, **8**, 807.
[237] B. V. K. S. R. A. Tilak and M. A. V. Devanathan, *J. Phys. Chem.*, 1969, **73**, 3582.

**Table 3** *Medium effects – origin and composition of pseudo-Frumkin isotherms in multi-component systems*

| Model | $\beta_1{}^*$ | $a_1{}^*$ | Relevance/remarks |
|---|---|---|---|
| (A) $\beta_2 c_2 \approx \theta_2$ | $\beta_1 \exp(2a_{12}\beta_2 c_2) = \beta_1{}^*$ | $a_1{}^* = a_{11}$ | Congruence retained; $a_{11}$ unaffected; the free energy of adsorption of 1 influenced by concentration $c_2$ of 2 and distorted by $\beta_2$; medium effect and competitive adsorption. |
| (B) $\beta_2 c_2 \theta_1 \approx \theta_2$ | $\beta_1 = \beta_1{}^*$ | $a_1{}^* = a_{11} + a_{12}\beta_2 c_2$ | Congruence lost through potential dependence (via $\beta_2$) of the pseudo-$a_1{}^*$; the 'interaction parameter' depends on the medium; $\beta_1$ is unaffected; adsorption of relevant counter-ion on adsorbate sites.[79] |
| (C) $\beta_2 c_2 \approx \theta_2/(1-\theta_1)$ | $\beta_1(1-\beta_2 c_2)\exp(2a_{12}\beta_2 c_2) = \beta_1{}^*$ | $a_1{}^* = a_{11} - a_{12}\beta_2 c_2$ | Congruence lost; both $\beta_1{}^*$ and $a_1{}^*$ are medium-dependent; applicable when $\theta_1$ is appreciable but $\theta_2$ small. |
| (D) Langmuir's law for 2 (exact) i.e., $\beta_2 c_2 = \theta_2/(1-\theta_1-\theta_2)$ | $\beta_1{}^* = \dfrac{\beta_1 \exp[2a_{12}B(E,c_2)]}{1+\beta_2 c_2}$ | $a_1{}^* = a_{11} - a_{12}B(E,c_2)$ | Congruence no longer valid $B(E,c_2) = \beta_2 c_2/(1+\beta_2 c_2)$ *cf.* the potential dependence. |

tors [see $b$(ii) of Section 22] can also be modelled, as a first step on this basis [see equation (8) in reference 79].

Other models that are more complicated than (A)—(D) of Table 3 would result in the parameters $\beta_1^*$ and $a_1^*$ becoming functions of $c_1$ (note that they *are* already functions of $c_2$) and thus violating the form of the FI [see also equation (4) of reference 210 for a similar elimination in a different context].

As repeatedly emphasized here and elsewhere,[45] a proper, non-empirical understanding must start with a realistic physical picture. A two-component model for ionic/neutral adsorption may be inadequate, since this can recognize the solvent as a continuum only, and the configurational freedom for the solvent (see Part I of this chapter) has to be suppressed. Similar to the approach in Section 19, wherein the problem of geometrical asymmetry is brought out by a *very* simple illustration, we shall now consider the problem of the adsorption isotherm of a molecule that has no permanent dipole moment, its polarizability being the same ($=\alpha$) as that of the solvent. The latter can be in any of two orientations (with equal and opposite dipole-moment vectors) whereas the adsorbate has one configuration only. The site requirements of the adsorbate and the solvent are assumed to be identical. Then it is easy to show that

$$\frac{\theta}{1-\theta} = \beta_0 c \left\{ \exp\frac{1}{kT}\left[-\phi_1 - \left(\frac{\mu}{K}\right)(4\pi\sigma + aR)\right] + \exp\frac{1}{kT}\left[-\phi_2 + \left(\frac{\mu}{K}\right)(4\pi\sigma + aR)\right]\right\}^{-1} \quad (17)$$

$$R = -(1-\theta)\tanh\left\{\frac{1}{2kT}\left[\phi_1 - \phi_2 + \frac{2\mu}{K}(4\pi\sigma + aR)\right]\right\} \quad (18)$$

When $\theta = 0$, equations (18) and (1) are identical. $\phi_1$ and $\phi_2$ are energetics of the metal–solvent states, with respect to that of the adsorbates; $\mu$ is the solvent dipole moment; and $K$ is a parameter that is dependent on the polarizability, co-ordination, and lattice spacing. Equation (18) can be inverted in the form, say, $R = \psi(\sigma, \theta, T)$, and substitution in equation (17) yields the isotherm that is clearly seen to be not of the Frumkin type.

But, consider now the approximation

$$R \approx R_0 + R_1 \theta \quad (19)$$

$$R_0 = -\tanh\left\{\frac{1}{2kT}\left[\phi_1 - \phi_2 + \frac{2\mu}{K}(4\pi\sigma + aR_0)\right]\right\} \quad (20)$$

(which is a solution of the 'solvent-only' model): $R_1$ can be shown to be

$$R_1 = -R_0\left[1 + \frac{\mu a}{KkT}(1 - R_0^2)\right]^{-1} \quad (21)$$

Even with equation (19), equation (17) does not straightaway reduce to the FI, but if $\theta$ is not too large it is possible to rewrite equation (17) approximately as

$$\beta^* c = [\theta/(1-\theta)]\exp(-2 A^* \theta)$$

where

$$\beta^* = \beta_0 \left\{ \exp\frac{1}{kT}\left[-\phi_1 - \left(\frac{\mu}{K}\right)(4\pi\sigma + aR_0)\right] + \exp\frac{1}{kT}\left[-\phi_2 + \left(\frac{\mu}{K}\right)(4\pi\sigma + aR_0)\right]\right\}^{-1} \quad (22)$$

and

$$2A^* = \left(\frac{\mu a R_1}{Kkt}\right)\tanh\left\{\frac{1}{2kT}\left[\phi_2 - \phi_1 - \left(\frac{2\mu}{K}\right)(4\pi\sigma + aR_0)\right]\right\} \quad (23)$$

Note that $R_0$ is also a function of $\sigma$ [equation (20)]. The 'interaction coefficient', $A^*$, has nothing to do with adsorbate–adsorbate interactions, which are, anyway, neglected in the above treatment, and $A^*$ is even a strong function of charge! Simplifications such as ignoring adsorbate–adsorbate interactions and assuming zero dipole moment for the adsorbate and equal polarizabilities have been relaxed, but the results will be reported elsewhere. The purpose of deducing the above result is only to indicate what one should expect in multi-component situations.

## 24 The Frumkin Isotherm, Congruence, and Electrosorption Valency

**The Frumkin Isotherm (FI).**—The anatomy of this isotherm

$$\beta c = \left(\frac{\Gamma}{\Gamma_m - \Gamma}\right) \exp(-2a\Gamma/\Gamma_m) \tag{24}$$

frequently employed[238–252] in the above form or in special forms such as the Langmuir[253–255] or the virial,[256–261] indicates the presence of a parameter to account for *each* of the broad categories of interactions, *viz.* substrate–adsorbate ($\beta$); adsorbate–external field (also in $\beta$; *cf.* the congruence principle); adsorbate–adsorbate, hard-core ($\Gamma_m$); and adsorbate–adsorbate, other interactions ($a$).

The *modus operandi* of an isotherm analysis, in practice, is usually to fit the data to the above equation and report the parameters $\beta(\sigma)$ or $\beta(E)$, $\Gamma_m$, and $a$, and also to correlate the interaction picture, whenever possible. In view of the several simplifications implicit in equation (24) (*cf.* Section 23), too much emphasis should not be placed on the quantitative (or sometimes even the qualitative; *e.g.*, the attractive or

---

[238] A. Abd-el-Nabey, A. de Battisti, and S. Trasatti, *J. Colloid Interface Sci.*, 1977, **60**, 67.
[239] K. G. Baikerikar and R. S. Hansen, *J. Colloid. Interface Sci.*, 1977, **61**, 239.
[240] A. de Battisti, V. Faggiano, and S. Trasatti, *J. Electroanal. Chem. Interfacial Electrochem.*, 1976, **73**, 327.
[241] F. Pulidori, G. Borghesani, R. Pedriali, and C. Bighi, *J. Electroanal. Chem. Interfacial Electrochem.*, 1976, **72**, 65.
[242] A. R. Alumaa and U. V. Palm, *Sov. Electrochem. (Engl. Transl.)*, 1978, **14**, 1192.
[243] M. A. Manvelyan and L. I. Boguslavskii, *Sov. Electrochem. (Engl. Transl.)*, 1976, **12**, 1636.
[244] M. A. Manvelyan, G. L. Neugodova, and L. I. Boguslavskii, *Sov. Electrochem. (Engl. Transl.)*, 1976, **12**, 1145.
[245] S. Minc, J. Jastrzebska, and M. Jurkiewicz-Herbich, *J. Electroanal. Chem. Interfacial Electrochem.*, 1975, **65**, 351.
[246] K. G. Baikerikar and R. S. Hansen, *J. Colloid. Interface Sci.*, 1978, **63**, 36.
[247] K. G. Baikerikar and R. S. Hansen, *J. Colloid. Interface Sci.*, 1975, **52**, 277.
[248] G. P. Gurina, M. Ya. Kats, G. A. Tedoradze, L. G. Feoktistov, and A. G. Gavlin, *Sov. Electrochem (Engl. Transl.)*, 1977, **13**, 1306.
[249] G. A. Tedoradze and Yu. A. Yuzbekov, *Sov. Electrochem. (Engl. Transl.)*, 1975, **11**, 234.
[250] L. E. Rybalka, B. N. Damaskin, and D. I. Lewis, *Sov. Electrochem. (Engl. Transl.)*, 1975, **11**, 7.
[251] N. B. Grigorev, V. P. Kuprin, Yu. M. Loshkarev, and R. V. Malaya, *Sov. Electrochem. (Engl. Transl.)*, 1975, **11**, 1315.
[252] S. L. Dyatkina, G. M. Rott, and B. B. Damaskin, *Sov. Electrochem. (Engl. Transl.)*, 1977, **13**, 807.
[253] F. Pulidori, G. Borghesani, R. Pedriali, A. de Battisti, and S. Trasatti, *J. Chem. Soc., Faraday Trans. 1*, 1978, **74**, 79.
[254] R. Parsons, R. Peat, and R. M. Reeves, *J. Electroanal. Chem. Interfacial Electrochem.*, 1975, **62**, 151.
[255] E. Boukaram, R. Bennes, and D. Bellostas, *J. Electroanal. Chem. Interfacial Electrochem.*, 1977, **84**, 21.
[256] N. B. Grigorev, V. B. Kuprin, and Yu, M. Loshkarev, *Sov. Electrochem. (Engl. Transl.)*, 1977, **13**, 93.
[257] D. A. De Vooys and J. H. A. Pieper, *J. Electroanal. Chem. Interfacial Electrochem.*, 1976, **72**, 129.
[258] R. V. Ivanova and B. B. Damaskin, *Sov. Electrochem. (Engl. Transl.)*, 1976, **12**, 532.
[259] U. V. Palm and E. K. Petyärv, *Sov. Electrochem. (Engl. Transl.)*, 1975, **11**, 126.
[260] E. K. Petyärv and U. V. Palm, *Sov. Electrochem. (Engl. Transl.)*, 1975, **11**, 295.
[261] M. G. Väärtnõu, E. K. Petyärv, and U. V. Palm, *Sov. Electrochem. (Engl. Transl.)*, 1975, **11**, 449.

repulsive nature as predicted by the sign of $a$) interpretations of these parameters. The nature of interactions and their origin (Are the adsorbate–adsorbate interactions substrate-induced or solvent-mediated? Has any agreement been obtained with the expected orders of energy assuming them to be of van der Waals type? Have pseudo-forms of interactions, resulting from the reduction of degrees of freedom in the model, been considered? *etc*.) and their plausibility (Can the particle–particle short-range lateral interaction show strong variations with the external field, perpendicular to the substrate? What are the various mechanisms that can give rise to the loss of congruence? Has the influence of the solvent been isolated to enable conclusions to be drawn about the adsorbates? *etc*.) need careful study.

Apart from the FI, and its discreteness version for the Coulombic effects, there is also a 'discreteness version' for non-Coulombic effects,[262] if there are any operating. The existence of a discreteness effect, having been recognized as the deviation of the actual potential at the adsorption site from that based on the mean-field-approximated value, suggests that a similar feature may be present for the non-Coulombic part too. From the two-state version,[45] it can be seen that, even in ideal situations, the Hamiltonian turns out to be the (spin $\frac{1}{2}$) Ising model with the external field and 'long-range' interactions. The known series expansion methods[262] suggest that, even if the Coulombic effects are ignored, a modified form, *e.g.*

$$\beta c = \left(\frac{\theta}{1-\theta}\right)\exp[-2(a_0\theta + a_1\theta^2 + a_2\theta^3 + \ldots)] \qquad (25)$$

should result, away from the MFA. In the simple case of the Ising model, *all* the parameters $a_0, a_1, \ldots$ are completely determined once the parameter denoting the strength of nearest-neighbour interactions is given.

A few heuristic models have been considered too. We may cite the 'generalized surface layer model',[263–266] whose isotherm is derived from a phenomenological postulate that is reducible to the form

$$\sigma = a + b\theta + c/(1 + n\theta - \theta) \qquad (26)$$

where $s$, $b$, and $c$ are constants. Apart from $n$, there is one more adjustable parameter $k$, and it involves $E_N$ (the shift of the potential of zero charge from $\theta = 0$ to $\theta = 1$) and $C_0, C_1$; $C_0$ and $C_1$ are the capacitances at $\theta = 0$ and $\theta = 1$ respectively. This leads to an isotherm like

$$\beta c = \frac{\theta}{1-\theta}\exp(-2a_0\theta)\exp\left[a_1 + \frac{a_2}{(1+n\theta-\theta)} + \frac{a_3}{(1+n\theta-\theta)^2}\right] \qquad (27)$$

where $a_0, a_1, a_2$, and $a_3$ are constants that are related to the potential $E, E_N, n, k$, *etc*. It will be interesting to know to what *physical* situation the above isotherm corresponds! Does this correspond to the hard-disc behaviour of the adsorbate or to the medium (solvent-induced) effects? At the moment, it seems to elude a satisfactory explanation. The values reported for $n$ and $k$ are invariably fractional;

---

[262] S. K. Rangarajan, *J. Electroanal. Chem. Interfacial Electrochem.*, 1975, **57**, 1.
[263] B. N. Afanas'ev and B. B. Damaskin, *Sov. Electrochem. (Engl. Transl.)* 1976, **12**, 308.
[264] B. N. Afanas'ev, B. B. Damaskin, G. I. Avilova, and N. A. Borisova, *Sov. Electrochem. (Engl. Transl.)*, 1975, **11**, 548.
[265] B. B. Damaskin, Yu. N. Kuryakov, and S. L. Dyatkina, *Sov. Electrochem. (Engl. Transl.)*, 1976, **12**, 1462.
[266] B. B. Damaskin and Yu. N. Kuryakov, *Sov. Electrochem. (Engl. Transl.)*, 1977, **13**, 79.

e.g., for mercury/aqueous $0.1M-NaF+n-C_4H_9OH$, $n=1.13$, $k=1.31$.[265] The $a_0$ values reported in such studies cannot be taken too literally as indicating the nearest-neighbour interaction or as any model for interaction.

Another heuristic attempt[267] to understand deviations from the FI is to model *mathematically* the free energy of adsorption, $\Delta G$, as a function of coverage, $\theta$. This method exploits the one characteristic that can be said *a priori* about the unknown function $\Delta G$, viz. $\Delta G \to \infty$ as $\theta \to 1$. The physical meaning of $\Delta G$ becoming infinity at saturation coverage is obvious. This suggests that $\Delta G$ may be split into two parts: (i) an analytic $\Delta G_a$, well behaved at $\theta = 1$, and (ii) a singular $\Delta G_s$ which tends to $\infty$ as $\theta \to 1$. The simplest way of modelling the functions $\Delta G_a$ and $\Delta G_s$ is to write

$$\Delta G_s = a\ln(1-\theta) + \frac{b}{1-\theta} + \frac{c}{(1-\theta)^2} + \ldots \qquad (28)$$

and

$$\Delta G_a = C_0 + C_1\theta + \ldots \qquad (29)$$

or

$$\Delta G_a \approx \frac{a_0 + a_1\theta}{b_0 + b_1\theta} \qquad (30)$$

where $a, b, c, a_0, \ldots$ are constants. Several known isotherms are shown to follow[267] the above pattern for a simple choice of values for $a_0, a_1, \ldots a, b, \ldots$! In particular, even the 'generalized surface layer model' mentioned above is a special case when the choice $a=1$, $b=c=\ldots=0$ is made in equation (28), together with equation (30).

Yet another phenomenological attempt is to describe the inner Helmholtz plane (IHP) as a plate of two series-connected capacitors,[268,269] reflecting the adsorbate-covered and the adsorbate-free parts of the IHP. This is claimed to be a *via media* between the *smeared out* IHP[12] and the *discrete* IHP[13] models.

Finally, we consider another model which is thermodynamically inspired but phenomenological in procedure. This is the 'non-ideal solution model' recently proposed by Mohilner.[189,270]

In deriving the FI, historically, the surface pressure was modelled as the osmotic pressure of a two-dimensional solution. Mohilner models the interface (ignoring ionic adsorption) as a two-component non-electrolyte solution ('surface solution'). In carrying out the anaylsis, two standard states (symmetrical and unsymmetrical) are considered for adsorbates.[189,270]

A theory of excess electrochemical Gibbs free energy of mixing, $\overline{\Delta G_E}$, is given, based on a power-series expansion in the surface mole fraction $x_A^{ads}$, the order of the power retained being *m*. By an elegant handling of the two types of standard states and a simple algebra, $\overline{\Delta G_E}$ is then constructed as a function of the control variables $x_A$ and the electrical state. It is claimed that analysis at the level of $\overline{\Delta G_E}$, rather than the usual isotherms, is more proper, especially because it is a direct measure of the

---

[267] S. K. Rangarajan, *J. Electroanal. Chem. Interfacial Electrochem.*, 1973, **45**, 283.
[268] Yu. V. Alekseev, Yu. A. Popov, and Ya. M. Kolotyrkin, *Sov. Electrochem. (Engl. Transl.)*, 1976, **12**, 841.
[269] Yu. V. Alekseev, Yu. A. Popov, and Ya. M. Kolotyrkin, *Sov. Electrochem. (Engl. Transl.)*, 1976, **12**, 924.
[270] D. M. Mohilner, H. Nakadomari, and P. R. Mohilner, *J. Phys. Chem.*, 1977, **81**, 244.

extent of non-ideality of the surface solution and does not contain any standard-state quantities.[270]

The experimental observations with 2-butanol indicate positive deviations from Raoult's law, thus suggesting a negative excess electrochemical entropy of mixing.[60] On this basis, the Flory–Huggins isotherm is considered to be incorrect.[270] Finally, an isotherm analysis is possible too. The FI is shown to be a special case if $n$, the number of solvent molecules displaced by an adsorbate, is assumed to be equal to unity, and $m=2$, i.e., the retention in the series for $\overline{\Delta G}_E$ of terms up to $(x_A^{ads})^2$ only. The interaction coefficient resulting therein is identified with the ratio of the activity coefficients of the adsorbate with symmetrical and unsymmetrical choice of standard states.

Though the above procedure seems novel and different, its closeness to well-known series expansion procedures in statistical mechanics must be noted. It is, therefore, not surprising that the FI, which is the MFA version,[45] must follow when we restrict the series for $\overline{\Delta G}_E$ to $\theta^2$ [N.B. $x_A^{ads} = \theta$ when $n=1$; misprints in equations (34) and (35) of reference 270 should be corrected].

The observation that one has to go beyond the FI can be taken to imply that MFA is inadequate. Translated into Mohilner's language, $m$ must be $>2$. Note also the discussion above, and especially equation (25).

The need to consider various states of water at the interface and their mutual equilibrium conditions, as well as the prevalent emphasis on 'finer' molecular information than gross thermodynamic descriptions, detract from the usefulness of the above approach.[271]

**Congruence.**—Recent reports indicate that the congruence hypothesis[22, 23] is still a talking point, with diverse approaches to this problem. Tests for congruence reveal the following trends:

(a) the notion of real and pseudo-congruence;[272, 273]
(b) experiments that support $E$-congruence only and those that satisfy $\sigma$-congruence only. We do not differentiate here, for this purpose, between $E$ and the inner-layer potential $\phi_{m2}$;[274, 275]
(c) arguments and experiments in support of the 'both or none principle', viz. either there is simultaneous congruence or no congruence;[253, 276]
(d) rejecting the hypothesis per se as unlikely;[289, 216, 270]
(e) the case of reluctant congruence, i.e., observance in one region of the charge axis and not in the other![274]

Mohilner et al.[189, 270] believe that the present approaches on congruence are 'unlikely to lead to correct results'. According to Fawcett,[216] 'one does not expect ionic adsorption isotherms to be congruent in either electrode potential or charge density on the basis of the electrostatic model for ionic adsorption'.

It may be noted that the dependence of parameters like $a$ and $\Gamma_m$ in equation (24) on potential/charge is a *manifestation* of the failure of congruence, not the *cause*.

---

[271] R. Amadelli, A. Daghetti, L. Vergano, A. de Battisti, and S. Trasatti, *J. Electroanal. Chem. Interfacial Electrochem.*, 1979, **100**, 379.
[272] S. K. Rangarajan, *J. Electroanal. Chem. Interfacial Electrochem.*, 1973, **41**, 279.
[273] A. Batana, R. C. Rocha, F. L. A. Avaca, and E. R. Gonzalez, *Electrochim. Acta*, 1980, in the press.
[274] A, de Battisti, B. A. Abd-el-Nabey, and S. Trasatti, *J. Chem. Soc., Faraday Trans. 1*, 1976, **72**, 2076.
[275] B. B. Damaskin and S. L. Dyatkina, *Sov. Electrochem. (Engl. Transl.)*, 1978, **14**, 128.
[276] K. Doblhofer and D. M. Mohilner, *J. Phys. Chem.*, 1971, **75**, 1698.

*Pseudo-congruence* has been noted as that leading to such conditions as[272]

$$\beta(E)c = f_1(\Gamma) \tag{31}$$

and

$$g(\sigma)c = f_2(\Gamma) \tag{32}$$

if there is a relationship of the form:

$$\Delta G_E^{\ominus} = \Delta G_\sigma^{\ominus} - RT \ln [f_1(\Gamma)/f_2(\Gamma)] \tag{33}$$

where $\beta(E) = \exp(-\Delta G_E^{\ominus}/RT)$ and $g(\sigma) = \exp(-\Delta G_\sigma^{\ominus}/RT)$. One of the two forms only is likely to be more 'physical' and direct, and thus may be preferred.[273]

*Simultaneous congruence* implies the existence of, say, $\beta(\phi_{m2})c_{OHP} = f_1(\Gamma)$, and the transformation of the variable $E \to \sigma$ should result in a form like $g(\sigma)c_{OHP} = f_2(\Gamma)$. The variable $c_{OHP}$ is the concentration at the OHP. This is unlikely, but may result under certain conditions, especially when a phenomenological law such as

$$\phi_{m2} = \sigma/K + F\Gamma/K_1 \tag{34}$$

holds, Table 4 is an illustration.

Based on parallel and series capacitance models, Damaskin[277] views this as an equivalence of

$$\frac{1}{C} = \frac{1-\theta}{C_0} + \frac{\theta}{C_1} \tag{35}$$

and

$$C = C_0(1-\theta) + C_1\theta \tag{36}$$

where $C_0$ is the value of $C$ when $\theta = 0$ and $C_1$ is the value when $\theta = 1$.

**Table 4** *Congruence and the $\phi_{m2} \to \sigma$ transformation*

| | $\ln\beta(\phi_{m2})$ | $K, K_1$ | $\ln\beta^*(\sigma)$ | $\sigma$-congruence | Frumkin isotherm | Remarks |
|---|---|---|---|---|---|---|
| (A) 1 | $\alpha\phi_{m2} + \beta$; $\alpha$ and $\beta$ are constants; linear in $\phi_{m2}$ | constants | $\alpha\sigma/K + \beta$; linear in $\sigma$ | Yes | Yes | Simultaneous congruence; an example of pseudo-congruence; $2a^* = 2a + \alpha F/K_1$ |
| 2 | $a\phi_{m2} + \beta$; linear in $\phi_{m2}$ | $K$ is a function of $\sigma$; $K_1$ is constant | $\alpha\sigma/K + \beta$; non-linear in $\sigma$ | Yes | Yes | As before |
| 3 | $\alpha\phi_{m2} + \beta$; linear in $\phi_{m2}$ | $K$ and $K_1$ are functions of $\sigma$ | $\alpha\sigma/K + \beta$; non-linear in $\sigma$ | No | Yes | $a^*$ varies with $\sigma$ |
| (B) | $\alpha(\phi_{m2} - \phi_{m2}^0)^2$; quadratic in $\phi_{m2}$ | constants | | No | No | see ref. 275 |

This Table illustrates the transformation from $E$-based isotherms to $\sigma$-based isotherms. It is assumed that $\phi_{m2}$-congruence holds with the isotherm $\beta(\phi_{m2})c_{OHP} = [\Gamma/(\Gamma_m - \Gamma)]\exp(-2a\Gamma/\Gamma_m)$ with the forms of $\beta(\phi_{m2})$ given in the first column. The transformation $\phi_{m2} = \sigma/K + F\Gamma/K_1$ [equation (34)] is employed to change the Frumkin isotherm [equation (24)] into the form $\beta^*(\sigma)c_{OHP} = \Gamma/(\Gamma_m - \Gamma) \exp(-2a^*\Gamma/\Gamma_m)$. More realistic equations and the conversion from $\sigma$ to $\phi_{m2}$ isotherms are not discussed above, for simplicity.

---

[277] B. B. Damaskin, *Sov. Electrochem. (Engl. Transl.)*, 1975, **11**, 395.

Equations (35) and (36) can strictly be true only if $C_0 = C_1$ [and hence is equal to $C(\theta)$, whatever be $\theta$!] It has been suggested that if the parameter $[(C_0/C_1)(C_0 - C_1)\theta]$ is small compared to unity (relatively low surface activity and relatively high limiting capacitance), simultaneous congruence may be 'observed'. (But note that the parameter cited[277] is $\sim 0.25$ even for acetonitrile,[274] and this cannot really be considered as small.)

The congruence principle has been shown to be valid for general two-state models (but not considering polarizability effects) if the background parameters are only weakly dependent on the electrical variables.[45] In this case, the potential $\phi_{m2}$ is 'chemical-potential-like'. More complex models may lead to a different conclusion, but the non-observance of congruence can be traced essentially to challenges posed by multi-component situations and the inability of the FI to respond to it. It must be noted that the two-component version of the congruence principle, for the isotherms of equations (4) and (5) in particular, is that $\beta_1$ and $\beta_2$ in equations (4) and (5) are functions of $\phi_{m2}$ or $\sigma$ only. Table 3 shows how, in reducing the two-component to pseudo-one-component systems, congruence may be lost. Also see equations (22) and (23) in this connection.

**Electrosorption Valency.**—The notion of covalent bond/chemisorption introduces the possibility of partial charge transfer. Originally this was due to Lorenz (for a duscussion of this question, especially the phenomenology and the connection with the theory of perturbations near and far from equilibrium, see the recent exhaustive survey by Lorenz and Salie[278]).

In simple situations, the electrosorption valency, $\gamma$, is defined as[279,280]

$$\gamma F = -(\partial \sigma / \partial \Gamma)_E = (\partial \mu / \partial E)_\Gamma$$

The valency $\gamma$ cannot be identified with the partial charge transfer on the adsorbate.[281–283] The composite parameter $\gamma_N$, i.e. the value of $\gamma$ at the potential of zero charge, is related to the chemisorption parameter, but the relationship involves discreteness of charge coefficients, the dipole terms of the adsorbate and water, the number of water molecules displaced, etc. With such a burden to carry, it is no wonder that $\gamma$ can be non-integral or even change sign!

Related issues like (i) how thermodynamic is $\gamma$; (ii) the propriety of describing an apparently (Faradaic) partial charge process through a formalism that is suitable for ideally polarizable systems; (iii) mixed adsorption and the variables to employ, and (iv) correlations of $\gamma$ with difference of electronegativities of the metal and the adsorbate, with ionic radii, and with the reciprocal diameter of solvent molecules have been discussed.[280,281,284–286] the relationship of $\gamma$ with the kinetics of reactions,

---

[278] W. Lorenz and G. Salie, *J. Electroanal. Chem. Interfacial Electrochem.*, 1977, **80**, 1.
[279] J. W. Schultze and K. J. Vetter, *Electrochim. Acta*, 1974, **19**, 913.
[280] K. J. Vetter and J. W. Schultze, *J. Electroanal. Chem. Interfacial Electrochem.*, 1974, **53**, 67.
[281] A. N. Frumkin, B. B. Damaskin, and O. A. Petrii, *Z. Phys. Chem. (Leipzig)*, 1975, **256**, 728; *Sov. Electrochem. (Engl. Transl.)*, 1976, **12**, 1.
[282] J. W. Schultze, *Trans. Soc. Adv. Electrochem. Sci. Technol.*, 1977, **12**, 305.
[283] J. W. Schultze, *Croat. Chem. Acta*, 1976, **48**, 643.
[284] A. N. Frumkin, B. B. Damaskin, and O. A. Petrii, *J. Electroanal. Chem. Interfacial Electrochem.*, 1974, **53**, 57.
[285] J. W. Schultze and F. D. Koppitz, *Electrochim. Acta*, 1976, **21**, 326.
[286] F. D. Koppitz and J. W. Schultze, *Electrochim. Acta*, 1976, **21**, 337.

the formation of monolayers or submonolayers, *etc.* have also been considered.[278,287–293]

More direct, even if only approximate, information on the partial charge transfer must come from quantum-chemical calculations. Two such studies have been reported.

Dohnert *et al.*[294] employ the extended Hückel method (EHM) in their calculations of bond length $L$ (measured in Å), bond energy $E_m$ (in kcal mol$^{-1}$), and the electron density $n$ (for mercury/X$^-$). Their results indicate that the $\pi$-bonds are favoured for both bond formation and charge transfer. The values reported for $L$, $E_m$, $n$, and the empirical charge-transfer coefficients $\lambda_{emp}$ are given below, together with the electronegativity differences $\Delta\chi$ (for comparison[294]):

| X$^-$ | $L$/Å | $E_m$/kcal mol$^{-1}$ | $n$ | $\lambda_{emp}$ | $\Delta\chi$ |
|---|---|---|---|---|---|
| Cl$^-$ | 2.4 | $-17$ | 1.00 | $-0.02$ | 1.16 |
| Br$^-$ | 2.55 | $-19$ | 1.03 | $-0.06$ | 0.96 |
| I$^-$ | 2.7 | $-33$ | 1.19 | $-0.27$ | 0.66 |

Agres *et al.*[295] report the *excess* charges on the chemisorbed halogen ions, calculated (at the PZC) as 0.469, 0.319, 0.276, and 0.177 for F$^-$, Cl$^-$, Br$^-$, and I$^-$, respectively. The method used is the MO LCAO one in the simple Hückel approximation (self consistency with respect to charge).

The above calculations must only be considered a beginning, and more *detailed* models for chemisorption[46,296] are required.

## 25 Perspectives

*'Would you tell me, please, which way I ought to go from here?'*
*'That depends a good deal on where you want to get to', said the Cat.*

This Report has recognized the reliability of the double-layer data for solid and liquid[297,298] electrodes, though there is no discussion of recent advances in the design of experiments, data acquisition, and analysis. The emphasis has been to put the questions in a sharper focus, with special reference to the origin and nature of interactions. The interfacial system, if it is to be cast into a pseudo-monocomponent mould, must recognize the pseudo-interactions too (*e.g.*, see the discussion of electrolyte theory in reference 299). This is even more true for substrate interactions where three-body effects are unavoidable. It is necessary to probe deeper into these

---

[287] G. Salie and W. Lorenz, *Z. Phys. Chem. (Leipzig)*, 1975, **256**, 386.
[288] G. Salie and W. Lorenz, *Electrochim. Acta*, 1975, **20**, 309.
[289] W. Lorenz, *Z. Phys. Chem. (Leipzig)*, 1976, **257**, 63.
[290] W. Lorenz and R. Siegemund, *Z. Phys. Chem. (Leipzig)*, 1975, **256**, 390.
[291] D. G. Wierse, M. M. Lohrengal, and J. W. Schultze, *J. Electroanal. Chem. Interfacial Electrochem.*, 1978, **92**, 121.
[292] J. W. Schultze and D. Dickertmann, *Ber. Bunsenges. Phys. Chem.*, 1978, **82**, 528.
[293] W. J. Lorenz, H. D. Herman, H. Wuthrich, and F. Hilbert, *J. Electrochem. Soc.*, 1974, **121**, 1167.
[294] D. Dohnert, J. Koutecky, and J. W. Schultze, *J. Electroanal. Chem. Interfacial Electrochem.*, 1977, **82**, 81.
[295] E. M. Agres, A. I. Altsybeeva, and S. Z. Levin, *Sov. Electrochem. (Engl. Transl.)*, 1976, **12**, 26.
[296] K. L. Sebastian and S. K. Rangarajan, *Mol. Phys.*, 1979, **38**, 1567.
[297] O. A. Petrii, *Russ. Chem. Rev. (Engl. Transl.)*, 1975, **44**, 965.
[298] U. V. Palm and V. Past, *Russ. Chem. Rev. (Engl. Transl.)*, 1975, **44**, 965.
[299] J. S. Hoye and G. Stell, *Faraday Discuss. Chem. Soc.*, 1977, No. 64, p. 16.

and the multi-component nature (cf. solvent effects) before making anything out of the 'accurately evaluated' isotherm parameters!

It is but natural that the study of statistical mechanics is the next one to be emphasized, in all its aspects – thermodynamic, fundamentals, and model analysis.[300-303] Such studies, coupled with investigations on the electronic description of surfaces, may be considered a beginning of what can ultimately resolve the doubts on conventional descriptions.[6,19—21,44,50]

More important can be the study of phase transitions. With such a manifold possibility of the choice of adsorbates, it should be possible to manipulate the physics of interactions *via* the chemistry of molecular structure, and a rich area of investigation of several pseudo-two-dimensional model systems[304] is in store. In particular, the study of critical exponents leading to 'invariant' laws relating capacitance with potential and concentration in the neighbourhood of transitions must prove to be valuable.

As mentioned in the beginning, contributions relating to experimental analysis, the study of interfaces other than those considered here, and phenomenological coupling of a more intricate nature (*e.g.* adsorption effects on electrode kinetics and electrocrystallization) are omitted, but it is hoped that they will be taken up in future volumes.

*Acknowledgement:*
The author expresses his thanks to W.R. Fawcett for providing him with preprints of his papers; he acknowledges his indebtedness to A.C. Ramamurthy for his enthusiasm and support in the preparation of this Report, to M.V. Sangaranarayanan for his help, and to others in the Theoretical Electrochemistry Group, Bangalore, for their interest.

---

[300] J. P. Badialli and J. Goodisman, *J. Electroanal. Chem. Interfacial Electrochem.*, 1978, **91**, 151.
[301] J. P. Badialli and J. Goodisman, *J. Electroanal. Chem. Interfacial Electrochem.*, 1975, **65**, 523.
[302] J. P. Badialli and J. Goodisman, *J. Phys. Chem.*, 1975, **79**, 223.
[303] W. I. Plesner and I. Micheli, *J. Chem. Phys.*, 1974, **60**, 3016.
[304] J. F. Nagle and J. C. Barner, *Ann. Rev. Phys. Chem.*, 1976, **27**, 291.

# Author Index

Abbey, K. M., 46
Abdel-Aziz, S. A. M., 116
Abd-el-Nabey, B. A., 240, 249, 252
Abe, S., 153
Abou-Romia, M. M., 232
Abraham, M. G., 130
Accomazzo, M. A., 134
Adametzova, H., 92, 93, 100
Adams, F., 97, 105
Adhikari, M., 70, 72
Afansev, B. N., 250
Agres, E. M., 255
Ahmad, F., 69
Ahmed, F., 59
Aihara, M., 231
Aikens, D. A., 67
Aizawa, M., 60, 61, 106, 111
Alabiso, G., 105
Alagova, Z. S., 66
Albery, J., 1
Alekseev, Yu. V., 251
Alexander, P. W., 105
Alfani, F., 143
Aliev, I. Y., 26
Alpatova, N. M., 2
Al-Sibaai, A. A., 62, 66, 104
Altsybeeva, A. I., 255
Alumaa, A. R., 249
Alzheimer, D. P., 145
Amadelli, R., 252
Amar, P., 139
Amjad, M., 2
Ammann, D., 63, 66, 86, 91, 94
Ampaya, J. P., 119
Amu, T., 115
Anderson, J. E., 47, 51, 134, 142
Anderson, J. L., 48, 54, 126
Anderson, K. P., 70
Anderson, W., 116
Andreev, V. N., 240
Andrieux, C. P., 3
Andronati, S. A., 10
Andrychuck, D., 88
Anfalt, T., 83, 99
Angerstein-Kozlowska, H., 232
Anghel, D. F., 67, 72
Anson, F. C., 208, 226, 231, 232
Aoki, J., 34
Aoki, N., 53
Aoyagui, S., 37
Appelbaum, J. A., 235
Aptel, P., 136
Arakelyam, R. A., 210

Archer, W. I., 191
Arita, S., 31
Arita, T., 118
Armand, J., 8
Armstrong, N. R., 8
Armstrong, P. A. M., 116
Armstrong, R. D., 166, 173, 177, 180, 183, 189, 190, 191, 192, 194, 197
Armstrong, W. McD., 100
Arneri, G., 142
Arvanitis, S., 122
Asada, M., 56
Asenov, I., 102
Ashbrook, A. W., 144
Asirvatham, M. R., 19
Asmus, K. D., 3
Astrom, O., 66
Augustowska, Z., 64
Avaca, L. A., 6, 252
Avilova, G. I., 250
Avrutskaya, I. A., 9, 12
Ayuzawa, M., 98
Axel, L., 49
Azumi, T., 133

Babieuskii, K. K., 9
Badialli, J. P., 256
Baer, H. J., 8
Bagat, M. R., 240
Baggaley, A. J., 23
Bagotskaya, I. A., 204, 211, 213, 215, 221
Bahr, G., 60, 122
Baikerikar, K. G., 232, 249
Bailey, P. L., 41, 69
Bailey, R. A., 67
Bairamov, F. G., 36
Baiulescu, G. E., 89, 90, 93
Baizer, M. M., 12, 16
Baker, R. W., 43, 53, 118, 139
Bamberg, E., 42
Band, D. M., 91
Bansal, I. K., 143, 144
Barba, F., 32
Barber, M. N., 210
Bard, A. J., 4, 5, 6, 29
Barker, G. C., 182
Barlow, C. A., 204
Barner, J. C., 256
Barone, G., 241
Barthel, J., 211
Basson, A. J., 84
Batana, A., 252

Bates, R. G., 72
Batrakov, V. V., 204
Baucke, F. G. K., 78
Baudras, A., 112
Bauer, R., 188
Bauerle, J. E., 177
Baugh, L. M., 241
Bauke, F. G. K., 100
Baumann, E. W., 65, 84, 105
Baxter, A. G., 144
Becerro, E. R., 138
Becher, H. M., 18
Beck, F., 7, 24
Becker, B. F., 33
Becker, J. Y., 24, 26, 35
Bednas, M. E., 143, 144
Beekmans, N. M., 187
Beer, E., 8
Beevers, C. A., 188
Beg, M. A., 53, 59, 61, 69
Behr, B., 241
Behret, H., 211
Beilis, J., 33
Beland, F. A., 18
Belfort, G., 141
Belikov, V. M., 9
Bell, A. T., 143
Bell, M. F., 180
Bellido, A., 138
Belling, J. R., 3
Bellostas, D., 232, 249
Bender, M., 54, 142
Bennes, R., 232, 249
Bennion, D. N., 122, 143
Benz, R., 42
Beran, S., 13
Berenblit, V. V., 36
Bergakad, F., 32
Berge, H., 36
Bergner, K., 78
Berndt, A. F., 98
Bertrand, C., 215
Berube, D., 16
Bettman, M., 188
Bewick, A., 41, 205
Bezuglyi, V. D., 9
Bhandari, V. M., 9
Bhattacharyya, D., 142, 144
Bhuiyan, L. B., 209
Bighi, C., 249
Binder, H., 117
Bindra, P., 182
Birch, B. J., 107, 108
Birraux, C., 89

Bisgaard, P., 74
Bissig, R., 66
Bixenman, W. R., 100
Black, D. B., 117, 118
Blackham, A. U., 31
Blaedel, W. J., 88, 111
Blais, P., 51, 142, 143
Blanc, G., 182
Bloch, R., 91, 117
Bloebaum, R. K., 143
Blokhra, R. L., 129
Bloomfield, J. J., 30
Blum, L., 209
Blum, Z., 24
Blumberg, A. A., 152
Bo, P., 141
Boari, G., 134, 145
Bobbitt, J. M., 102
Bobilliart, F., 22
Bockris, J. O'M., 203, 210, 216, 218, 232
Bodkin, J. B., 99
Boeck, C., 34
Boehm, R. E., 233
Boel, T., 98
Boesen, C. E., 141
Bogatskii, A. N., 10
Boger, Z., 54
Boguslavskii, L. I., 249
Boitsov, V. G., 213
Boles, J. H., 81
Bollenbeck, P. H., 58, 145
Bonjouklian, R., 54, 142
Bontempelli, G., 2, 19, 34
Bordner, J., 10
Borghesani, G., 249
Borisova, L. V., 134
Borisova, N. A., 250
Borkowska, Z., 216
Borodkin, S., 118
Borovaya, N. A., 246
Borowitz, I. R., 66
Booker, H. E., 109
Bos, M., 81
Boto, K. G., 1, 4
Botre, C., 59
Boukaram, E., 232, 249
Boulares, L., 8
Boyd, G. E., 52
Boyd, J. W., 113
Boyett, J. D., 145
Brabec, V., 231, 238
Bradley, J. N., 157
Bragg, W. L., 188
Brand, M. J. D., 85
Brandt, E. S., 139
Branica, M., 238
Breant, M., 2
Bresler, E. H., 49, 50
Breslow, R., 2, 23
Brett, A. C., 218
Brettle, R., 22, 23
Brich, W., 3
Brick, E. B., 205

Britz, D., 3
Brohult, S., 188
Broun, G., 54
Brousse, C. I., 139
Brown, G., 108
Brown, G. M., 37
Brown, H. M., 87
Brown, O. R., 9
Brown, W., 115
Bruckenstein, S., 2
Brun, J. T., 145
Brun, P. F., 71
Bruner, L. J., 234
Buck, R. P., 42, 47, 62, 65, 80, 81, 92, 97, 123, 134, 183
Buff, F. P., 210
Buffle, J., 70, 98
Buldini, P. L., 97
Bulvestre, G., 145
Bunce, S. C., 67
Bunzl, K., 132
Burgers, C., 95
Burghoff, H. G., 51
Burman, J. O., 102
Burnham, R. A., 183, 190
Burns, D. T., 90, 91, 97, 103, 107
Byrd, L. R., 35
Byrn, S. R., 56
Bystrikova, E. F., 12

Cabasso, I., 117, 136
Cabon, J. Y., 2
Cahn, R. P., 118, 119
Cameron, R. G., 131
Cammann, K., 62, 64
Camoes, G. F. C., 79
Campbell, B. H., 1
Canepa, P., 139, 143
Caplan, S. R., 51, 135
Caracciolo, F., 56
Carmack, G. D., 75, 121, 134
Carr, J. D., 71, 80
Carr, P. W., 110
Castagnola, A., 156
Cattrall, R. W., 75, 77
Caullet, C., 8
Cauquis, A., 20, 33, 36
Caza, J., 16
Cekavicius, B., 33
Chabaud, B., 33
Chagelishvilli, V. A., 210
Challard, N., 136
Chambers, J. Q., 36
Chammas, S., 121
Chang, F. C., 71, 99
Chang, H., 161
Chang, I. F., 4
Chang, T. M. S., 52
Chang, Y. G., 54
Chang, Y. J., 150
Channabasappa, K. C., 138
Chantrey, G., 53
Chao, E. E., 89

Chapurlat, R., 139
Chawla, A. S., 52
Chen, C. T., 54, 150
Chen, Y. W., 118
Cheng, K. L., 71, 89
Cheung, J. Y., 116
Cheung, M. T., 67
Chian, E. S. K., 138, 139, 141, 142
Chiolle, A., 139
Chitumbo, K., 115
Chkir, M., 29
Chlanda, F. P., 133
Chmielewski, M. E., 152
Cho, T., 5
Choi, K. M., 143
Choung, J. Y., 144
Choy, E. M., 56
Christian, G. D., 93
Christian, S. D., 231, 238
Christoffersen, G. R. J., 87
Christova, R., 100, 102
Chum, H. L., 2
Chung, L. L., 5
Cichy, M., 87
Cikurel, H., 54
Ciocan, N., 67, 72, 90
Clark, L. C., 41
Clarke, D. E., 107, 108
Clarke, R., 1
Clay, J. R., 210
Clermont, L. P., 122
Clysters, H., 97, 105
Cocu, F. G., 89
Coetzee, C. J., 84
Cogoni, G., 34
Colaruotolo, J. F., 102
Cole, K. S., 80
Comi, R., 27
Comper, W. D., 58
Comtat, M., 112
Connolly, J. F., 6
Conway, B. E., 203, 205, 232
Cook, R. L., 144
Cooney, D., 42
Cooper, I. L., 204, 217
Copas, A. L., 143
Cornish, D. C., 62
Cosofret, V. V., 89, 90, 93
Costich, P. M., 54
Covington, A. K., 62, 69, 70, 79, 80
Cowan, D. O., 37
Craggs, A., 67, 86
Creason, S. C., 182
Credali, L., 139
Cristescu, C., 93
Crombie, D. J., 88, 105
Cserfalvi, T., 104, 105
Cullen, L. F., 114
Cummings, A. L., 70
Cuny, J., 136
Cuong, N. H., 215
Curran, P. F., 48

# Author Index

Cusbert, P. J., 97
Cussler, E. L., 56
Czaban, J. D., 82

Daghetti, A., 252
Dagner, D., 11
D'Alkaine, C. V., 213, 215
Dalrymple-Alford, P., 223
Damadian, R., 59
Damaskin, B. B., 204, 210, 212, 213, 215, 216, 218, 221, 236, 237, 239, 240, 241, 246, 249, 250, 252, 253, 254
Dana, G., 8
Danchik, R. S., 94
Danesi, P. R., 156
Dariel, M. S., 153
David, A., 54
Davis, D. G., 150
Davis, J. A., 144
Davis, V. J., 36
Davydov, G. A., 36
De Battisti, A., 240, 249, 252
Decoodt, P., 152
Deelstra, H., 82
Deffner, U., 31
Degner, D., 25
Deinzer, M., 117
Dejong, J., 50
De Korosy, F., 123
Delahay, P., 204
De Levie, R., 44, 46, 174, 208, 234, 235
Demange-Guerin, G., 2
Demisch, H. U., 137
Den Boef, G., 70, 88, 89
Dencks, A., 68
Dennison, C., 144
De Nobriga, R. M., 218
D'Epenoux, B., 65
Déportes, C., 187
Deronzier, A., 20
Desbene-Monvernay, A., 12
DeSousa, R. C., 124
Deutsch, D., 99
Devanathan, M. A. V., 204, 246
De Vooys, D. A., 259
Dewolfs, R., 82
Dezelic, G., 138
Dhar, H. P., 232
Diadov, V., 102
Dickel, G., 55
Dickertmann, D., 255
Dickinson, T., 177, 182, 189, 190, 192, 194, 197
Dickson, J. M., 142, 143, 144
Dietmar, W., 7
Dinwiddie, D. E., 88
Discala, V. A., 50
Doblhofer, K., 181, 252
Dohner, R. E., 72
Dohnert, D., 255
Dohno, R., 133
Dokunikhin, N. S., 18

Dolnakov, P., 36
Dolphin, D., 37
Dorabialska, A., 52
D'Orazio, P., 70
Dorfman, Ya. A., 226
Dorset, D. L., 126
Doubrow, M., 118
Drakesmith, F. G., 36
Dresner, L., 140
Drew, D. M., 75
Drioli, E., 143
Drobnik, J., 118
Dryhurst, G., 13, 231, 238
Dryon, L., 100
Dubois, R., 12, 82, 152, 153
Dubova, L. M., 213, 221
Dubovik, G., 5
Duburs, G., 33
Duda, J. L., 145
Dudnik, S. S., 133
Duer, W. C., 88
Duffy, J. E., 52
Duff, D. J., 72
Duff, E. J., 98, 100
Dufresne, J. C., 5
Dunsch, L., 32
DuPont, J. T., 15
Dupre La Tour, O., 182
Durliat, H., 112
Durst, R. A., 89
D'yachenko, A. I., 16
Dyatkina, S. L., 212, 237, 240, 249, 250, 252
Dzhafarov, E., 36
Dzhaparidze, D. I., 210
Dzumedzei, N. V., 14

Eagan, M. L., 82
Eberson, L., 24, 25
Ebra-Lima, O. M., 144, 150
Eckfeldt, E. L., 79
Edström, K., 111
Edwards, G. J., 4, 20
Efstathiou, C. E., 103
Ehara, R., 43
Eijsermans, J. C., 147
Eisenbach, W., 37
Eisenman, G., 41, 46, 81
Ekel, V. A., 9
Elbakai, A. M., 72
Elia, V., 241
Elliott, C. M., 231
Elliot, I. W., 10
El Murr, N., 38
El Taib Heakal, F., 232
El-Taras, M. F., 89
Endoh, R., 139
Englar, J. R., 98
Engovatov, A. A., 2, 23
Enkelmann, V., 123
Eørland, T., 46
Epelboin, I., 182
Epling, G. A., 27
Erabi, T., 21

Eremisini, C., 79
Ermakov, A. N., 134
Ershlirand, A. B., 212
Essig, A., 49, 124, 125
Evans, D. F., 56
Evans, W. H., 83
Eyrich, T. L., 91
Ezuddin, K. H., 50

Fabiani, C., 67, 156
Faggiano, V., 249
Fahey, D. R., 37
Falk, M., 117
Fang, H. H. P., 138, 139, 141, 142
Farcy-Bravacos, J., 182
Farrington, G. C., 190
Farwell, S. O., 18
Fawcett, W. R., 210, 211, 212, 213, 216, 218, 240
Fedorovich, N. V., 236, 237
Feher, F., 72, 108
Felton, R. H., 37
Fenyo, J-Cl., 133
Feoktistov, L. G., 16, 249
Ferguson, D. A. jun., 113
Ferles, M., 7
Fernandez-Prini, R., 52
Ferra, M. I. A., 79
Ferris, C. D., 41
Ferry, J. D., 53
Fetters, L. J., 53
Fiedler, U., 74, 85, 90, 94, 95
Filho, H. B., 74
Filomena, M., 79
Fioshin, N. Ya., 9, 12
Fischer, H., 3
Fischer, W., 188
Fishman, H. M., 126
Flanagan, O. H., 118
Fleet, B., 63, 85, 99, 105
Fleischmann, M., 1, 3, 41, 182, 205
Fleury, D., 16
Fligier, J., 74
Flynn, G. L., 42, 118
Fodor, L., 108
Fogg, A. G., 66, 90, 91, 95, 97, 103, 104, 107
Fogt, E. J., 70
Force, R. K., 71
Ford, A., 145
Ford, W. T., 2
Forgacs, C., 125
Formicheva, M. G., 2
Foulatier, P., 8
Fozzard, H. A., 82
Franceschetti, D. R., 166
Franck, U. F., 43
Frank, S., 231
Franz, R. L., 15
Frazer, J. W., 63, 72, 83, 89
Frehland, E., 126
Frei, R. W., 99

# Author Index

Freiberg, E., 32
Freiser, H., 74, 75, 77, 121, 134
French, R. J., 47
Friauf, R. J., 161
Fried, A., 54, 142
Friedman, M., 118
Fritz, H. P., 33
Fritz, I., 108
Frommer, M. A., 116
Frumkin, A. N., 204, 212, 215, 216, 218, 221, 236, 240, 241, 246
Fry, A. J., 5, 16
Fueno, T., 25
Fuerst, W., 117
Fuginaga, T., 74
Fujita, H., 55
Fujita, I., 37
Fukada, E., 123
Fukushima, H., 66
Fung, K. W., 89
Fung, Y. S., 89
Funke, K., 201
Furukawa, J., 56

Gable, R. J., 124
Gabrielli, C., 182
Gal, J. Y., 2
Gale, R. M., 53, 117
Galey, W. R., 145
Galus, Z., 3, 10
Gangopadhyay, D., 72
Gara, W. B., 3, 34
Garbett, K., 85
Garbuzova, N. V., 67
Garcia, R. A., 54
Garcin, M., 119
Gardner, C. L., 221
Gardner, C. R., 135
Garmon, W. E., 119
Garrison, K. A., 144
Gary-Bobo, C. M., 53
Gasalbore, G., 36
Gassee, J. P., 152
Gassman, P. G., 32
Gassner, S., 54
Gavach, C., 65
Gavlin, A. G., 249
Gedheim, L., 24
Geer, R. D., 18
Geissler, M., 93
Geller, S., 157
Geminova, M. V., 66
Genies, M., 31, 33
Germano, G., 192
Gesser, H. D., 56
Geyer, R., 66
Ghaly, H. A., 232
Ghosh, D., 70
Gibson, K., 113
Gilbert, B. L., 4
Gilbert, J. C., 29
Gileadi, E., 1
Gilgen, P., 2

Ginsberg, O. G., 12
Gircud, G., 187
Gittens, G. J., 115
Glavaski, S., 205
Glick, M., 188
Glueckauf, E., 140, 141
Gluskoter, H. J., 99
Goddard, J. D., 42
Goel, N. S., 210
Goelz, J. F., 205
Gol'din, M. M., 16
Goldsmith, M., 59
Goldstein, D. A., 50
Goldstein, W. E., 153
Golubev, V. N., 133
Gomer, R., 208
Gonzalez, E. R., 241, 252
Goodfellow, G. I., 78
Goodhue, E. C., 143
Goodin, R. D., 29
Goodisman, J., 256
Goodman, D., 102
Goodridge, F., 9
Goossens, I., 142
Gordievskii, A., 66, 68, 71, 87, 94, 103
Gorton, L., 95
Gottesfeld, S., 205
Gottfried, C., 188
Gough, K. M., 56
Gourney, J. G., 16
Govindan, K. P., 132
Grachev, V. L., 36
Graetzel, M., 3
Grafov, B., 182
Graneli, A., 83
Grant, R. J., 190
Grasser Bauer, M., 101
Grazan, A. M., 59
Green, M. E., 58, 126
Greene, P. P., 157
Gregor, H. P., 111, 139
Gregorowicz, Z., 74
Grekovich, A. L., 67, 91
Greter, F. L., 70
Griepink, B., 68
Grieves, R. B., 142, 144
Grigoryev, N. B., 204, 211, 215, 221, 249
Grimshaw, J., 19
Grivgorjeva, M. G., 66
Grodzinsky, A. J., 153
Gross, M., 3, 38
Grossman, G., 132
Grubb, W. T., 91
Grypta, R. D., 6
Gryte, C. C., 139
Gueggi, M., 66
Guerout, D., 8
Guggi, M., 84, 86, 90
Guilbault, G. G., 74, 82, 104, 105, 108, 109, 110, 112, 113, 114
Guillou, M., 145

Gulick, W. M., 15
Gul'tyai, V. P., 5, 10
Gurevich, Yu. Ya., 238
Gurina, G. P., 249
Gurney, M., 99
Gusarskaya, N. L., 18
Gutch, C. F., 43
Gutknecht, W. F., 71, 100
Gutman, R., 116

Haar, W., 188
Haase, R., 47
Habib, M. A., 203, 204, 210, 218, 232
Haddadin, E. S., 152
Hadjiioannou, T. P., 103
Haerdi, W., 70, 89, 98
Hagen, G., 205
Haikala, E., 78
Hakozaki, S., 23
Halligam, J. E., 58
Hallsworth, A. S., 99
Hamaguchi, H., 31, 34
Hamann, D. R., 235
Hamano, Y., 5
Hamelin, A., 211
Hamilton, I. C., 75
Hammond, S. M., 91
Hanaoka, K., 154
Hanji, K., 56
Hansen, E. H., 69, 72, 73, 74, 89, 102, 108, 110, 205, 232, 249
Haque, A., 53, 61
Hara, S., 139
Hargens, R. D., 71
Hargis, L. G., 37
Harigome, S., 50, 141
Harrington, M. G., 46
Harrison, J. A., 9, 204, 217
Harshman, R. C., 144
Hartman, R. B., 145
Hashimoto, K., 43
Hashimoto, T., 34
Haslam, J. L., 109
Hassan, S. S. M., 97, 99, 102
Hasslinger, B. L., 15
Hatayama, T., 5
Hawley, M. D., 19
Hawlicka, E., 52
Hayakawa, T., 89
Hayasi, J., 23
Hazama, S., 6
Hazard, R., 33
Hazeldine, R. N., 36
Hazelrigg, M. J., 6
Hazemoto, N., 64, 106, 107
Headridge, J. B., 83
Heijne, G. J. M., 71, 88
Heilmann, S. M., 19
Heineman, W. R., 205
Heiss, J., 24
Helgee, B., 25
Helmy, A. K., 59

# Author Index

Henderson, D., 209
Hendra, P. J., 4, 205
Henry, F., 65
Herbig, H., 11
Herlem, M., 22
Herman, H. B., 231
Herman, H. D., 255
Hermans, H. B., 95
Herscovici, G., 138
Heyde, M. E., 51, 142
Heyne, L., 187
Hibino, K., 34
Hiro, K., 67, 96, 106
Hilbert, F., 255
Hildebrandt, W. A., 90
Hilgen, H., 50
Hills, G. J., 3, 204
Hirai, N., 31
Hirata, H., 98
Hirsch, R. F., 67
Hirsch-Ayalon, P., 60, 122
Hiura, H., 2
Hjemdahl-Monsen, C. E., 112
Hlewaty, J., 10
Ho, A. Y. W., 105
Hochhauser, A., 56
Hodge, I. M., 190
Hoffman, J. F., 50
Hofton, M. E., 72
Homola, A., 99
Hooper, A., 190
Hopirtean, E., 62, 64, 74, 82, 90
Hor, D., 59
Horigome, S., 120
Horiguchi, K., 56
Horikoshi, I., 54
Horner, L., 3, 11, 14
Horsfall, G. A., 56
Horvai, G., 72, 89
Hoye, J. S., 255
Hrabeczy-Pall, A., 99
Hsieh, S., 204
Hsiung, C. P., 112
Hsiung, K. P., 74
Hubbard, A. T., 2
Huffman, W. J., 144
Hughes, D. A., 36
Hughes, W. B., 37
Hulanicki, A., 64, 65, 67, 85, 86, 87, 88, 89, 102
Humphreys, M. W., 205
Humphreys, T. W., 205
Hung, C. C., 142
Hung, G. W. C., 116
Hünig, S., 13
Hurwitz, H. W., 215, 237
Hussein, W. R., 82
Hutchins, C. S., 5
Hwang, S. T., 42

Ianova, M., 102
Iberall, A. S., 150
Ibi, N., 2
Ibrisagic, Z., 8

Ichijo, N., 156
Ichise, M., 182
Ide, H., 6
Ihara, H., 139
Ikeda, A., 23
Ikeda, B. M., 211
Ikeda, K., 139
Ikeda, S., 101, 102
Ikenberry, L. D., 114
Il'yasov, A. V., 13
Imura, Y., 152
Inada, K., 6
Ingram, M. D., 190
Inno, S. K., 240
Inouye, A., 82
Ioffe, N. T., 2, 23
Ishibashi, N., 65, 66, 102
Isotani, T., 6
Itoh, M., 31, 35
Ivanov, V. F., 211, 212
Ivanova, R. V., 213, 249
Iwasaki, T., 15
Iwase, A., 113
Iwashita, H., 6
Izekov, V. P., 105
Izmestev, A. A., 236

Jaber, A. M. Y., 84
Jackson, H. W., 47
Jaffé, G., 161
Jagner, D., 83, 85, 99
Jagur-Grodzinski, J., 43, 54, 117, 136, 151
Jain, A. K., 127, 129, 130
James, R. O., 99
Janacek, K., 42
Janda, M., 28
Jansson, R. E. W., 1
Japan, P., 26
Jasinski, R., 77, 88
Jastrzebska, J., 249
Jeanmaire, D. L., 41, 205
Jehring, H., 237
Jeminet, G., 16
Jenard, A., 215, 237
Jenkins, R. A., 111
Jensen, B. S., 5
Jensen, J. B., 70
Jensen, S., 5
Jeroschewski, P., 2
Johansen, E. S., 87
Johansson, G., 77, 102, 110, 111
Johnson, A. R., 143
Johnson, C. C., 143
Johnson, J. M., 113
Johnson, J. S. jun., 143, 144
Johnson, R. W., 23
Johnston, H. K., 141
Johnston, W. V., 192
Jones, I. W., 188
Jones, S. R., 20
Jonsson, G., 141
Joshi, K. M., 240
Jozefonvicz, J., 136

Juday, R. E., 7
Jugel, W., 33
Jumawan, A. B. jun., 144
Junghans, K., 5
Juni, K., 118
Jurkiewicz-Herbich, M., 249
Jyo, A., 65

Kadis, V., 13
Kadish, K. M., 38
Kaempf, K., 2
Kahr, G., 62, 63, 91
Kakabadse, G. J., 72
Kakiuchi, T., 244
Kalyuzhnaya, A. M., 211, 221
Kambara, T., 71, 75, 106
Kammermeyer, K., 42
Kamo, N., 46, 60, 61, 64, 67, 106, 107
Kanbe, T., 25
Kaneko, N., 144
Karachentseva, Yu. M., 105
Kardos, A. M., 1
Kargin, Yu. M., 13
Kariv, E., 3
Karlberg, B., 77, 78
Karube, I., 106, 111
Kataoka, M., 71, 75
Katchalsky, A., 48
Kato, S., 60, 61
Kats, M. Ya., 249
Katto, T., 124
Kaufmann, F., 37
Kawaguchi, M., 150
Kawahara, A., 67, 96, 106
Kawamata, F., 134
Kawashima, T., 110
Kawata, T., 4
Kazarinov, V. E., 240
Keddam, M., 182
Kedem, O., 47, 124, 132, 150, 153
Keidel, F. A., 24
Keil, L., 86
Keleyan, A. P., 210
Kemula, W., 241
Kenforce, R., 80
Kent, P., 67
Kergreis, A., 145
Kessler, M., 41, 86
Kessler, Yu. M., 2
Ketzinel, Z., 54
Kharkats, Yu. I., 238
Khayat, M. N., 72
Khmel'nitskaya, E. Yu., 18
Kiang, C. A., 114
Kikkawa, N., 59
Kim, M. H., 231
Kimmerle, F. M., 16, 226, 240
Kimura, S., 43
Kina, K., 66, 102
Kinoshita, H., 238
Kipling, B., 51, 52
Kirihara, H., 99

Kirowa-Eisner, E., 1
Kirsch, N. N. L., 66
Kissel, T. R., 111
Kitaoka, M., 152
Kivalo, P., 78, 84, 89
Klahre, P., 238
Klehr, M., 22
Klein, E., 47
Klein, R., 54, 142
Kleinschmager, H., 188
Kleitz, M., 187
Klimova, T. A., 16
Kloow, G., 115
Klump, B., 120
Klyagul, T. A., 10
Klyuer, B. L., 7
Kmetec, E., 113
Knittel, D., 3
Knizakova, E., 92
Knolle, J., 29
Kobatake, Y., 55, 60, 61, 64, 67, 106, 107
Kobayashi, H., 156
Kobayashi, R., 119
Kobayashi, S., 156
Kobuke, Y., 56
Koch, V. R., 2
Kock, K., 139
Kodama, K., 99
Koebel, M., 105
Koenhen, D. M., 139
Koh, W., 126
Kojima, H., 100
Kojima, T., 182
Kolb, D. M., 205
Kolm, W., 235
Kolotyrkin, Ya. M., 205, 251
Kondo, Y., 53
Kononowa, O., 144
Kontoyannakos, J., 101
Kooijnan, D. J., 182
Kopecek, J., 121
Koppitz, F. D., 254
Köppel, H., 33
Korngold, E., 132
Kornyshev, A. A., 236
Koryta, J., 41, 62
Kosovich, J. K., 106
Kostin, M. D., 47
Kotok, L. A., 9
Kotyk, A., 42
Koupparis, M. A., 103
Koutecky, J., 255
Kraicer, P. F., 117
Kraig, R. P., 94
Krasne, S. J., 46
Kratochvil, J., 91
Kraus, M. A., 116
Kravath, R. E., 144
Krawchuk, B., 56
Kray, A. M., 72, 83, 89
Krayushkin, M. M., 16
Krebs, A., 23
Krishnan, V., 12

Krishtalik, L. I., 213
Krygowski, T., 10
Krylov, V. S., 212, 234, 236
Kryuchkova, E. I., 35
Krznaric, D., 238
Kuan, S. S., 74, 112, 114
Kuhn, A., 1
Kuklin, R. N., 236
Kumar, J. L., 90
Kumar, N., 87
Kumar, R., 131
Kumar, S., 69
Kummer, J. T., 157
Kunimatsu, K., 205
Kunst, B., 138, 142
Kuprin, V. P., 249
Kurihara, K., 67
Kurihara, M., 139
Kuriyakov, Yu. N., 212, 250
Kuroda, R., 70
Kurosawa, Y., 154
Kusakari, K., 134
Kusumoto, K., 132, 139
Kuta, J., 237
Kutowy, O., 138
Kuznetsova, L., 213
Kwak, J. C. T., 60
Kwak, S., 31

Labbe, M., 133
Lacan, M., 1, 8, 9
Lacaze, P. C., 12
Lakshmanan, S., 241, 244
Lakshminarayanaiah, N., 41, 44, 45, 58, 61, 121, 128
Lamaze, C. E., 114
Lambert, F. L., 15
Lambert, H. J., 118
Lambert, P. A., 91
Lambert, P. P., 152
Landelout, H. G., 52
Landry, J.-Cl., 89
Lang, N. D., 235
Lange, Y., 53
Lanza, P., 97
Lapatnik, L. N., 97
Lardy, H. A., 87
Laren, E., 10
Larsen, N. R., 69, 74, 108, 110
Larson, G., 38
Lauger, P., 42, 48, 125
Laurent, E., 29
Laviron, E., 38
Lavitaya, P. I., 105
Lavrinovics, E., 13
Lazorenko-Manevich, R. M., 205
Lebl, M., 7
Leblanc, O. H. jun., 91
Lechner, J. F., 68, 83, 93, 97, 99
Ledemezet, M., 2
Lee, C. H., 48, 50, 60, 82, 108, 116, 139
Lehmkuhl, H., 37

Leibovitz, J., 125
Leibzon, V. N., 16
Leitz, F. B., 134, 135
Lelandais, D., 29
Lemoine, P., 38
Lengton, W., 81
Leonelli, E., 139
Lepore, V., 241
Lerche, D., 58, 125
Lessard, J., 16
Le Vanda, C., 37
Levart, E., 181, 182
Levin, S. Z., 255
Levin, Ya. A., 13
Levine, S., 209, 210, 218, 231
Levitin, I. Ya., 39
Levitt, D. G., 48, 49
Lewandowski, R., 67, 89, 102
Lewenstam, A., 65, 88
Lewis, D. I., 249
Lewis, S. B., 92
Lexa, D., 38
L'Her, M., 2
Li, J. H., 49, 124, 125
Li, N. N., 56, 118, 119
Liberti, A., 114
Lidorenko, N. S., 236
Light, T. S., 79
Lightfoot, E. N., 41
Lih, M. M., 41
Lim, R., 72
Lin, T. K., 118
Lindner, E., 63
Lindstrom, M., 2
Linhart, F., 13
Lion, H., 3
Lipkowski, J., 3
Lipsztajn, M., 10
Lisberg, W., 58
Liss, A. R., 41
Lister, K., 9
Litan, A., 122
Liteanu, C., 64, 72, 74, 90
Liu, K., 133
Llenado, R. A., 82, 86
Lobel, S., 91, 117
Loeb, S., 140
Loginova, N. F., 14
Lohrengal, M. M., 255
Lokhande, H. T., 137
Long, G. D., 83
Lonsdale, H. K., 43, 118, 141, 143
Lopez, M., 51, 52
Lopez-Gonzalez, J. D., 54
Lorenz, W., 254, 255
Loshkarev, Yu. M., 249
Loughi, P., 127
Loutfy, R. O., 213
Lovering, E. G., 117, 118, 119
Lowe, B. M., 79
Lowenstam, A., 89
Lubbers, D. W., 41
Lubrano, G. J., 109, 110

# Author Index

Luca, C., 65
Lund, H., 14, 17
Lundquist, S., 235
Lyle, I. G., 131
Lynch, S., 47

Mackey, M. D., 213
Mackie, J. S., 114
McAllister, D. L., 13
Macaskill, J. B., 72
McCadless, F. P., 145
McCallum, C., 58, 131, 132
McCarlhy, J. M., 142
McCarrick, T. A., 240
McDonald, A. C., 218
Macdonald, J. R., 161, 204
Macdonald, R. C., 46, 123
McGregor, R., 50
McInnes, D. A., 44
McIntyre, J. D. E., 205
McKeever, L. D., 1
McKenzie, L. R., 83
McKinney, R. jun., 143
McNeely, R. L., 231
McNelis, E., 103
McQuaker, N. R., 99
McQuillan, A. J., 4, 205
McRae, W. A., 134
McSpadden, S. K., 119
Madsen, R. F., 42
Magini, M., 156
Magno, F., 2, 19
Magouyrk, D. W., 95
Mahenc, J., 112
Mainville, C. A., 119
Mairanovskii, S. G., 5, 16, 19
Mairanovskii, V. G., 2, 14, 23
Maj, M., 102
Makarochkina, S. M., 7
Malaterre, P., 182
Malaya, R. V., 249
Malik, W. U., 69
Malissa, M., 101
Malone, D. M., 48, 54
Malone, T. L., 93
Maloy, J. T., 6, 110
Malyugina, N. I., 18
Mamadieva, G. R., 66
Manahan, S. E., 70, 99
Mandai, T., 29
Manning, D. L., 48, 95
Mano, E. B., 40
Manousek, O., 9, 10
Manvelyan, M. A., 249
Marcenaro, G., 105
Margalit, R., 56
Margaretha, P., 25
Margolis, J. A., 81
Marinyuk, V. V., 205
Marr, D. H., 25
Marsh, H. C., 139
Marshall, E. A., 47
Martelli, M., 38
Martigny, P., 29

Martin, F. E., 54
Martinchek, G. A., 2
Martinez, F. A., 32
Martire, D. E., 233
Maruyama, M., 18
Mascini, M., 59, 70, 79, 93, 96, 99, 108, 114
Mason, E. A., 49
Massart, D. L., 99, 100
Masson, M. R., 71
Mastragostino, M., 36
Materova, E. A., 66, 67, 91
Mathieson, J. G., 232
Mathis, D. E., 62, 92, 102, 123, 183
Matsubera, M., 3
Matulevicius, E. S., 56
Matsumoto, K., 15
Matsumoto, N., 6, 134
Matsuura, T., 51, 142, 143, 144
Matsumura, Y., 31, 34
Mattson, J. S., 4
May, G., 188
Mayeda, E. A., 28
Mayoral, J., 124
Mazur, S., 4
Mazzocchin, G. A., 2, 19
Meares, P., 40, 41, 42, 58, 114, 132
Mechanik, G. F., 236
Medvedev, I. G., 236
Meier, P. C., 72
Meier, T., 72
Meites, L., 1
Melcher, J. R., 153
Mell, L. D., 110
Mellor, J. M., 4, 20
Mel'nikov, B. V., 13
Memoli, A., 59
Menard, H., 226, 240
Mendkovich, A. S., 16
Merli, C., 134
Mertens, J., 98, 99
Merti, C., 145
Mesplede, J., 100
Metayer, M., 54
Metcalfe, A. A., 173, 180, 192
Meyer, E., 14
Meyer, J. P., 47
Meyer, T. J., 37
Meyerhoff, M., 107, 112
Mezentseva, G. A., 18
Michaels, A. S., 117, 142
Michaeli, I., 122
Michel, M. A., 14, 17, 29
Micheli, I., 256
Midgley, D., 63, 78, 88, 96
Middleman, S., 143
Middleton, E., 152
Mikhailov, V. S., 5
Millat, H., 36
Miller, A. D., 137
Miller, I. F., 124
Miller, J. H., 54

Miller, L. L., 2, 3, 24, 26, 35
Millero, F. J., 88
Mills, J. L., 15
Mills, K. L., 36
Min, S., 145
Minc, S., 249
Minturn, R. E., 143, 144
Mison, P., 20
Misono, A., 5
Mitani, M., 56
Mitoff, S. P., 190
Mitzlaff, M., 29
Miyazaki, Y., 56
Miyazima, S., 224
Miyoshi, M., 15
Mizutani, Y., 132, 139
Mocanu, A., 72
Mohan, M. S., 70, 72
Mohan, R. R., 111
Mohilner, D. M., 223, 237, 240, 251, 252
Mohilner, P. R., 223, 237, 251
Moiroux, J., 16
Momenteau, M., 38
Monk, D. W., 116
Monnier, D., 98
Monzie, P., 144
Moody, G. J., 62, 67, 84, 86, 88, 96, 105
Moon, J. K., 54, 142
Moore, G. E., 144
Moore, J., 188
Moore, W. M., 6, 18
Mor, E., 105
Moreira, H., 44
Morel, G., 136
Morf, W. E., 44, 62, 63, 91, 122
Mori, H., 81
Mori, K., 139
Morie, G. P., 83
Morita, S., 154
Morita, Z., 119
Moriyama, T., 143, 152
Morozova, I. D., 13
Moses, P. R., 36
Mossa, G., 145
Motomura, H., 119
Motonaka, J., 102
Motsonelidze, E. P., 105
Mousa, A. A., 232
Mousset, G., 14
Muchovikov, V. V., 66
Mukhtarov, A. Sh., 13
Mukhtarov, V. A., 36
Mulder, M. H. V., 139
Muller, H. R., 10
Muller, R. H., 205
Mumallah, N. A. A., 85
Munari, S., 139, 143
Murakami, H., 102
Murakami, K., 18
Muratore, E., 144
Muratsugu, M., 67
Murayama, N., 124

Murray, R. W., 231
Mussini, T., 127
Myall, C. J., 24
Myamlin, V. A., 234

Naby, J. B., 3
Nadjo, L., 3
Naegele, K., 4
Nagasawa, S., 55
Nagasawa, M., 53, 153, 154
Nagasubramanian, K., 133
Nagayanagi, Y., 182
Nagle, J. F., 256
Nagy, G., 72, 82, 108
Nagy, O. B., 3
Nahmijuz, D., 9
Nakabayashi, N., 43
Nakadomari, H., 223, 237, 240, 251
Nakagawa, G., 89, 99, 102
Nakagawa, T., 43
Nakajima, A., 142
Nakajima, T., 106
Nakamoto, Y., 111
Nakamura, K., 124
Nakano, M., 81, 118
Nakano, T., 4
Nakayama, Y., 56
Nanjo, M., 104, 109, 110
Napoli, A., 70, 96
Narasimham, K. C., 1
Naryanan, P. K., 132
Natale, I. M., 59
Nath, K. C., 9
Nechepurnoi, V. D., 7
Nedea, C., 65
Neeb, R., 68
Neel, J., 136
Nefedova, G. Z., 133
Neihof, R., 54
Nekrasov, L. N., 10
Nelson, K. G., 118
Nemas, M., 116
Nemes, M., 208
Neshkova, N. T., 105
Neugodova, G. L., 249
Neumcke, B., 126
Neves, E. A., 226
Nicchia, M., 139, 143
Nicholson, C., 94
Nielsen, H. J., 102
Nien, T., 37
Nikol'skii, B. P., 91
Nilles, J. L., 137
Nilsson, A., 25
Nishiguchi, I., 23, 32
Nishimura, M., 139
Nishimura, T., 139
Nishimura, Y., 31
Nohe, H., 7, 10
Noma, T., 139
Nomura, T., 102
Nota, G., 96
Notter, R. H., 145

Novikov, S. S., 16
Novikov, V. T., 9
Novkirishka, M., 100, 102
Novozamsky, I., 104
Nurnberg, H. W., 231, 238
Nyberg, K., 24

Oakes, J., 108
O'Brien, R. N., 125
O'Connor, P. E., 46
Oelschläger, H., 10
O'Dea, J. J., 16
Oehme, M., 86, 90, 93
Oei, H. A., 4
Ogawa, S., 6
Oglesby, G. B., 88
Ogren, L., 110, 111
Ogumi, Z., 3
Ogura, S., 123
Ohkubo, K., 4
Ohya, H., 143, 152
Oishi, N., 6
Okada, Y., 82
Oikawa, M., 23, 61
Oikawa, T., 124
Okazaki, S., 74
Okoh, E., 1
Okonogi, N., 43
Okubo, K., 6
Okuyama, M., 123
Olderman, G. M., 67
Oldfield, J. W., 182
Oldham, K. B., 223
Oleinik, A. V., 18
Olson, V. K., 71
Oren, Y., 122
Oshima, K., 100
Oskina, O., 240
Osswald, H. F., 72
Østergaard-Jensen, J. P., 85
Osterhoudt, H. W., 54, 114
Osteryoung, R. A., 2
Ostis, E. K., 9
Østvold, T., 46
Outhwaite, C. W., 209
Owen, J. D., 87
Owens, B. B., 192

Paatsch, W., 41
Paciotti, J. D., 136, 144
Pageau, L., 138, 143
Pak, C. M., 15
Palecek, E., 238
Palin, M. J., 115
Pallozzi, F., 93
Palm, U. V., 212, 213, 218, 240, 249, 255
Palmer, J. L., 31
Papariello, G. J., 114
Paparone, D., 241
Papastathopoulos, D. S., 108, 112, 113
Papay, M. K., 101, 105
Parker, V. D., 3, 5, 25, 26

Parkinson, B. A., 208
Parmar, M. L., 129
Parrini, P., 139
Parsons, R., 203, 204, 205, 211, 216, 217, 220, 223, 241, 249
Parthasarathy, N., 98
Pasek, W., 9
Passino, R., 134, 145
Past, V., 255
Patane, I., 2
Paterson, R., 131
Pathan, A. S., 70, 91, 97, 107
Pathy, M. S. V., 1
Patlak, C. S., 50
Patridge, B. F., 83
Paul, D. R., 119, 136, 144, 148, 150
Paul, R. L., 205
Payne, R., 204
Peat, R., 211, 249
Pedriali, R., 249
Pefferkorn, E., 55
Pemberton, J. P., 87
Penciner, J., 1
Periasamy, N., 4
Perone, S. P., 99
Perram, J. W., 210
Perry, E., 48
Perry, M., 91
Persin, M., 2
Peter, F., 3
Peterlin, A., 148
Peters, C., 188
Peters, C. R., 142
Peters, D. G., 6, 18
Petranek, J., 92, 93
Petres, J. J., 138
Petrii, O. A., 1, 204, 212, 240, 254, 255
Petrosyan, V. A., 10
Petyarv, E., 212, 249
Philipp, B., 43
Philipp, M., 52
Phillips, A. W., 113
Pichon, M., 144
Pickett, C. J., 38
Pieper, J. H. A., 249
Pierre, G., 36
Pilla, A. A., 181
Pilloni, G., 38
Pistoia, G., 192
Pitti, C., 22
Plashkin, V. S., 36
Plesch, P. H., 14
Pleskov, Yu. V., 205, 211
Plesner, W. I., 256
Pletcher, D., 1, 2, 22, 24, 38
Plieth, W. J., 4
Plonka, A., 52
Poirier d'Ange d'Orsay, E., 181
Polos, L., 101
Polyanovskaya, N. S., 240, 241
Pool, K. H., 90, 103
Popescu, I. C., 62, 64, 72

# Author Index

Popicol, W. J., 85
Popov, A., 211
Popov, Yu. A., 251
Porter, G. B., 115
Porter, M. C., 138
Porter, R. D., 81
Porthault, M., 100
Pospisil, L., 237
Potter, E. C., 216
Povarov, Yu. M., 211
Powers, R. W., 190
Pragst, F., 33, 34
Prasolova, O. D., 134
Prather, R., 117
Pretsch, E., 66, 84, 86, 90, 94, 122
Preuss, I., 66
Prishchepa, Yu. A., 205, 211
Proctor, W. E., 79
Proskurovskaya, C. V., 19
Proskurovskaya, I. V., 5
Pui, C. P., 75
Pulidori, F., 249
Pungor, E., 63, 72, 84, 89, 99, 101, 105, 108
Purin, B., 133
Purz, H. J., 43
Pusch, W., 43, 47, 51, 132, 134, 137, 141, 143, 145

Quentin, J. P., 139
Quinn, J. A., 48
Quinn, R. K., 8

Rach, R., 2
Radic, Nj., 99
Ragupathy, K., 12
Rajaput, A. R., 75
Ramachandran, V., 26
Ramakaeva, R. F., 9
Ramamurthy, A. C., 237
Ramirez, W. F., 58, 145
Ramp, F. L., 121
Randles, J. E. B., 204
Rangarajan, R., 98, 143
Rangarajan, S. K., 206, 212, 235, 241, 244, 250, 251, 252, 255
Rastogi, R. P., 127
Rattee, I. D., 53
Rauch, K., 138
Rauf, P. W., 111
Rausch, M. D., 37
Rautenbach, R., 138
Rauzen, F. V., 133
Razzaq, A., 59
Rechnitz, G. A., 62, 64, 70, 82, 95, 98, 101, 105, 107, 108, 110, 112, 113
Reed, P. R., 87
Reed, R. C., 17
Reeves, R. M., 203, 220, 249
Reichman, B., 205
Reich, S., 122

Reidhammer, T. M., 2
Reilley, C. N., 139
Reitano, M., 27
Ren, K., 67, 94
Renkin, E. M., 150
Retter, U., 237
Reymann, E., 74
Reynaud, J. A., 10
Rhee, C. K., 53
Rhodes, R. K., 183
Rice, G., 77
Richter, G., 2, 43, 132
Rigdon, L. P., 63
Riley, D. S., 2
Riley, M., 69
Riley, R., 141
Ritcher, B., 43
Roberts, B. P., 34
Roberts, J. L., 1
Robertson, P. M., 2
Robillard, Y., 12
Robinson, J., 205
Robinson, K., 210, 218, 231
Robinson, R. A., 69
Rocha, R. C., 252
Rock, P. A., 91
Rohm, T. J., 74, 104
Roling, P. V., 37
Ronlan, A., 5, 25, 26
Roseman, T. J., 42
Ross, J. W. jun., 97
Ross, M. A. S., 188
Ross, S. D., 11
Rotenburg, Z. A., 205, 211
Rott, G. M., 240, 249
Rouchouse, A., 100
Rouse, J. D., 54
Rowe, M. L., 118, 119
Rozhkov, I. N., 26
Rozinek, R., 28
Rubinshtein, A. I., 236
Rubinskaya, T. Ya., 19
Rubinstein, I., 133
Rudich, S. W., 50
Rushing, J. F., 114
Ruzicka, J., 72, 73, 74, 89
Ryan, D. E., 67
Ryan, T. H., 63, 85
Ryba, O., 92, 93
Rybalka, L. E., 249
Ryvkina, I. P., 10

Sachs, S. B., 138
Saeki, H., 134
Saier, H. D., 139
Saji, T., 37
Sakai, T., 43
Salajegheh, A., 18
Salaun, J. P., 8
Salaun-Bouix, M., 8
Salem, M. R., 223
Salie, G., 254, 255
Salomon, A., 89
Salve, M. A., 212, 213, 218, 240

Salvi, A. S., 137
Sammon, D. C., 43
Samokhvalov, G. I., 23
Sand, H. J. S., 65
Sandifer, J. R., 80, 134
Sansoni, B., 132
Santhanam, K. S. V., 4
Sara, R., 99
Sarangapani, S., 237
Sapio, J. P., 102
Sasaki, K., 26
Sastri, V. S., 144
Sasty, K. S., 9
Sathyanarayana, S., 237
Sato, N., 127
Satoh, I., 111
Sausins, A. E., 33
Savéant, J. M., 3, 15
Savvin, N. I., 66, 68
Sawada, K., 81
Sawyer, D. J., 1
Scanio, C. J. V., 28
Schaaf, D. P., 144
Schäfer, H. J., 22, 29
Schafer, O. F., 68
Scheide, E. P., 89
Schejter, A., 56
Scherfig, J., 141
Scheutzow, D., 13
Schiavon, G., 34
Schier, E. J., 7
Schiffer, D. K., 56
Schiffrin, D. J., 204
Schindler, A. M., 145, 150
Schleifer, A., 114
Schlogl, R., 47, 55
Schmithals, F., 211
Schnabel, H., 29
Schofield, W. M., 143
Scholer, R. P., 82
Schouler, E., 187
Schreiber, B., 99
Schrøder, K. H., 71
Schuette, R. W., 116
Schugerl, K., 119
Schultz, F. A., 72, 102
Schultz, J. S., 42
Schultze, J. W., 254
Schumann, D., 182, 215
Schwabe, K., 77
Schwager, F., 2
Scibona, G., 156
Scott, W. J., 1
Scotto, V., 105
Scrosati, B., 192
Scuppa, B., 156
Seabach, D., 4
Sease, J. W., 17
Sebastian, K. L., 212, 255
Sederel, W. L., 50
Seevers, R., 117
Sehring, R., 18
Seidah, N. G., 44
Sekerka, I., 68, 83, 97, 99

265

Seko, M., 6
Selegny, E., 133
Selig, W., 63, 71, 72, 83, 89, 93, 103
Sellan, J. B., 211, 240
Semenescu, Gh., 65
Semler, M., 92, 93
Sen, J., 70
Senda, M., 244
Senguen, F. I., 10
Seno, M., 139
Sequaris, J. M., 10
Sergievskii, V. V., 103
Serkerka, I., 93
Serve, D., 20, 32
Settineri, W. J., 1
Sevastyanov, E. S., 211, 241
Sevast'yanova, V. V., 16
Severova, T. A., 213
Shah, A. C., 118
Sharp, M., 66, 69
Shatkay, A., 63
Shasholia, V. E., 134
Shchori, E., 151
Shean, G. M., 60
Sheina, N. M., 105
Shelepakov, A. V., 241
Shepard, V. R., 97
Sherwood, W. G., 2
Shibata, N., 83, 100
Shigei, T., 35
Shigi, M., 25
Shijo, Y., 100
Shin, M., 71
Shinaberger, J. H., 54
Shinagawa, Y., 45
Shinbo, T., 117
Shinoda, K., 142
Shimbo, T., 59
Shiraishi, N., 99
Shono, T., 1, 23, 31, 34
Shorr, J., 135
Shterman, V. S., 66, 68, 71, 94
Shults, M. C., 106
Shyam, R., 61
Siddiqi, F. A., 53, 61
Siegel, T. M., 35
Siegemund, R., 255
Sigan, A. L., 39
Silver, I. A., 41
Simeonov, V., 102
Simon, W., 41, 44, 62, 63, 66, 72, 82, 84, 86, 90, 91, 93, 94, 122
Simonet, J., 14, 16, 17, 29
Simons, R., 50
Singleton, D., 182
Singh, J., 127
Singh, K., 127
Singh, S. P., 53, 61
Singhal, T. C., 129
Skaletz, D. H., 11
Skevin, D., 138
Skorobogatova, M. S., 13

Slovetakii, V. I., 10
Sloan, C. H., 83
Sluyters, J. H., 157
Sluyters, M., 182
Smid, J., 56
Smionov, K. M., 16
Smirnov, S. K., 7
Smirnov, Yu. D., 7
Smit, J. A. M., 43, 147
Smith, A. L., 218
Smith, C. A., 4
Smith, C. Z., 22
Smith, D. E., 182
Smith, D. G., 79
Smith, G. R., 102
Smith, L. R., 15
Smith, W. H., 4, 5
Smolders, C. A., 139, 142
Smoler, I., 237
Smul'skii, S. P., 10
Snakin, V. V., 71
So, Y. H., 24, 35
Soignet, D. M., 37
Sokolov, S. V., 36
Sokolov, Yu. M., 210
Soler, A., 32
Sollner, K., 41, 43, 51, 54
Solomin, A. V., 35
Solunov, H., 124
Sonobe, N., 111
Sorrentino, M. H., 101
Sourirajan, S., 51, 138, 142, 143, 144
Spacek, P., 118
Spiegler, K. S., 125, 136
Sproule, R. C., 121
Srivastava, M. L., 131
Srivastava, R. C., 127, 129, 130
Srivasthava, S. K., 69
Srogl, J., 28
Stannett, V., 141
Stark, G., 42
Staude, E., 43
Staverman, A. J., 43, 147
Stearns, R. I., 98
Steele, R. D., 58
Stefaniga, E., 74, 82
Stell, G., 255
Stelting, K. M., 70, 99
Stenin, E. V., 236
Stenina, E. V., 237
Stermitz, F. R., 26
Stern, P., 7
Stevenson, J. F., 116
Stibor, I., 28
Stine, J., 54, 142
Stokbro, W., 108
Stokely, J. R., 95
Stormberg, H. P., 117
Storz, D. G., 152
Stoupel, E., 153
Stover, F. S., 47, 62
Stradins, J., 13, 33
Strathmann, H., 139, 142

Stringer, B., 43
Stroka, J., 241
Stuart, J. L., 72, 98, 100
Stulik, K., 98
Stunic, Z., 9
Styer, S., 91
Suchadevo, S. R., 42
Suchanski, M. R., 4
Suemmermann, W., 31
Sugiura, S., 59, 117, 153
Sugawara, M., 106
Su Khac Bink, 15
Sun, T. I., 4
Sundholm, G., 2, 84
Sunu, W. G., 122
Surana, P., 231
Suschicital, C., 123
Sutherland, J. O., 36
Sutton, J. R., 22, 23
Suvorov, N. N., 12
Suzuki, A., 31, 43
Suzuki, S., 60, 61, 106, 111
Suzuki, T., 36
Svanholm, U., 3
Swartz, J. L., 79
Sweetsur, A. W. M., 102
Syrchenkov, Ya. A., 66, 103
Szczepaniak, W., 67, 94

Tabakovic, I., 1
Taits, S. Z., 5
Tajima, M., 4
Takahashi, Y., 6, 31
Takai, N., 135
Takaishi, T., 209
Takashima, S., 133
Takatori, Y., 43
Takeda, A., 26
Takeguchi, N., 54
Takehara, Z., 3
Takeshita, K., 31
Takeshita, M., 127
Taketani, Y., 139
Takizawa, A., 150
Taksishvili, O. G., 105
Tallec, A., 33
Tamura, K., 102
Tanaka, H., 29
Tanaka, M., 2, 81
Tanaka, S., 541
Tanaka, T., 67, 96, 106, 139
Taniguchi, T., 150
Taniguchi, Y., 50, 120, 141
Tanikawa, S., 99
Tankersley, R., 145
Tanny, G., 150
Tardivel, R., 20
Tarle, M., 1
Tarp, M., 108
Tasaka, M., 53, 55, 153, 154, 156
Tatarinov, V. I., 210
Tateda, A., 102
Taylor, K., 194, 197

# Author Index

Taylor, P. J., 113
Taylor, P. N., 220
Tedoradze, G. A., 1, 210, 249
Temerk, Y. M., 231, 238
Tendulkar, S. R., 240
Tennant, H. G., 3
Terada, Y., 134
Tereshchenko, M. N., 133
Tessier, D., 3
Thain, J. M., 70
Thayer, W. L., 138
Theeuwes, F., 53
Thiebault, A., 22
Thirsk, H. R., 192
Thoma, A. P., 122
Thomas, D., 54
Thomas, F. G., 1, 14
Thomas, J. jun., 99
Thomas, J. D. R., 62, 67, 84, 86, 88, 96, 105
Thomas, R. C., 90
Thompson, D. E., 94
Thompson, H., 112
Thornalla, M., 29
Tien, C., 142
Tien, H. T., 41
Tilak, B. V. K. S. R. A., 246
Tiravanti, G., 134
Tissot, P., 25
Tobolsky, A. V., 54, 150
Tochigi, K., 150
Tock, R. W., 116, 144
Todd, D. K., 143
Tokuda, M., 31, 35
Tokuyama, A., 6
Tombalakian, A. S., 54
Tomilov, A. P., 7, 16
Tomita, M., 43
Tong, S. L., 113
Topol, L. F., 192
Torii, H., 29
Torikai, M., 65
Toth, K., 63, 72, 84, 89, 99, 101, 105, 108
Touboul, E., 8
Toyoda, H., 59
Toyoshima, Y., 55
Trachtenberg, I., 77, 88
Tran-Minh, C., 108
Trasatti, S., 204, 215, 220, 240, 249, 252
Tredgold, R. H., 121
Treichel, I., 106
Treichel, P. M., 106
Tribuzio, S., 77
Trkovnik, M., 9
Trocha-Grimshaw, J., 19
Trojanek, A., 24
Troll, T., 12
Tronjanowicz, M., 85, 86, 87
Trska, P., 7
Tsai, H. T., 99
Tseng, P. K. C., 71, 100
Tsien, R. Y., 234

Tucker, F. E., 118
Turner, J., 189
Turovskii, A. A., 14
Turovskii, N. A., 14
Tyas, D., 72

Uchimura, K., 119
Udo, K., 5
Udupa, H. V. K., 12
Uiltman, J., 54
Uldrikjis, J., 33
Umezawa, K., 135
Unterker, D. F., 2
Updike, S. J., 106
Urabe, Y., 15
Urusov, Yu. I., 66, 68, 71, 87, 94, 103
Ushakova, Z. N., 211, 212
Utley, J. H. P., 6

Väärtnõu, M. G., 212, 213, 240, 249
Vacik, J., 121
Vadura, R., 100
Vafina, A. A., 13
Vajnaht, Z., 142
Valcher, S., 36
Valenta, P., 231, 238
Valenzuela, C. C., 54
Valette, G., 211
Vallance, C. R., 2
Vallo, F., 99
Van Bruggen, J. T., 145
Van Bueren, A. L., 145
Van de Leest, R. E., 68
Van den Winkel, P., 98, 99
Vandeputte, M., 100
Vanderborgh, N. E., 8
Van der Linden, W. E., 70, 71, 88, 89
Van der Meer, J. M., 70, 88, 89
Van der Velden, P. M., 139, 142
Van Duyne, R. P., 4, 205
Vanel, P., 215
Van Haute, A., 142
Van Heuven, J. W., 143
Van Leeuwen, H. P., 182
Van Osch, G. W. S., 68
Van Riemsdijk, W. H., 104
Vansant, E. F., 82
Varadi, M., 72
Varoqui, R., 55
Vartires, I., 5
Vasina, S. Ya., 240
Vassilev, T., 124
Vecchio, C. D., 59
Veis, A., 58
Venkateswaran, A., 122
Venkateswarlu, P., 99
Verbeek, F., 97
Vereecken, J., 98
Vergano, L., 252
Verhoff, F. H., 153
Verniory, A., 152

Vesely, J., 98, 101
Vetter, K. J., 254
Vetterl, V., 237, 238
Vickery, B., 99
Vickery, M. L., 99
Vieil, E., 20
Vigo, F., 139, 143
Vikhlyaev, Yu. I., 10
Vincent, F., 20
Virtanen, R., 84, 89
Vita, O. A., 83
Vitanov, T., 211
Viviani-Nauer, A., 122
Vlad, R., 90
Vofsi, D., 43, 54, 117, 136
Voigt, R., 43
Volke, J., 9, 10, 13
Volkel, W., 119
Vol'pin, M. E., 39
Volodina, T. A., 12
Von Deak, M. A., 116
Von Mylius, U., 139
Voronovskaya, M. N., 87
Vorotyntsev, M. A., 236
Vrentas, J. S., 145

Wada, C., 154
Wada, H., 89
Wade, J. B., 50
Walch, A., 141
Walker, J. F., 131
Wallace, R. A., 119, 124
Walz, D., 47
Wandrey, C., 119
Wang, C. C., 234
Wang, C. L., 205
Wanninen, E., 99
Ward, R. M., 144
Warren, G. B., 30
Watanabe, A., 43
Waters, W. L., 7
Watkins, B. F., 3
Wawro, R., 101, 113
Wawzonek, S., 19
Wayt, H., 145
Weatherell, J. A., 99
Weaver, M. J., 208, 232
Webber, H. M., 78
Weber, F. G., 33
Weddigen, G., 188
Wegner, G., 123
Wei, L. Y., 47
Weinberg, M., 116
Weinberg, N. L., 24, 25
Weinreb, S. M., 27
Weinstein, J. N., 135
Weinstein, J. W., 135
Weintraub, M., 140
Weiss, L., 66
Wellisch, E., 116
Wendt, R. P., 47, 49, 50
West, A. R., 190
West, J., 188
Whitfield, M., 72

# Author Index

Whitfield, R., 182, 192
Whyte, J. N. C., 98
Wichterle, O., 118
Wickstrom, K., 84, 89
Wiedersich, H., 192
Wiedner, G., 47, 121, 150
Wiersdorff, W. W., 11
Wierse, D. G., 255
Wijnen, M. D., 181
Wikby, A., 77
Wiley, A. J., 143, 144
Wilhelm, G., 150
Williams, D. R., 36
Willis, P. M., 177, 189, 190, 192
Wilson, A. G., 102, 103
Wilson, C. N., 137
Wilson, M., 84, 89
Wilson, M. F., 78
Woermann, D., 47, 121
Wojtkowski, W., 100
Woermann, D., 120, 121
Wolf, F., 144
Wolf, H., 58, 125
Wolff, H. J., 145
Wong, K. H., 56
Wong, P. S. L., 117
Wong, S. G., 60

Woo, B. Y., 47
Wu, C. N., 24, 25
Wu, S. C., 99
Wuhrman, P., 44, 122
Wuthrich, H., 255
Wydeven, T., 143

Yagi, K., 56
Yahara, S., 36
Yalkowsky, S. H., 42, 118
Yamabe, T., 135
Yamagishi, H., 5
Yamaguchi, A., 6
Yamaishi, T., 56
Yamamoto, H., 32
Yamamoto, M., 4
Yamamoto, Y., 144
Yamana, M., 4
Yasuda, H., 114, 139, 148
Yeager, H. L., 51, 52, 205
Yijan, A., 141
Yonezawa, K., 224
Yoo, K. S., 66
Yoshida, K., 25
Yoshida, M., 6
Yoshida, Sh., 135
Yoshida, T., 123

Yoshikuni, N., 70
Yoshizawa, S., 3
Young, P. N. W., 83
Yu, P., 36
Yuriado, E. K., 212
Yurinskaya, V. E., 91
Yuzbekov, Yu. A., 249

Zabusova, E. E., 2
Zagatto, E. A., 72
Zalogina, E. A., 18
Zaporozheto, E. V., 9
Zarins, P., 13
Zarinskii, V. A., 134
Zecchin, S., 34
Zelman, A., 145
Zhukov, A. F., 66, 68, 71, 87, 94, 103
Zhukov, M. A., 133
Zimmerman, R. L., 123
Zipper, J. J., 99
Zisner, E., 138
Zoroatskaya, E. I., 13
Zotti, G., 38
Zuman, P., 1

te Due

JAN 12 1981